MOLECULAR TARGETS IN PROTEIN MISFOLDING AND NEURODEGENERATIVE DISEASE

ELSEVIER  *science & technology books*

Companion Web Site:

http://booksite.elsevier.com/9780128001868

Molecular Targets in Protein Misfolding and Neurodegenerative Disease
Pierfausto Seneci, *Author*

Resources:

• All figures from the book available as both Power Point slides and .jpeg files

• Links to web sites carefully chosen to supplement the content of the book

• Contact the author with questions and/or suggestions

ELSEVIER

AP
ACADEMIC PRESS

MOLECULAR TARGETS IN PROTEIN MISFOLDING AND NEURODEGENERATIVE DISEASE

PIERFAUSTO SENECI
Universitá degli Studi di Milano,
Dipartimento di Chimica,
Milan, Italy

AMSTERDAM • BOSTON • HEIDELBERG
LONDON • NEW YORK • OXFORD • PARIS
SAN DIEGO • SAN FRANCISCO • SINGAPORE
SYDNEY • TOKYO
Academic Press is an Imprint of Elsevier

Academic Press is an imprint of Elsevier
32 Jamestown Road, London NW1 7BY, UK
225 Wyman Street, Waltham, MA 02451, USA
525 B Street, Suite 1800, San Diego, CA 92101-4495, USA

Notice
No responsibility is assumed by the publisher for any injury and/or damage to persons
or property as a matter of products liability, negligence or otherwise, or from any use or
operation of any methods, products, instructions or ideas contained in the material herein.
Because of rapid advances in the medical sciences, in particular, independent verification
of diagnoses and drug dosages should be made

British Library Cataloguing-in-Publication Data
A catalogue record for this book is available from the British Library

Library of Congress Cataloging-in-Publication Data
A catalog record for this book is available from the Library of Congress

ISBN: 978-0-12-800186-8

For information on all Academic Press publications
visit our website at http://store.elsevier.com/

Typeset by Thomson Digital

Printed and bound in United States of America

Working together
to grow libraries in
developing countries

www.elsevier.com • www.bookaid.org

Dedication

Plaques, fibrils, and tangles could not conceal your beautiful soul, that still shines. Ti vorrò sempre bene, mamma!

Contents

6. Assembly and Disassembly of Protein Aggregates

Abbreviations

4EBP1	4E-binding protein 1
AA	amino acid
AAA+	ATPases associated with various cellular activities
Aβ	amyloid β
ABIN	A20 binding inhibitor of NF-kappaB
ACD	α-crystallin domain
AChE	acetylcholinesterase
AD	Alzheimer's disease
ADI	Alzheimer's Disease International
Adrm1	adhesion regulating molecule 1
AFM	atomic force microscopy
Ag	aggregate
AgD	argyrophilic disease
AgR	aggrephagy receptor
AgS	aggrephagy scaffold
Aha1	activator of Hsp90 ATPase
AIM	Atg8-interacting motif
Akt	protein kinase B
ALFY	autophagy linked FYVE protein
ALR	autophagic lysosomal reformation
ALS	amyotrophic lateral sclerosis
AMBRA1	autophagy/beclin 1 regulator 1
AmF	amyloid fiber
AMPA	α-amino-3-hydroxy-5-methyl-4-isoxazolepropionic acid
AMPK	AMP-activated protein kinase
AN	aggregation nuclei
AMSH-LP	associated molecule with the SH3 domain of STAM-like protein
AP	autophagosomes
APD	atypical parkinsonism disorder
APOE	apolipoprotein E
APP	amyloid precursor protein
ARA54	AR-associated protein 54
ARCA	autosomal recessive cerebellar ataxia
ARIH	ariadne homolog
Atg	autophagy-related

AV	autophagic vacuoles
BACE1	β-APP cleaving enzyme 1
BAG	Bcl-2-associated athanogene
BARD1	BRCA1–associated RING domain protein 1
Barkor	beclin 1-associated autophagy-related key regulator
BDNF	brain-derived neurotrophic factor
BEACH	PH-Beige and Chediak-Higashi
BH	Bcl-2-homology
BIF-1	BAX-interacting factor-1
Bnip-3	Bcl-2/E1B-19 kDa interacting protein 3
BRCA1	breast cancer type 1
BRMS1	breast cancer metastasis suppressor 1
CA1	cornus ammonis 1
Car	cargo
CASA	chaperone-assisted selective autophagy
Cath A	cathepsin A
CBD	corticobasal degeneration
CD4	cluster of differentiation 4
cdc37	cell division cycle 37 homolog
Cdk	cyclin-dependent kinase
CFTR	cystic fibrosis transmembrane regulator
Ch	cholesterol
CHIP	C-terminus of Hsc70 interacting protein
CHMPB2	charged multivesicular body protein B2
cIAP	cellular Inhibitor of Apoptosis Protein
CK2	casein kinase 2
Clp	caseinolytic protease
CMA	chaperone-mediated autophagy
CMT	Charcot–Marie–Tooth disease
CNS	central nervous system
CP	core particle
CRL	cullin RING ligases
CTE	chronic traumatic encephalopathy
Cvt	cytosol-to-vacuole transport
CYLD	cylindromatosis
Cyp40/Cpr6	cyclophilin 40/cytoplasmic ribosomal protein-6
Cyt	cytosolic
D	dimer
DA	disordered aggregate
DAP1	death-associated protein 1
DAPK	death-associated protein kinase
dAV	degradative autophagic vacuoles
DBM	dynein binding motor

Ddi1	DNA damage-inducible 1
DHMN	distal hereditary motor neuropathy
DLB	dementia with Lewy bodies
DLS	dynamic light scattering
DM1	type I myotonic dystrophy
DNTC	diffuse neurofibrillary tangles with calcification
DRAM	damage-regulated autophagy modulator
DS	Down syndrome
Dsk2	dominant suppressor of kar2
DUB	de-ubiquitinating enzyme
DYN	dynein motors
E1	UBQ-activating enzyme
E2	UBQ-conjugating enzyme
E3	UBQ ligase
E6AP	E6-associated protein
EC	entorhinal cortex
ECD	evolutionary conserved domain
eIF4E	eukaryotic translation initiation factor 4 epsilon
EM	electron microscopy
ER	endoplasmic reticulum
ERAD	endoplasmic reticulum-associated degradation
ERGIC	ER-Golgi intermediate compartment
ERK	extracellular-regulated signal kinase
ES	endosomes
ESCRT	endosomal sorting complex required for transport machinery
ESI	electron spray ionization
FAT	fast axonal transport
FATC	FRAP, ATM, TRRAP C-terminal
FBD	familial British dementia
FDD	familial Danish dementia
FIP200	focal adhesion kinase family-interacting protein
FKBP	FK-binding protein
FL	full length
FRB	FKBP–rapamycin binding
FTDP-17	frontotemporal dementia and parkinsonism linked to chromosome 17
FTLD	frontotemporal lobar degeneration
FUS	fused in sarcoma
FYCO	FYVE and coiled-coil domain containing 1
FYVE	Fab1-YotB-Vac1p-EEA1
G2E3	G2/M-phase-specific E3 UBQ ligase
G3BP1	Ras-GTPase-activating protein SH3 domain-binding protein 1

GABA	γ-aminobutyric acid
GABARAP	GABA receptor-associated proteins
Gad	gracile axonal dystrophy
Gd-PDG	Guadeloupean-parkinsonism dementia complex
GGT	globular glial tauopathies
GOF	gain-of-function
GR	glucocorticoid receptor
GSK-3	glycogen synthase kinase 3
GSS	Gerstmann–Sträussler–Scheinker disease
H	heparin
HACEI1	HECT domain and ankyrin repeat-containing E3 ubiquitin-protein ligase 1
HbYX	hydrophobic Tyr-X
HD	Huntington's disease
HDAC	histone deacetylase
Hdj1	DnaJ protein homolog 1
HDM2	human double minute 2 homolog
HEAT	Huntingtin, elongation factor-3, protein phosphatase 2A, tor1
HECT	homologous to the E6AP carboxyl terminus
HEK293	human embryonic kidney 293
HERC	HECT and RCC1-like domains
HIP	Huntingtin-interacting protein
HMGB1	high mobility group box 1
HMN	hereditary motor neuron
HOIL-1	heme-oxidized IRP2 ubiquitin ligase 1
HOIP	HOIL-1-interacting protein
HOP	Hsp70–Hsp90 organizing *protein*
HOPS	homotypic vacuole fusion and vacuole protein sorting
HP	hyperphosphorylation, hyperphosphorylated
Hsc	heat shock constitutive
HSF1	heat shock factor 1
HSP	heat shock protein
HSPG	heparan sulfate proteoglycan
HTS	high throuput screening
HUWEI1	HECT, UBA, and WWE domain-containing protein 1
IAPP	islet amyloid polypeptide
iAV	initial autophagic vacuole
IBMPFD	inclusion body myopathy, Paget disease of the bone, and frontotemporal dementia
IBR	in-between RING
IDP	intrinsically disordered protein

IDR	intrinsically disordered region
IκBα	nuclear factor of kappa light polypeptide gene enhancer in B-cells inhibitor, alpha
IKK	inhibitor of NF-κB kinase
IM	isolation membrane
IMP2B	integral membrane protein 2B
IP$_3$R	inositol-1,4,5-triphosphate receptor
IPOD	inclusion protein deposit
ISG15	interferon-stimulated gene 15
JAMM	JAB1/MPN/Mov34
JIP1	JNK-interacting protein 1
JNK	c-jun N-terminal kinase
JUNQ	juxtanuclear quality control
kDa	kilodalton
Keap-1	kelch-like ECH-associated protein 1
KIR	keap1-interacting region
KO	knockout
LAMP	lysosomal-associated membrane protein
LC3	light chain 3
LIR	LC3-interacting region
LOF	loss-of-function
LPF	long protofibrils
LRRK2	leucine repeat-rich kinase 2
LS	lysosomes
LUBAC	linear UBQ chain assembly complex
MA	macroautophagy
MALDI	matrix-assisted laser desorption ionization
MAP	microtubule-associated protein
MAPK2	mitogen-activated protein kinase 2
MCPIP1	monocyte chemotactic protein-induced protein 1
MDM2	mouse double minute 2 homolog
MEFs	murine embryonic fibroblasts
MEKK	mitogen-activated protein kinase kinase kinase 3
MF	mature fibrils
MK2	mitogen-activated protein kinase-activated protein kinase 2
mLST8	mammalian lethal with Sec13 protein 8
MM	monomer, monomeric species
MPP$^+$	1-methyl-4-phenylpyridinium
MS	mass spectrometry
MSA	multiple system atrophy
mSIN1	mammalian stress-activated protein kinase interacting protein
MT	microtubule

MTBR	MT-binding repeat
MTOC	MT-organizing center
mTORC	mammalian target of rapamycin complex
MVB	multi-vesicular bodies
MW	molecular weight
NAE	NEDD8-activating enzyme
NBD	nucleotide-binding domain
NBIA	neurodegeneration with brain iron accumulation
NBR1	neighbor of BCRA gene
NDD	neurodegenerative disease
NDP52	nuclear dot protein 52
NEDD4	neural precursor cell expressed, developmentally down-regulated 4
NEF	nuclear exchange factor
NEMO	NF-κB essential modulator
NES	nuclear export system
NF-κB	nuclear factor kappa-light-chain-enhancer of activated B cells
NFs	neurofilaments
NFT	neurofibrillary tangles
NFTPD	neurofibrillary tangle-predominant dementia
NLS	nuclear localization systems
NMJ	neuromuscular junction
NMR	nuclear magnetic resonance
NPC	Niemann–Pick type C disease
NPM	nucleophosmin
Nrf2	NF-E2-related factor 2
NSF	N-ethylmaleimide-sensitive factor
O	oligomer
OPN	optineurin
ORP1L	oxysterol-binding protein-related protein 1 L
OTD	ovarian tumor domain
p23	progesterone receptor complex 3
p70S6K	p70 ribosome S6 kinase
PAS	phagophore assembly site
PB1	phox and bem 1p domain
PCNA	proliferating cell nuclear antigen
PD	Parkinson's disease
PDB	Paget's disease of bone
PDC	Parkinsonism-dementia complex of Guam
PE	phosphatidylethanolamine
PEP	post-encephalitic parkinsonism
PF	protofibril
PH	pleckstrin homology

PHD	plant homeo domains
PHF	paired helical filament
PI	phosphatidylinositol
PI3P	phosphatidylinositol-3-phosphate
PiD	Pick's disease
PIKfyve	FYVE finger-containing 1-phosphatidylinositol-3-phosphate 5-kinase
PIKK	phosphoinositide kinase-related kinase
PINK1	PTEN-inducible kinase 1
PKA	protein kinase A
PKAN	pantothenate kinase-associated neurodegeneration
PNS	peripheral nervous system
PolyQ	polyglutamine
PP2A	protein phosphatase 2A
PPI	protein–protein interaction
PPIase	peptidyl-prolyl isomerase
PQC	protein quality control
PR	protein-rich regions
PRAS40	proline-rich Akt substrate p40
Protor	protein observed with rictor-1
PrP	prion protein
PRP5	proline-rich protein 5
PrPCAA	prion protein cerebral amyloid angiopathy
Pru	pleckstrin-like receptor for ubiquitin
Prx	peroxiredoxins
PS1	presenilin-1
PSP	progressive supranuclear palsy
PTEN	phosphatase and tensin homolog
PTM	post-translational modification
R&D	research and development
Rabs	Ras-related in brain
Rad23	radiation-sensitive 23
Rag	Ras-related small GTP binding protein
Raptor	rapamycin-sensitive scaffolding protein of mTOR
Ras	rat sarcoma
RBCK	RanBP-type and C3HC4-type zinc finger containing 1
RBR	RING-in-between-RING
RCC1	regulator of chromosome condensation 1
REDD1	regulation of DNA damage response 1 protein
Rheb	Ras homolog enriched in brain
Rictor	rapamycin-insensitive companion of mTOR
RILP	Rab7-interacting lysosomal protein
RING	really interesting new gene

RIP-1	receptor-interacting protein 1
RLD	RCC1-like domain
RNF	ring finger protein
ROCK	Rho-associated coiled-coil kinase
ROS	reactive oxygen species
RP	regulatory particle
Rpn	regulatory particle non-ATPase
Rpt	regulatory particle triple A
Rsp5	reverses SPT-phenotype protein 5
Rubicon	RUN domain and cysteine-rich domain containing
SAR	selective autophagy receptors
SBD	substrate-binding domain
SBMA	spinal and bulbar muscular atrophy
SCA-3	spinocerebellar ataxia type-3
SEC	size exclusion chromatography
SG	stress granule
SGK	serum and glucocorticoid-inducible kinase
SHARPIN	SH3 and multiple ankyrin repeat protein-associated RBCK1 homology domain-interacting protein
sHsp	small heat shock protein
SIK	salt-inducible kinase
siRNA	small interference RNA
Sirt	silent information regulator
SKD1	Suppressor of K^+ transport growth defect 1
SMIR	SOD1 mutant interaction region
SNAP-29	synaptosomal-associated protein 29
SNARE	soluble N-ethylmaleimide-sensitive factor (NSF) attachment protein receptor
SOD-1	superoxide dismutase 1
SP	senile plaque
SPF	short protofibrils
SSPE	subacute sclerosing pan-encephalitis
STAT	signal transducer and activator of transcription
STC	substrate translocation channel
Stx17	syntaxin 17
SUMO	small ubiquitin-like modifier
TAK1	TGFβ-activated kinase 1
TANK	TRAF family member-associated NF-kappa-B activator
TAR	trans-active response
TAT	trans-activator of transcription
TBK1	TANK-binding kinase 1
TBS	TRAF6 binding site
TEM	transmission electron microscopy

TG	transgenic
TDP-43	TAR DNA-binding protein 43
TGFβ	transforming growth factor-β
TIEC1	TGFβ-inducible early growth response protein 1
TLS	translated in liposarcoma
TNF	tumor necrosis factor
TPPP-1	tubulin polymerization-promoting protein 1
TPR	tetratricopeptide
TRAF6	TNF receptor associated factor 6
Tregs	T-regulatory cells
TREM2	triggering receptor expressed on myeloid cells 2
TRIM50	tripartite motif-containing 50
TRIP12	thyroid hormone receptor interactor 12
TSC	tuberous sclerosis complex
UBA	ubiquitin-associated
UBAN	ubiquitin binding in ABIN and NEMO
UBD	ubiquitin-binding domain
UBL	ubiquitin-like
UBQ	ubiquitin
UBZ	ubiquitin-binding zinc finger
UCH	ubiquitin C-terminal hydrolase
UIM	ubiquitin-interacting motif
ULK	UNC-51-like kinase
UPS	ubiquitin–proteasome system
UPR	unfolded protein response
USP	ubiquitin-specific protease
UVEAG	UV irradiation resistance-associated gene
VAMP8	vesicle-associated membrane protein 8
VCP	valosin-containing protein
VCPIP	VCP (p97)/p47 complex interacting protein 1
VDAC1	voltage-dependent anion-selective channel protein 1
Vps34	vacuolar protein sorting protein 34
Vt1b	vesicle transport through interaction with t-SNAREs homolog 1B
WT	wild type
ZnF-UBP	zinc finger ubiquitin-specific protease

CHAPTER

1

Protein Misfolding, Neurodegeneration and Tau
The Main Players, or the Usual Suspects?

1.1 THE NEURODEGENERATION SCENARIO

If one looks at life expectancy, we seem to be moving in the right direction [1]. Life expectancy at birth did not significantly vary from the Neolithic ($\approx$20 years) to Rome (between 20 and 30 years), to medieval Britain ($\approx$30 years), and even to early 20th century Britain ($\approx$31 years). Death at birth was a huge negative factor, as 10-year-old boys in Rome could expect to reach $\approx$47 years of age, while 21-year-olds from Middle Age Britain could as an average reach 64 years of age. Improved sanitary conditions, disease prevention (e.g., vaccinations) and treatment (e.g., antibiotics) have significantly increased life expectancy, up to the 67.2-year value in 2010 [2]. One should not forget about regional differences due to country development (compare the 82.6-year value in Japan with the 49.4-year value in Swaziland), or to local bursts of otherwise curable diseases (a life expectancy value of 41.2 years in South Africa around 2010 that would increase to 69.9 years if HIV did not exist [3]). Nevertheless, the average lifespan is steadily increasing.

Eternal life is the mankind dream, but one should carefully look at the details before signing any Faustesque contract with the devil's associates. Living longer a poor life does not represent anyone's dream, but that's exactly what we're facing now: and that's due to neurodegeneration. A few facts will better define and explain it.

Alzheimer's Disease International (ADI) estimates in its 2013 report [4] that there are more than 35 million people with dementia worldwide as of 2010, that the number will double by 2030, and triple to 115 million by 2050. The risk of Alzheimer's disease (AD) increases with age, so unless new AD treatments are launched, this number will grow sharply as the baby boomer generation reaches old age. In industrialized countries, the

Molecular Targets in Protein Misfolding and Neurodegenerative Disease. DOI: 10.1016/B978-0-12-800186-8.00001-8

prevalence of Parkinson's disease (PD) is about 1% for people over 60, with estimates of up to 4% for people in the highest age groups [5]. Numbers further increase when one takes into account all neurodegenerative diseases (NDDs). AD and related disorders currently affect over 7 million people in Europe, and this figure is expected to double every 20 years as the population ages (16% of the European population is over 65 now, and this figure is expected to reach 25% by 2030 [6]). In the US, an estimated 5.2 million Americans of all ages have AD in 2013 [7]. The estimated annual incidence (rate of developing disease in one year) of AD increases dramatically with age, from ≈53 new cases per 1000 people aged 65 to 74, to 170 new cases between 75 and 84, to 231 new cases at age 85 and older. Because of the increasing number of people age 65 and older in the US, the annual number of new cases of AD and other dementias is projected to double by 2050 [8].

Treatment strategies for NDDs are inadequate. Limited benefits come from compensation for neuronal loss by increasing levels of corresponding neurotransmitters in the central nervous system (CNS), without directly slowing or halting neurodegeneration. Acetylcholinesterase (AChE) inhibitors raise acetylcholine levels in the cortex of AD patients, partially compensating for loss of cholinergic neurons [9]. L-DOPA increases dopamine levels in the brains of PD patients, temporarily compensating for loss of dopaminergic neurons [10]. Tetrabenazine reduces hyperkinetic movement disorders (chorea) in Huntington's disease (HD) patients through depletion of monoamines, and dopamine in particular, in presynaptic neurons [11]. Such symptomatic therapies offer temporary relief, but the ultimate NDD outcome does not change. Riluzole [12] and memantine [13] reduce basal levels of glutamate excitotoxicity in the CNS, marginally slowing the progression of amyotrophic lateral sclerosis (ALS) and AD, respectively. The former adds a few months to the expected lifespan of ALS patients [14], while the latter has small effects on the rate of cognitive decline in moderate to severe AD patients [15]. Today, there is no effective disease-modifying treatment for any NDD, so a question must be asked: Is it good that our life expectancy steadily increases, if existing treatments for NDD/aging diseases only treat the symptoms, rather than addressing the cause and eradicating it, or at least halting disease progression?

We often hear that pharmaceutical research and development (R&D) dealing with CNS, and in particular with NDDs, is extremely risky and expensive. There's no question about that, but other costs should also be considered to fully evaluate the financials of NDDs. ADI estimated that for 2010 the global cost of neurodegeneration, including medical costs and cost of formal (e.g., nursing homes and skilled nurses) and informal (e.g., relatives) care, exceeded $600 billion (about 1% of world gross domestic product), with disproportionately high costs in wealthy countries [16]. The cost of providing care for AD patients in the US was ≈$200 billion per year in 2012, projected to grow to $1.1 trillion per year by 2050 [17]. A recent estimation [6] set to ≈€130 billion the yearly cost of providing care for demented people across Europe. A detailed analysis [18] considered the global cost of so-called "diseases of the brain" in Europe at €798 billion in 2010 (37% direct healthcare costs, 23% direct non-medical costs, 40% indirect costs associated with patients' production losses). The cost includes dementia (€105.2 billion) and PD (€13.9 billion). As to PD, an estimation dated 2007 sets the total cost in the US at $10.78 billion per year [19]. Are we really sure, then, that preclinical and clinical research is not affordable for mankind? Isn't the conservative figure of ≈$1 trillion—what we should spend in 2050 in global care for NDD patients—enough to stimulate public funding agencies and the public opinion to steadily invest in R&D?

I do not want to overstate the emotional motivation that each of us has when a loved one—my mother for me—is wasted by neurodegeneration: that should be *the* main motivation to target NDD treatments, and has been mine to choose such a challenging area for my efforts.

1.2 PROTEIN FOLDING: PHYSIOLOGICAL BENEFITS AND PATHOLOGICAL CONSEQUENCES

NDDs are a heterogenic set of diseases, and multiple therapeutic intervention strategies can be conceived. This book focuses on disease-modifying therapies, aiming to halt neurodegeneration or, better, to cause its remission—symptomatic treatments [20,21] are not covered. This book focuses on interfering with the development and/or the progression of NDDs with small molecules—immunotherapy-based approaches [22,23], although relevant, are not covered either.

Disease-modifying pathways, which should prove beneficial in the treatment of several NDDs, include oxidative [24] and nitrosative [25] stress, endoplasmic reticulum (ER) stress [26], mitochondrial injuries [27], impaired protein degradation [28], chaperone malfunctioning [29], inflammatory responses [30], and heavy metal accumulation in the brain [31]. This book focuses on a mechanism shared by most of the ≈600 characterized NDDs that overlaps with, influences, and is influenced by most of the mentioned disease-modifying pathways: *the aggregation and precipitation of misfolded amyloidogenic proteins*. The resulting insoluble polymeric protein aggregates accumulate in the cytosolic and/or in the nuclear space of affected brain cells, or in the extracellular CNS space, in a NDD- and protein-specific manner [32,33].

Unfolded proteins are synthesized by the ribosome, and require proper folding to assume unique three-dimensional structures to act, *inter alia*, as enzymes, membrane receptors, and molecular scaffolds [34]. Folding takes advantage of components of the protein quality control (PQC) machinery, such as chaperones, and proceeds in parallel with protein synthesis [35]. The high protein concentration ($\approx$300 mg/mL) in cells [36] could cause protein aggregation due to aspecific interactions between unfolded or partially folded proteins. The PQC network prevents aspecific interactions and ensures an efficient protein folding in physiological conditions [37].

The energy stabilization of properly folded *vs.* unfolded proteins does not exceed a few kCal/mol, and depends on the so-called hydrophobic effect [38]. The hydrophobic effect initiates protein folding by packing its hydrophobic protein core, and directing water molecules to the higher entropy liquid phase of water. Many proteins, reacting to external stimuli, dynamically oscillate between folded and unfolded conformations to perform different functions. Conformer switching may expose hydrophobic side chains, decreasing solvation and increasing the risk for aggregation unless specialized chaperones bind and protect such side chains [39].

Ribosomal protein synthesis is an error-prone process that even in physiological conditions, in addition to the large population of correctly translated proteins (**P1**, Figure 1.1), creates a small population of incorrectly translated proteins (**p2**, Figure 1.1) [40]. Non-native and aggregation-prone states are accessible to incorrectly translated proteins, and even to unfolded native proteins. Thus, PQC-driven troubleshooting acts under *basal/physiological* conditions through a process depicted in Figure 1.1.

Kinetic competition between single copy-folding leading to functional proteins (**a**, Figure 1.1), single copy-misfolding leading to non-functional protein copies (**b**, Figure 1.1), aggregation leading to soluble oligomeric complexes (**c**, Figure 1.1), and to insoluble aggregates (**d**, Figure 1.1) is strongly biased towards functional proteins/**a**. The vast majority of the large **P1** population folds correctly (**a**, Figure 1.1), with a small percentage of protein copies experiencing kinetically disfavored partial misfolding (**b**, Figure 1.1). Their refolding is assisted by holding and folding–unfolding chaperones [29]. The former family of holdases-small heat shock proteins (sHsps) holds partially folded or misfolded protein copies. sHsps switch from an oligomeric state to smaller subunits that expose hydrophobic sequences [41]. These hydrophobic sHsp sequences bind similar sequences from partially misfolded proteins, and block their misfolding/aggregation tendency during the folding process [42]. Proteins "on hold" are then rescued by ATP-dependent folding–unfolding chaperones [43] that work by preferentially binding misfolded protein substrates [44]. Once bound to a misfolded protein copy, an unfoldase uses energy from ATP binding and/or hydrolysis to unfold misfolded proteins (**b**[1], Figure 1.1) [45]. The unfolded intermediates are now ready for proper refolding (**a**, Figure 1.1). Very few unfolded or misfolded protein copies escape respectively proper folding and chaperone-assisted refolding, and aggregate into soluble oligomeric complexes (**c**, Figure 1.1) and insoluble

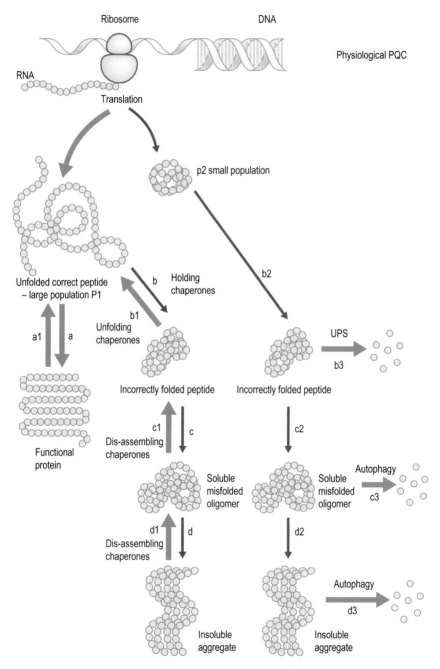

FIGURE 1.1   Tau and protein quality control: basal/physiological conditions.

aggregates (d, Figure 1.1). Then, chaperones with disaggregating activity [46] drive their disassembly/unfolding (c^1 and d^1, Figure 1.1) and lead them once more towards proper refolding (a, Figure 1.1) [45]. Dynamic unfolding–refolding (a^1–a, Figure 1.1) is needed by functional proteins to reach specific cellular compartments or to perform specific functions, but the physiological abundance of properly folded protein copies is assured by the PQC machinery [47]. The small $p2$ population of incorrectly translated proteins is intrinsically dysfunctional, and is disposed of either as soluble misfolded copies (from b^2, Figure 1.1) *via* the ubiquitin–proteasome system (UPS) (b^3, Figure 1.1) [48], or as soluble oligomers (from c^2, Figure 1.1) and/or insoluble aggregates (from d^2, Figure 1.1) *via* autophagy (respectively c^3 and d^3, Figure 1.1) [49]. The former mechanism entails soluble misfolded protein labeling with ubiquitin (UBQ), a 76-mer protein [50], followed by recognition and degradation of UBQ-protein copies by the proteasome, a multi-subunit proteolytic complex [51]. The latter mechanism degrades insoluble protein aggregates specifically (aggrephagy [52]), or aspecifically together with other cellular components (macroautophagy [53]). Insoluble aggregates are enclosed in degradative autophagic vacuoles (dAVs) [54], where they are degraded by lysosomal proteases at strong acidic pH [55].

Under *pathological* conditions, inherited toxic mutations, and/or a decrease in efficiency for the PQC machinery, lead to the translation of increasingly large populations of folding-deficient proteins ($p2$ to $P2$, Figure 1.2), with an increasingly small population of correctly folded/functional proteins ($P1$ to $p1$, Figure 1.2).

During age-dependent loss of efficiency for the PQC machinery, single protein copy-folding leading to functional proteins (a, Figure 1.2) slowly decreases. Single protein copy-misfolding leading to non-functional protein copies (b, Figure 1.2) increases with similar speed [56]. The dynamic folding–unfolding equilibrium (a^1 and b^1, Figure 1.2) contributes to the increase of $P2$/decrease of $p1$ populations. Once misfolded proteins exceed the capacity of holding and folding–unfolding chaperones, and of the UPS system (b^2, Figure 1.2) [57], soluble misfolded proteins aggregate first into soluble oligomeric complexes (c, Figure 1.2), then into insoluble aggregates (d, Figure 1.2). Once soluble oligomers and insoluble aggregates exceed the capacity of autophagy (respectively c^2 and d^2, Figure 1.2) [58], the proteostasis equilibrium inexorably shifts towards the accumulation of insoluble protein aggregates. The process takes years to transition from a benign/symptomless phase to an overt pathological/proteopathic phase in NDD/aging diseases [29]. The proteostasis equilibrium is more rapidly shifted towards pathological aggregation when inherited genetic mutations cause the translation of a large $P2$ population of folding-deficient proteins [59]. Their folding deficiency prevents chaperone-driven protein refolding, but chaperones still bind the mutated protein copies.

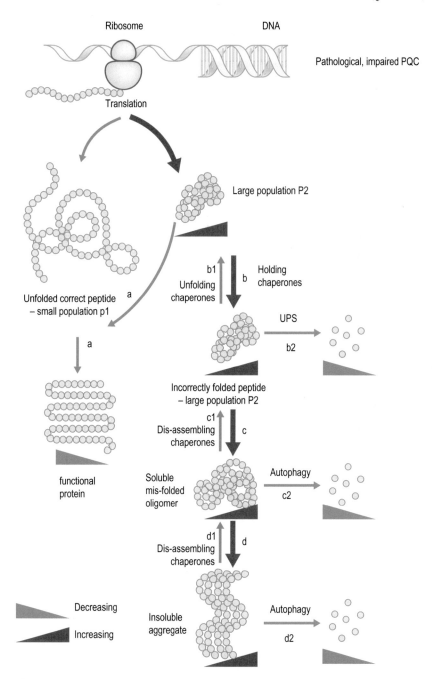

FIGURE 1.2 Tau and protein quality control: pathological conditions.

Thus, the overall capacity of the PQC machinery is rapidly affected [60]. Accumulation of misfolded protein copies, of soluble oligomers, and insoluble aggregates (respectively **b**, **c** and **d**, Figure 1.2) happens faster, while the beneficial activity of the PQC machinery (respectively b^1, c^1 and d^1, Figure 1.2) is rapidly impaired, and the capacity of UPS and autophagy (respectively b^2, c^2 and d^2, Figure 1.2) is fast exceeded [61].

Aging-dependent processes negatively influence the efficiency of cellular PQC, and increase the risk of protein aggregation [62,63]. Stress-responsive pathways are activated when the PQC capacity of cells and tissues is impaired [64,65], but even stress-induced PQC may not be able to rescue advanced proteinopathies. Pathological effects of systemically expressed amyloidogenic proteins are often prevented by the cellular turnover in tissues and organs. Conversely, post-mitotic tissues, and neuronal tissues in particular, cannot regenerate [66]. Thus, proteinopathies caused by their aggregation and accumulation are often restricted to CNS.

Soluble neuronal proteins are slowly converted into insoluble, filamentous amyloidogenic polymers with crossed-β-pleated sheet structures as depicted in Figure 1.2. Aggregates accumulate in a disease-, CNS compartment-, cellular compartment-, and protein-specific manner. NDDs develop over the lifetime of an individual, but usually become symptomatic late in life—a sign of their slow progression (Figure 1.2), and of the high damage tolerance of brain regions and functions. The earlier age of disease onset in the mutated protein/familial NDD scenario, compared to its wild-type protein/sporadic NDD counterpart, depends on their accelerated progression rate (Figure 1.2).

Aggregation-prone neuronal proteins are the core of NDDs. The same protein aggregate may determine the insurgence of several NDDs. Conversely, a single NDD may entail the simultaneous presence of more than one protein aggregate. A thorough description of disease-modifying approaches targeted against >600 known NDDs, and consequently focused onto the physiopathological features of a large number of aggregation-prone neuronal proteins, would largely exceed the length of any book. The following chapters provide a brief survey on the relevance for each selected target/mechanism, and on the activity for each described compound, on most common NDDs and aggregation-prone neuronal proteins. Selected targets/mechanisms are chosen for their therapeutic relevance in a subclass of NDDs caused by a specific neuronal protein. This protein is preferentially used as an example to study the protein aggregation–NDDs scenario throughout this book.

Figure 1.3 shows the relationships between major proteinopathies (large colored circles) and NDDs (small circles). It also highlights the main protein aggregate for each NDD (color coded; only AD has a hybrid color, due to equal relevance of two protein aggregates), and the presence of secondary protein aggregates (connecting lines).

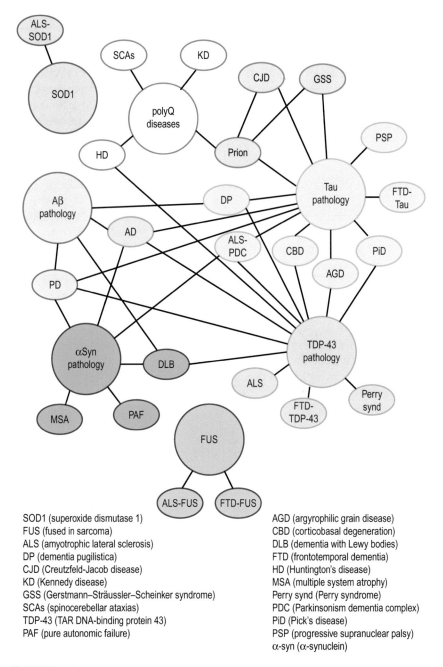

SOD1 (superoxide dismutase 1)	AGD (argyrophilic grain disease)
FUS (fused in sarcoma)	CBD (corticobasal degeneration)
ALS (amyotrophic lateral sclerosis)	DLB (dementia with Lewy bodies)
DP (dementia pugilistica)	FTD (frontotemporal dementia)
CJD (Creutzfeld-Jacob disease)	HD (Huntington's disease)
KD (Kennedy disease)	MSA (multiple system atrophy)
GSS (Gerstmann–Sträussler–Scheinker syndrome)	Perry synd (Perry syndrome)
SCAs (spinocerebellar ataxias)	PDC (Parkinsonism dementia complex)
TDP-43 (TAR DNA-binding protein 43)	PiD (Pick's disease)
PAF (pure autonomic failure)	PSP (progressive supranuclear palsy)
	α-syn (α-synuclein)

FIGURE 1.3   The neurodegenerative proteinopathy network: main players and disease connectivity.

Extracellular senile plaques (SPs) in the AD brain consist of β-*amyloid* (*Aβ*) [67], a family of amyloidogenic peptides resulting from the cleavage by β- and γ-secretase of the amyloid precursor protein (APP) [68]. Amyloid plaques are observed in familial and sporadic AD. Prevention of Aβ formation and of its aggregation, reduction of Aβ neurotoxicity, and degradation of Aβ species are hotly pursued as AD treatments [69,70]. Intracytoplasmic protein inclusions in familial and sporadic PD, in dementia with Lewy bodies (DLB), and in multiple system atrophy (MSA) contain α-*synuclein* [71]. α-Synuclein is a small protein found predominantly in neuronal tissue, which becomes aggregation-prone either when mutated (familial PD) [72], or after post-translational modifications (PTMs) (sporadic PD, DLB, MSA) such as phosphorylation [73], oxidative modification [74], and proteolytic cleavage [75]. Intraneuronal protein inclusions in nine polyglutamine repeat (polyQ) diseases [76], such as HD, contain polyQ-containing proteins such as *polyQ-huntingtin* [77]. PolyQ sequences between 10 and 36 residues increase the fibrillization tendency of mutated proteins [33]. Therapeutically relevant aggregation-prone proteins/ NDD couples also include *superoxide dismutase 1 (SOD1* [78])/ALS; *TAR DNA-binding protein 43 (TDP-43* [79])/ALS; *fused in sarcoma (FUS* [80])/ ALS; and the *prion protein (PrP* [81])/prion disease. Any of them could be representative enough of aggregation-prone neuronal proteins, as it causes clinically relevant NDDs. The same is true for *tau* and *tauopathies*, respectively the protein and the NDDs (including AD) chosen as a focus for this book.

1.3 TAU: AN INTRINSICALLY DISORDERED, FLEXIBLE, AND AGGREGATION-PRONE PROTEIN

Tau is a highly soluble microtubule-associated protein (MAP) discovered in 1975 [82] that promotes microtubule (MT) assembly. Tau is mostly expressed in neurons in general, and axons in particular, ensuring their structural integrity [83]. The almost total absence of secondary and tertiary structural elements in tau makes it an intrinsically disordered protein (IDP) [29]. IDPs exist as dynamic ensembles, rather than unique 3D structures, and are highly abundant in nature [84]. The conformational flexibility of IDPs allows their folding through adaptation to varying cellular environments (e.g., interaction with other proteins, nucleic acids, and membranes) [85]. Target-induced rearrangement for a given IDP may vary between substantially disordered and tightly folded states [86]. The switch of IDPs between folding states dynamically regulates their interaction with multiple partners through high-specificity/low-affinity interactions, and modulates many cellular processes and signaling pathways [87].

Chaperones and aggregation-prone neuronal proteins often are IDPs [29]. They physiologically adapt their conformation to external stimuli (e.g., binding to an exposed hydrophobic protein sequence for a hold-ase chaperone, or MT binding for tau), but may pathologically start the aggregation/proteinopathy process due to their susceptibility to NDD-specific neurotoxic events (e.g., abnormal PTMs, radical, and oxidative insults). As to tau, a dynamic MT–tau interaction network is established through a set of low-energy interconverting tau structures in solution [88]. Tau–MT binding–unbinding events control the stabilization or destabilization of MT segments to regulate neuritic growth and promote axonal transport [89].

The single copy human tau gene *MAPT*, composed by 16 exons, is located on chromosome 17 [90]. Alternative splicing produces up to 30 tau isoforms [91], six of which are expressed in CNS (Figure 1.4). The longer human brain tau isoform (2N4R, 441 amino acids—AAs) contains a basic C-terminal domain (AAs 244–441), including four MT-binding repeats (MTBRs) that modulate MT–tau interactions [92]. A basic middle domain (AAs 151–243), containing two proline-rich regions (PR), contributes to MT binding and binds the multifunctional protein actin [93].

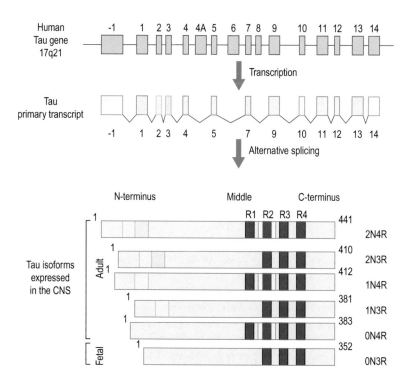

FIGURE 1.4   Structure of the *MAPT* gene, and of tau isoforms present in the CNS.

An N-terminal domain (AAs 1–150), containing two acidic inserts (MeI), interacts with plasma membrane [94] and the kinase Fyn [95].

Tau isoforms result from alternative splicing of exons 2, 3 (N-terminal domain, 29 AAs in each domain, isoforms 2N, 1N, and 0N), and 10 (C-terminal domain, 31 AAs, isoforms 4R and 3R). Six CNS isoforms are observed because exon 3 is expressed only in presence of exon 2 [96]. The shortest, 352 AA-containing tau (0N3R) is the only fetal tau isoform [97], while the adult tau pool in human brains encompasses all six isoforms (2N4R, 441 AAs; 1N4R, 412 AAs; 2N3R, 410 AAs; 0N4R, 383 AAs; 1N3R, 381AAs; 0N3R, 352 AAs).

Mutations in *MAPT*, resulting in familial tauopathies, are known. Missense mutations cause amino acid variations in tau, while silent mutations switch the physiological 4R:3R isoform ratio [98]. Alternative splicing of *MAPT* is regulating tau functions and its stability as a monomer. Exon 10 contains an MTBR, so that 4R isoforms show stronger binding to MTs than 3R isoforms. At embryonic stage, the shortest 3R isoform weakens MT–tau interactions and allows the growth of immature neurons [97]. Adult tau shows a ≈1:1 4R:3R ratio, a compromise between strong MT cohesion to secure neuronal integrity, and morphological plasticity needed by dynamic MT–tau complexes. An abnormal 4R:3R ratio in adult tau pools is invariably associated with tauopathies [99]. Isoforms 2N, 1N, and 0N influence axonal membrane [100] and dynactin binding [101]. Their relative abundance is connected with physiological and pathological events in neuronal functions [102]. Each tau isoform behaves and localizes differently in developmental and adult neuronal subpopulations [103]. They may bind MTs and/or other proteins at different sites, causing a shortage of available binding sites and increase in soluble, aggregation-prone tau in case of any imbalance of tau isoform ratio [104].

The basic-polar nature of tau supports its interaction with acidic MTs [105], and favors *PTMs* on tau [106,107]. PTM patterns, and phosphorylation in particular, heavily influence the conformational stability, the interaction network, and the physicochemical properties (including aggregation propensity) of tau [106].

Phosphorylation of tau has a strong impact on its functions [108]. Out of 85 Ser, Thr, and Tyr residues, more than 30 are phosphorylated in non-diseased brains, around 15 in both physiological and pathological conditions, and almost 30 are phosphorylated only in AD brains [106]. A dynamic phosphorylation balance is kept by the interplay between tau kinases and phosphatases [109]. Tau varies its phosphorylation state depending on its localization [110] and on developmental stage, as fetal human tau is more phosphorylated than adult tau [111]. Hyperphosphorylation (HP) in adult tau is a marker for tau aggregation, and a risk factor for tauopathies [112]. HP fetal tau is highly soluble and perfectly functional [97]. The HP pattern in human brain tissues from AD patients is different from human fetal tau. Namely, residues S202, T212, S214, T217, T231, S262, S396, S404, and S422 are linked to adult HP [113] as early, intermediate, or late-stage tauopathy-specific hyper-phosphorylated epitopes [114–118].

Figure 1.5 depicts the impact of phosphorylation and of other PTMs on the dynamic equilibrium between native tau (folded, soluble species) and HP-tau (aggregation-prone species). Pro-aggregation PTMs acting on HP-tau and promoting the formation of insoluble tau aggregates (neurofibrillary tangles, NFTs [119]) are also shown.

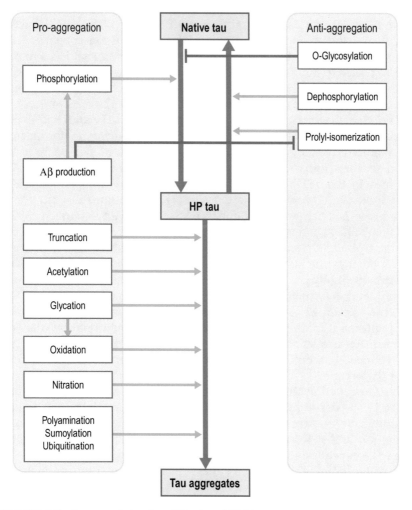

FIGURE 1.5   Post-translational modifications (PTMs) on tau: impact on tau aggregation.

O-glycosylation with N-acetylglucosamine (*O-GlcNAcylation*) is essential for the regulation of proteins in physiological and pathological states [120], including their phosphorylation [121]. O-GlcNAcylation negatively regulates tau phosphorylation [122], prevents tau aggregation and the risk for tauopathies. The O-GlcNAcylation-dependent inaccessibility of some phosphorylation sites for tau kinases may be due to direct enzyme competition on the same sites, or to O-GlcNAcylation-induced conformational changes of tau [121]. A decreased level of tau O-GlcNAcylation is observed in brains from AD patients, as is correlation between tau hypo-GlcNAcylation and HP-tau [123].

Cis and *trans* conformations of the X-Pro amide bond strongly influence the folding of Pro-rich protein regions [124]. *Peptidyl prolyl isomerization* [125] of one or more X-Pro amides in a protein may influence the accessibility of a domain to PTM enzymes, with functional consequences [126]. The middle region of tau contains two Pro-rich domains and 21 Pro residues. The X-Pro amide bond isomerization on the pT231–P232 sequence shows beneficial effects on tau PTM patterns, and decreases its aggregation tendency and the risk of tauopathies [127].

Tau acetylation reduces the affinity of tau for MTs and has a pro-aggregation effect [128]. It is observed in most patients suffering from tauopathies [128,129], while acetylated tau is not observed in primary cultured neurons [130]. Acetylation mostly takes place on Lys274 and Lys280 residues in the MTBR [130]. The latter epitope is strongly associated with sporadic AD and other tauopathies [131]. Tau itself has tau acetyltransferase activity, which is enhanced by phosphorylation on Ser262 and Ser356 [132] and may further drive tau towards aggregation.

The *proteolytic cleavage of tau* [133] is promoted by its IDP nature, which ensures the access of large tau regions to brain proteases [134]. Tau processing by proteases may control the cellular abundance of soluble tau, and is influenced by conformational changes imposed by PTM patterns. Epitope-specific phosphorylation promotes tau processing [135], although inhibition of tau cleavage by phosphorylation of a specific residue is also observed [136]. Physical shielding of tau epitopes, and reduced susceptibility to protease cleavage, may also result from PTM patterns.

Tau is cleaved *in vitro* by trypsin, chymotrypsin, and endogenous proteases [137]. Tissue samples from NDD patients contain tau fragments generated by caspases [138] and calpains [139]. In particular, caspases cleave tau at Asp421 in tissues from AD patients [140], producing the aggregation-prone and neurotoxic tau ΔC fragment [141].

Tau glycation [142], nitration [143], polyamination [144], sumoylation [145], and oxidation [146] have effects on tau functions, misfolding and aggregation, and may in future be fully validated as disease-modifying therapeutic options against tauopathies [106].

1.4 TAUOPATHIES: AGGREGATION-PRONE TAU IN NEURODEGENERATIVE DISEASE (NDD)

Each NDD showing intracellular accumulation of filamentous tau inclusions (paired helical filaments (PHF) [147], straight filaments [148], or random filaments [149]) into NFTs, and disease-dependent brain dysfunctions, is defined as a *tauopathy* [150,151]. The identification of multiple *tau* mutations in frontotemporal dementia and parkinsonism linked to chromosome 17 (FTDP-17) [152,153] proves that dysfunctions in tau cause NDDs without the need for any other neuropathological factor.

At least 27 tauopathies (Table 1.1) are known, including largely diffuse and extremely rare NDDs. Tau aggregates are either the only neuronal abnormality, or one of the main disease factors, or secondary inclusions associated with other main pathologies. Tauopathies are divided into familial and sporadic NDDs, and into five classes depending on the neuropathological observation of tauopathy-specific tau aggregates (tau isoforms, phosphorylation levels) corresponding to all or to some of the six CNS tau isoforms [154].

1.4.1 Class 0 Tauopathies

Class 0 tauopathies are characterized by a lack of neuropathological hallmarks of tau [154]. *Frontal lobe degeneration, non-AD, non-Pick* [155] is the only tauopathy in this class. It shows reduced tau expression without any tau aggregate, and neurodegeneration in the frontal area (frontal and temporal cortex). It causes speech and behavior disturbances, and it eventually leads to dementia [156].

1.4.2 Class I Tauopathies

Class I tauopathies are defined by the presence in insoluble HP-tau aggregates of three major tau bands at 60, 64, and 69 kDa, and of a less intense band at 72–74 kDa [154]. The 60-kDa band corresponds to the 0N3R isoform; the 64-kDa band corresponds to the 1N3R and 0N4R isoforms; the 69-kDa band corresponds to the 2N3R and 1N4R isoforms; and the 72–74-kDa band corresponds to the longest 2N4R tau isoform (usually present in low amounts in the adult human brain). Both 3R and 4R tau isoforms are found in HP insoluble tau aggregates, and in the soluble tau fraction [151].

AD is by far the most prevalent class I tauopathy, and the most prevalent NDD [157,158]. It is characterized by extracellular SPs made by β-amyloid deposits, and by intraneuronal NFTs made by insoluble HP-tau aggregates. SPs are spread in the cerebral cortex and in subcortical structures [159]. Their concentration increases with disease progression,

TABLE 1.1   List of Characterized Tauopathies

Tauopathy	Inclusions	Origin[d]	Class
AD	Main/Aβ	S, F	I
Amyotrophic lateral sclerosis/parkinsonism-dementia complex of Guam (ALS/PDC)	Main/TDP-43	S	I
Argyrophilic grain disease	Tau only	S	II
Corticobasal degeneration	Tau only	S, F	II
Dementia pugilistica/Chronic traumatic encephalopathy	Main/TDP-43, Aβ	S	I
Diffuse neurofibrillary tangles with calcification	Main/TDP-43, α-syn	S	I
Down's syndrome	Main/Aβ[a]	S	I
Familial British dementia	Main/Aβ-like, TDP-43	F[e]	I
Familial Danish dementia	Main/Aβ-like	F[e]	I
Frontal lobe dementia, non-AD, non-Pick	Lack of tau	S	0
Frontotemporal dementia and parkinsonism linked to chromosome 17 caused by MAPT mutations	Main/TDP-43, FUS[b]	F	I, II, III
Frontotemporal lobar degeneration caused by C9ORF72 mutations	Support[c]/TDP-43, p62	F[e]	I
Gerstmann–Sträussler–Scheinker disease	Support[c]/prion	S	I
Globular glial tauopathies/White matter tauopathy with globular glial inclusions	Tau only	S	II
Guadeloupean parkinsonism with dementia	Tau only	S	I
Guadeloupean PSP	Tau only	S	II
Multiple system atrophy	Support[c]/α-syn	S	I
Myotonic dystrophy	Tau only[a]	F[e]	IV
Neurodegeneration with brain iron accumulation/Hallevorden–Spatz disease/Pantothenate kinase-associated neurodegeneration	Support[c]/TDP-43, α-syn	F[e]	I
Neurofibrillary tangle-predominant dementia	Tau only	S	I
Niemann–Pick disease, type C	Tau only[a]	F[e]	I
Pick's disease	Tau only	S	III
Postencephalitic parkinsonism	Main/Aβ	S	I
Prion protein cerebral amyloid angiopathy	Support[c]/prion	F[e]	I
Progressive supranuclear palsy	Tau only	S	II
SLC9A6-related mental retardation	Tau only[a]	S	II
Subacute sclerosing panencephalitis	Tau only[a]	S	I

[a]The main disease cause is not related to the proteinopathy/tauopathy.
[b]Each pathology is typical of FTDP-17 subclasses, no overlapping proteinopathies.
[c]Tau pathology has lesser importance than other proteinopathies.
[d]S = sporadic, F = familial.
[e]Other proteins than tau are mutated.

although they can be found in aged non-demented controls. NFTs initially appear in the hippocampus and spread to the anterior, inferior, and mid-temporal cortex (early asymptomatic AD, non-demented aged controls [160]). They then expand to other cortex areas (mid-stage AD, first clinical manifestations [161]), and finally to primary motor and sensory areas (late-stage AD and demented patients). The progressive neurodegenerative pattern has been thoroughly described [162]. $A\beta$/SP and tau/NFT pathologies are clearly connected, and the former may be a trigger/potentiation factor for the latter in AD [154]. Familial AD cases are a small minority, and show mutations in genes encoding for $A\beta$ pathway-involved APP and presenilins (transmembrane proteins that are part of the γ-secretase intramembrane protease complex) [163–165]. Sporadic AD is by far predominant, and some genetic risk factors are known (the apolipoprotein 4 (APOE-4) allele [166,167], the triggering receptor expressed on myeloid cells 2 (TREM2) protein [168,169]). Mutations of tau and tau association with genetic risk factors in AD have not been observed.

Neurofibrillary tangle-predominant dementia (NFTPD) [170] shows NFT/tau inclusions resembling AD-type NFTs at a moderate stage of AD development, while SPs are almost absent—hence the disease name. NFTPD is a milder version of AD in terms of cognitive impairment and duration [171]. Tau tangles contain 4R and 3R tau isoforms, as in AD, and morphologically match their structure. A larger number of extracellular/ghost tangles is observed in NFTPD with respect to AD [172]. *PD* is not considered a tauopathy, although dementia and tau aggregation are sometimes observed in PD patients [173]. Atypical parkinsonism disorders (APDs), though, are characterized by dementia and tau abnormalities [174]. Among them, *amyotrophic lateral sclerosis/parkinsonism-dementia complex of Guam (ALS/PDC)* [175] is an extremely rare tauopathy/APD clinically undistinguishable from sporadic ALS. ALS/PDC is triggered by environmental factors present on the island of Guam. NFTs from ALS/PDC and AD patients are similar, but the former are found both in cortical and subcortical areas [176]. TDP-43-positive, tau-negative aggregates are found in the frontotemporal and hippocampal regions of ALS/PDC cases [177].

Frontotemporal lobar degeneration caused by C9ORF72 mutations (FTLD-C9) [178] is an APD caused by an expanded GGGGCC repeat insertion in a non-coding promoter region of open reading frame 72 (C9ORF72) in chromosome 9. FTLD-C9 is described as a TDP-43- and p62-driven proteinopathy, characterized by anxiety and agitation, memory impairment and motor neuron disease [179]. Tau pathology in limbic regions is also observed in FTLD-C9 [180], defining its tauopathy/APD nature.

Guadeloupean Parkinson-dementia complex (Gd-PDC) [181] is one of two frequent APDs on Guadeloupean islands, likely due to the consumption of local plants containing a neurotoxic combination of natural products [182]. Gd-PDC resembles other PD-dementia diseases, but dementia intervenes earlier than in PD patients; cognitive symptoms start with memory loss, rather than behavioral changes as in FTDP-17; and fluctuations in cognition, attention, and alertness that are characteristic of DLB are not observed in Gd-PDC [183].

Multiple system atrophy (MSA) [184] is an APD showing autonomic dysfunctions, parkinsonism, and ataxia in various proportions. MSA shows glial inclusion bodies, mostly constituted by α-synuclein [185]. Non-HP-tau is present in the same bodies [186], while granular accumulation of HP 4R and 3R tau isoforms is observed in α-synuclein-negative glial granules in MSA [187].

Postencephalitic parkinsonism (PEP) [188] is a chronic APD complication of *encephalitis lethargica*. PEP leads to progressive deterioration of motor and autonomic functions [189]. It shows scarce Aβ pathology, and completely lacks Lewy bodies and α-synuclein inclusions when compared to PD [190]. NFTs-containing tau triplets are abundant in cortical and subcortical regions, including AD-spared motor cortex and basal ganglia [191].

Subacute sclerosing panencephalitis (SSPE) [192] is a rare tauopathy affecting children even years after an infection with measles. SSPE starts with mental and behavioral changes, and progressively leads to neurovegetative state and death in 1 to 10 years [193]. Tau NFTs appear in SSPE patients with longer survival periods [194], although their relevance in the development of the pathology is questionable (probably they are a secondary effect of disease-relevant damage of neuronal tissues [193]).

Diffuse neurofibrillary tangles with calcification (DNTC) [195] is an extremely rare tauopathy observed mostly in Japan, which entails the presence of tau, α-synuclein, and TDP-43 aggregates with partial co-localization in the neocortex and limbic system [193]. DNTC patients show overlapping AD and frontotemporal lobar dementia (FTLD) symptoms [196], calcification, and parallel accumulation of Pb in calcified brain tissues [197].

Dementia pugilistica/Chronic traumatic encephalopathy (CTE) [198] is a long-term NDD initiated by repeated concussions that lead to cognitive, psychiatric, and behavioral disturbances observed among boxers and American football players. AD-like NFTs are abundantly found in CTE, but their distribution is more irregular due to the varying nature of physical traumas suffered by CTE patients [199]. Extensive SPs made by Aβ deposits [200], and TDP-43 accumulation [201], are also observed in CTE.

Down syndrome (DS) [202] shows cognitive impairment and somatic dysfunctions due to the trisomy of chromosome 21, leading to dementia around 50 years of age. DS patients show SPs/Aβ (starting at 15 years of age, increasing with age) and NFTs/tau (starting at 35 years of age, expanding NFT-affected brain areas with age [203]) similarly to AD cases [204].

Familial British dementia (FBD) [205] is a rare disease caused by a point mutation in the integral membrane protein 2B (*ITM2B*, *BRI2*) gene. FBD is characterized by progressive dementia and ataxia, amyloid plaque deposition and neurofibrillary degeneration [206]. AD-type NFTs are observed together with amyloid plaques made of ABri, a 22-mer peptide from abnormal cleavage of the BRI2 protein [207]. TDP-43 inclusions are co-localized with NFTs [208].

Another point mutation in the *BRI2* gene causes *familial Danish dementia (FDD)* [209]. Its clinical phenotype is similar to FBD, where amyloid plaques are made by ADan, a 34-mer neurotoxic peptide [210]. A transgenic (TG) mouse model of FDD shows enhanced tauopathy, tau truncation at the Asp421 residue, and synaptic loss prior to ADan deposition in the brain of TG mice. A role for soluble ADan peptides in the induction/acceleration of tauopathies can be inferred [211].

Gerstmann–Sträussler–Scheinker disease (GSS) is an extremely rare genetic transmissible spongiform encephalopathy [212]. Difficulty in speaking, progressive ataxia and dementia are among its symptoms [213]. Prion plaques are the dominant biochemical anomaly, and determine the clinical outcome of GSS [213]. AD-like NFTs containing class I-specific tau triplets are observed in some GSS patients [214].

Another rare tauopathy with dominant prion protein pathology is *prion protein cerebral amyloid angiopathy (PrPCAA)* [215]. PrPCAA is caused by a single point mutation on the gene coding for the prion protein [215]. It progressively leads to diffuse atrophy of the cerebrum, and to severe dementia [216]. PrPCAA is characterized by amyloid deposits of prion protein in cerebral vessels, and AD-like NFTs in the brain (especially in the hippocampus) [216].

Several NDDs share the accumulation of iron in the basal ganglia of the brain, and are collectively known as *neurodegeneration with brain iron accumulation (NBIA)* [217]. Among them, pantothenate kinase-associated neurodegeneration (PKAN, also known as Hallevorder-Spatz disease) is due to mutations in pantotenate kinase 2. α-Synuclein, tau, and TDP-43 pathologies are observed in PKAN patients [218], although some question the connection between tau and PKAN [219]. Tau NFTs are observed also in closely related PLA2G6-associated neurodegeneration (PLAN) and mitochondrial membrane protein-associated neurodegeneration (MPAN) [217].

Niemann–Pick type C disease (NPC) [220] is a rare lysosome storage disorder caused by mutations in Niemann–Pick C1 and C2, two proteins that transport cholesterol to various organelles in the cell [221]. Symptoms include clumsiness, ataxia, vertical supranuclear gaze palsy, seizures, and psychomotor retardation [222]. Abundant AD-like NFTs are observed in various brain areas of NPC patients [223].

1.4.3 Class II Tauopathies

Class II tauopathies are defined by the presence of two major tau bands at 64 and 69 kDa, and of a less intense band at 72–74 kDa in insoluble

HP-tau aggregates [154]. This profile corresponds to aggregates composed in large part by 4R tau isoforms, as class II tau aggregates are heavily stained by exon 10-specific antibodies. The soluble tau fraction contains all six CNS tau isoforms, and privileged aggregation of predominant 4R tau isoforms must be inferred [151].

SLC9A6-related mental retardation [224] is a familial disease, entailing ataxia, epilepsy, and autism. It is associated with neuronal and glial tau inclusions in cortical and subcortical regions. Tau inclusions contain both 3R and 4R isoforms, but the latter is more abundant than the former—hence the disease is a class II tauopathy. While the mutated sodium/proton exchanger SLC9A6 characterizes the disease, tau pathology contributes to the progressive neurodegeneration observed in its patients [224].

Argyrophilic grain disease (AGD) [225] is characterized by argyrophilic grains, neuronal inclusions other than PHFs or NFTs. The grains are mostly located in the hippocampus and other limbic areas, and contain predominantly 4R tau isoforms [226]. AGD is a tau-only dementia causing behavioral disturbances, memory and cognitive impairments [227]. AGD is the only common tauopathy lacking aggregation-prone tau acetylation [228].

Corticobasal degeneration (CBD) [229] is a late-onset, slowly progressive tau-only tauopathy leading to moderate dementia. Glial astrocytic plaques and tau inclusions are found in the white matter, while ballooned neurons are found in cortical and subcortical brain regions [230]. Immunohistochemistry shows only exon 10-expressing tau isoforms as NFT components in CBD [231]. Although CBD is a sporadic disease, tau polymorphism is a genetic risk factor [232]. A single tau mutation in exon 13 (leading to an increase in 4R isoform-selective alternative splicing of the MAPT gene) is observed in a CBD patient [233].

Global glial tauopathies (GGT) [234] are a recently characterized subset of diseases showing motor neuron disease and frontotemporal dementia symptoms. The heterogeneous clinical features stem from a common neuropathology, consisting of globular glial inclusions and non-fibrillary tau pathology [235]. 4R tau isoforms mainly compose the inclusions, but an ≈35-kDa tau peptide and other lower-size fragments are observed [235,236].

Progressive supranuclear palsy (PSP) [237] is, together with MSA, the most common APD. PSP is characterized by postural instability and unexplained falls, by cognitive impairment and dementia [238]. 4R isoform-predominant NFTs are accompanied by PSP-specific tau inclusions in astrocytes (tufted astrocytes) and in oligodendroglia (coiled bodies) [239]. Abnormal tau inclusions (strong 64- and 69-kDa bands, and a minor 72-kDa band) are first observed in subcortical areas, then move to cortical areas and cause dementia [240]. Atypical, tau-based PSP variants include a single example of familial PSP [241].

Guadeloupean PSP-like syndrome (Gd-PSP) [181] is the second common APD on Guadeloupean islands, due to environmental factors and local food diets [182]. Gd-PSP shows hallucinations, a higher frequency of sleep disorders, and a varied panel of oculomotor abnormalities when compared to "classical" PSP [183].

1.4.4 Class III Tauopathies

Class III tauopathies are defined by the presence of two major tau bands at 60 and 64 kDa, and of a less intense band at 69 kDa in insoluble HP-tau aggregates [154]. This corresponds to aggregates composed in large part by 3R tau isoforms, as class II tau aggregates do not stain in the presence of exon 10-specific antibodies. The soluble tau fraction still contains the six tau isoforms, and privileged aggregation of predominant 3R tau isoforms must be inferred [151].

Pick disease (PiD) [242] is the only known class III tauopathy. PiD is a rare frontal lobe dementia entailing mood disturbances, language impoverishment, and pre-senile dementia [243]. PiD is characterized by round, argyrophilic intraneuronal tau inclusions (Pick bodies) made by random coiled and straight filaments of tau. 3R isoform-containing Pick bodies are found in cortical and subcortical brain regions [244]. Less abundant glial inclusions are mostly composed of 4R tau isoforms [245]. Tau inclusions from PiD patients do not stain in presence of antibodies specific for the pathological pSer262/pSer356 epitopes, which are phosphorylated in most other tauopathies [246].

1.4.5 Class IV Tauopathies

Class IV tauopathies are defined by the presence of a single major tau band at 60 kDa, and of two less intense bands at 64 and 69 kDa in insoluble HP-tau aggregates [154]. This profile corresponds to aggregates composed in large part by the smallest tau isoform 0N3R [151].

Type I myotonic dystrophy (DM1) [247] is the only class IV tauopathy. DM1 is a complex, multisystemic disease strongly affecting the CNS (cognitive and psychiatric impairments). DM1 is a genetic disease caused by the expansion of a CTG trinucleotide motif in the myotonic dystrophy protein kinase gene (*dmpk*), located on chromosome 19 [248]. The 60-kDa tau band is exclusively observed in the entorhinal cortex and the temporal pole of DM1 patients. Both the 60- and 64-kDa bands are present in the hippocampus and the temporal amygdala. NFTs from the amygdala of DM1 patients do not stain in the presence of exon 2- and exon 10-specific antibodies [249]. The exon 2/3-exclusion mechanism leading to the 0N3R tau isoform in the alternative splicing of tau in DM1 depends on the *dmpk* mutation [250] and differs from the generation of the same tau isoform during tau development. A recent study shows that the reduced inclusion of exon 10-containing tau isoforms is not common to all DM1 patients [249].

1.4.6 Tau Mutations

At least 57 tau mutations are identified on chromosome 17 in patients affected by NDDs [251]. Different mutations of tau determine neuro-pathological scenarios typical of class I–III tauopathies [151]. The muta-tions cause dementia and affect the frontotemporal region, so that they are collectively responsible for *FTDP-17*, the only familial tauopathy [252,253]. Forty-four mutations modify the amino acidic sequence of tau, almost exclusively in the MTBR, while 13 are located in the intronic regions flanking exon 10 [251]. They are depicted in Figure 1.6, using a color code in accordance with their effects on exon 10 splicing and on MT affinity.

Exonic mutations usually reduce the affinity of tau for MTs, and de-crease its promotion of MT assembly. Mutations on exons 9, 11, 12, and 13 reduce the MT affinity of one of the MTBRs [254], with the exception of MT–tau affinity enhancer/MT polymerization inducer Q336R and V363A mutations on exon 12 [255,256]. Mutations on exon 1 induce a

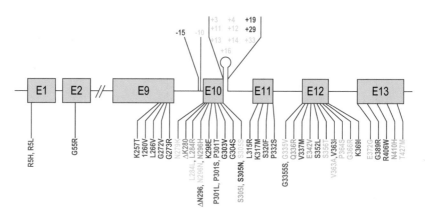

Color codes:
■ red for reduction of MT-tau affinity, no effect on exon 10 inclusion
▨ orange for reduction of MT-tau affinity, increase of exon 10 inclusion
■ pink for no effect on exon splicing
  yellow for no effect on MT-tau affinity, increase of exon 10 inclusion
▨ light blue for increase of MT-tau affinity, no effect on exon 10 inclusion
■ blue for increase both of MT-tau affinity and exon 10 inclusion
▨ light brown for no effect on MT-tau affinity, decrease of exon 10 inclusion
■ dark brown for no effect both on MT-tau affinity and exon 10 inclusion
▨ light green for reduction of MT-tau affinity
■ dark green for increase of MT-tau affinity
■ purple for increase of exon 10 inclusion
■ black for no reported data on these mutations
▨ light gray for MT tau affinity reduction and decrease of exon 10 inclusion

FIGURE 1.6   Tau mutations in FTDP-17.

conformational change that leads to reduced tau–MT binding [257]. Most intronic mutations close to exon 10 increase the 4R:3R ratio by destabilizing the stem–loop structure of the mRNA splice site, causing exon 10 inclusion [151]. Intronic +19 and +29 mutations decrease the 4R:3R ratio and cause exon 10 exclusion by altering a splicing silencer sequence [258]. Exon 10 mutations have multiple effects on tau and MT. They either reduce MT–tau affinity due to AA substitution (ΔN296, P301L, P301S [259]), or increase the 4R:3R ratio/exon 10 inclusion by mRNA splice site destabilization (N279K and G303V [260,261]) or by modifying splicing regulator sequences (L284L/R, N296N, S305I/S [262–264]). The same exon 10 mutation sometimes influences mRNA splicing and tau–MT affinity. Reduced MT–tau affinity and increased 4R:3R ratio are caused by the N296H mutation [265]. The ΔK280 and S305N mutation have opposite effects, as the former lowers the strength of tau–MT binding and causes an increased 4R:3R ratio [254,266], while the latter increases MT–tau affinity and 4R:3R ratio [267,268]. The E342V mutation on exon 12 reduces the affinity of tau for MTs, increases the 4R:3R ratio, and decreases the inclusion of exons 2/3 [269]. G55R (the only observed mutation on exon 2) does not influence MT–tau affinity or splicing ratios, but its 4R G55R tau isoform increases MT nucleation when compared with wild-type (WT) 4R tau [270]. Conversely, the 3R G55R tau isoform is equal to WT 3R tau in terms of MT nucleation.

Some mutant tau proteins display different *in vitro* aggregation properties when compared with WT tau isoforms [256,271–274]. G272V, G303V, L315R, and S320F mutations increase the nucleation rate of tau in an arachidonic acid-induced tau polymerization assay. They assemble into a larger number of shorter tau filaments than WT tau, possibly through an increased β-strand character for the mutated proteins. Conversely, R5L, P301L, and S352L mutations decrease the nucleation rate and lead to fewer, longer tau filaments through various molecular mechanisms. ΔN296, V337M, and E342V mutations do not significantly change the number and the length of tau filaments [271]. A heparin-induced tau polymerization assay shows a significant increase in fibrillization/aggregation for 4R P301L, P301S, and P364S isoforms, a moderate fibrillization increase for the 4R G272V isoform, and similar fibrillization/aggregation for 4R K257T, ΔK280, S305N, V337M, and R406W isoforms when compared with WT 4R tau. As to 3R isoforms, 3R K257T, G272V, and V337M isoforms show a moderate increase in fibrillization, and the R406W isoform shows no effect when compared with WT 3R tau [272–274]. Interestingly, V363I and V363A mutations show a decrease in tau fibrillization, but lead to the formation of soluble and potentially more neurotoxic tau oligomers [256].

The heterogeneity of FTDP-17 tau mutations in terms of MT assembly, MT–tau affinity, and aggregation propensity is mirrored by a large variety of clinical symptoms. Dementia-predominant and parkinsonism-predominant phenotypes are observed in FTDP-17 patients from hundreds of families in Europe, North America, Asia, and Australia [275]. Different symptoms may be observed in patients bearing the same mutation, and sometimes even in the same family [276], but correlations between tau mutation and clinical outcomes can be made [150,151]. Class I-like tau triplets/bands, PHF/tau and SP/Aβ inclusions, and AD-like symptoms are observed in the presence of G272V [153], V337M [277], and R406W [152] mutations. Class II-like tau doublet/bands, tau filaments, and PSP/CBD-like symptoms are observed in the presence of N279K [278], L284L [279], S305N [267], S305S [280], and +3/+13/+14/+16/+33 [152,281] intronic mutations. Class III-like tau doublet/bands, Pick bodies-like tau inclusions, and PiD-like symptoms are observed in the presence of K257T [282] and G389R [283] mutations. Multiple symptoms are observed in the presence of P301L (dementia/PSP/CBD [152,284]); P301S (dementia/CBD [285]); and K317M (ALS/PSP/CBD [286]) mutations. The *premortem* diagnosis, for these and other mutations, is often (not always) in agreement with *postmortem* neuropathological observations [287].

Alternative splicing, mutations, and PTMs have a strong influence on tau and tauopathies. Thus, various approaches target either of these areas to provide therapeutic benefits against tauopathies [106,107]. Tauopathies invariably lead to the weakening of tau–MT interactions, to the aggregation of tau, and to a number of neurotoxic consequences. Other proteins involved in proteinopathy-driven NDDs become aggregation-prone through similar mechanisms [288,289]. Aggregation and aggregation-dependent mechanisms are thus the common pathway leading to neurotoxicity for misfolded and mutated neuronal proteins.

The next chapters will thoroughly describe protein misfolding, aggregation into soluble oligomers and then into insoluble protein inclusions. The process will be dissected into pathological molecular mechanisms leading to proteinopathies, and promising points of intervention/targets for small molecule modulators/leads and candidates against proteinopathies. Tau, tauopathies in general, and AD in particular will be used as privileged examples of aggregation-prone proteins and proteinopathies.

References

1. http://en.wikipedia.org/wiki/Life_expectancy
2. https://www.cia.gov/library/publications/the-world-factbook/rankorder/2102rank.html
3. http://www.un.org/esa/population/publications/wpp2002/WPP2002-HIGHLIGHT-Srev1.PDFUN
4. http://www.alz.co.uk/research/WorldAlzheimerReport2013ExecutiveSummary.pdf
5. de Lau, L. M.; Breteler, M. M. Epidemiology of Parkinson's disease. *Lancet Neurol.* **2006**, *5*, 525–535.
6. http://www.neurodegenerationresearch.eu/about/why/
7. Hebert, L. E.; Weuve, J.; Scherr, P. A.; Evans, D. A. Alzheimer's disease in the United States (2010–2050) estimated using the 2010 Census. Neurology. Available at www.neurology.org/content/early/2013/02/06/WNL.0b013e31828726f5.abstract. Published online before print, Feb. 6, 2013.

8. Hebert, L. E.; Beckett, L. A.; Scherr, P. A.; Evans, D. A. Annual incidence of Alzheimer disease in the United States projected to the years 2000 through 2050. *Alzheimer Dis. Assoc. Disord.* **2001**, *15*, 169–173.

9. Small, D. H. Acetylcholinesterase inhibitors for the treatment of dementia in Alzheimer's disease: do we need new inhibitors? *Expert Opin. Emerg. Drugs* **2005**, *10*, 817–825.

10. Lees, A. J. L-dopa treatment and Parkinson's disease. *Q. J. Med.* **1986**, *59*, 535–547.

11. Ondo, W. G.; Tintner, R.; Thomas, M.; Jankovic, J. Tetrabenazine treatment for Huntington's disease-associated chorea. *Clin. Neuropharmacol.* **2002**, *25*, 300–302.

12. Cheah, B. C.; Vucic, S.; Krishnan, A. V.; Kiernan, M. C. Riluzole, neuroprotection and amyotrophic lateral sclerosis. *Curr. Med. Chem.* **2010**, *17*, 1942–1999.

13. Molinuevo, J. L.; Llado, A.; Rami, L. Memantine: targeting glutamate excitotoxicity in Alzheimer's disease and other dementias. *Am. J. Alzheimers Dis. Other Demen.* **2005**, *20*, 77–85.

14. Bellingham, M. C. A review of the neural mechanisms of action and clinical efficiency of riluzole in treating amyotrophic lateral sclerosis: what have we learned in the last decade? *CNS Neurosci. Therap.* **2011**, *17*, 4–31.

15. Schneider, L. S.; Dagerman, K. S.; Higgins, J. P.; McShane, R. Lack of evidence for the efficacy of memantine in mild Alzheimer disease. *Arch. Neurol.* **2011**, *68*, 991–998.

16. http://www.alz.co.uk/research/world-report

17. http://www.alz.org/downloads/facts_figures_2012.pdf

18. Gustavsson, A.; Svensson, M.; Jacobi, F.; Allgulander, C.; Alonso, J.; Beghi, E., et al. Cost of disorders of the brain in Europe 2010. *Eur. Neuropsychopharmacol.* **2011**, *21*, 718–779.

19. http://dbt.consultantlive.com/display/article/1145628/1488329

20. Riordan, K. C.; Hoffman Snyder, C. R.; Wellik, K. E.; Caselli, R. J.; Wingerchuk, D. M.; Demaerschalk, B. M. Effectiveness of adding memantine to an Alzheimer dementia treatment regimen which already includes stable donepezil therapy: a critically appraised topic. *Neurologist* **2011**, *17*, 121–123.

21. Linazasoro, G. Recent failures of new potential symptomatic treatments for parkinson's disease: causes and solution. *Movement Disorders* **2004**, *7*, 743–754.

22. Delrieu, J.; Ousset, P. J.; Caillaud, C.; Vellas, B. Clinical trials in Alzheimer's disease: immunotherapy approaches. *J. Neurochem.* **2012**, *120*, 186–193.

23. Hutter-Saunders, J. A.; Mosley, R. L.; Gendelman, H. E. Pathways towards an effective immunotherapy for Parkinson's disease. *Exp. Rev. Neurother.* **2011**, *11*, 1703–1715.

24. Qureshi, G. A., Parvez, S. H., Eds. *Oxidative Stress and Neurodegenerative Disorders*; Elsevier: Amsterdam, The Netherlands, 2007.

25. Pacher, P.; Beckman, J. S.; Liaudet, L. Nitric oxide and peroxynitrite in health and disease. *Physiol. Rev.* **2007**, *87*, 315–424.

26. Salminen, A.; Kauppinen, A.; Suuronen, T.; Kaarniranta, K.; Ojala, J. ER stress in Alzheimer's disease: a novel neuronal trigger for inflammation and Alzheimer's pathology. *J. Neuroinflamm.* **2009**, *6*, 41.

27. Moreira, P. I.; Zhu, X.; Wang, X.; Lee, H. G.; Nunomura, A.; Petersen, R. B., et al. Mitochondria: a therapeutic target in neurodegeneration. *Biochim. Biophys. Acta* **2010**, *1802*, 212–220.

28. Matsuda, N.; Tanaka, K. Does impairment of the ubiquitin-proteasome system or the autophagy-lysosome pathway predispose individuals to neurodegenerative disorders such as Parkinson's disease? *J. Alzheimers Dis.* **2010**, *19*, 1–9.

29. Uversky, V. N. Flexible nets of malleable guardians: intrinsically disordered chaperones in neurodegenerative diseases. *Chem. Rev.* **2011**, *111*, 1134–1166.

30. Amor, S.; Puentes, F.; Baker, D.; Van Der Valk, P. Inflammation in neurodegenerative diseases. *Immunology* **2010**, *129*, 154–169.

31. Bush, A. I. Metals and neuroscience. *Curr. Opin. Chem. Biol.* **2000**, *4*, 184–191.

32. Bossy-Wetzel, E.; Scharzenbacher, R.; Lipton, S. A. Molecular pathways to neurodegeneration. *Nat. Med.* **2004**, *10*, S2–S9.

33. Forman, M. S.; Trojanowski, J. Q.; Lee, V. M.-Y. Neurodegenerative diseases: a decade of discoveries paves the way for therapeutic breakthroughs. *Nat. Med.* **2004**, *10*, 1055–1063.
34. Debès, C.; Wang, M.; Caetano-Anollés, G.; Gräter, F. Evolutionary optimization of protein folding. *PLoS Comput. Biol.* **2013**, *9*, e1002861.
35. Junker, M.; Besingi, R. N.; Clark, P. L. Vectorial transport and folding of an autotransporter virulence protein during outer membrane secretion. *Mol. Microbiol.* **2009**, *71*, 1323–1332.
36. Ellis, R. J.; Minton, A. P. Protein aggregation in crowded environments. *Biol. Chem.* **2006**, *387*, 485–497.
37. Young, J. C.; Agashe, V. R.; Siegers, K.; Hartl, F. U. Pathways of chaperone-mediated protein folding in the cytosol. *Nat. Rev. Mol. Cell. Biol.* **2004**, *5*, 781–791.
38. Kuntz, I. D., Jr.; Kauzmann, W. Hydration of proteins and polypeptides. *Adv. Protein Chem.* **1974**, *28*, 239–345.
39. Ellis, J. Proteins as molecular chaperones. *Nature* **1987**, *328*, 378–379.
40. Drummond, D. A.; Wilke, C. O. Mistranslation-induced protein misfolding as a dominant constraint on coding sequence evolution. *Cell* **2008**, *134*, 341–352.
41. Horwitz, J. Alpha-crystallin can function as a molecular chaperone. *Proc. Natl. Acad. Sci. U.S.A.* **1992**, *89*, 10449–10453.
42. Stamler, R.; Kappé, G.; Boelens, W.; Slingsby, C. Wrapping the alpha-crystallin domain fold in a chaperone assembly. *J. Mol. Biol.* **2005**, *353*, 68–79.
43. Rothman, J. E. Polypeptide chain binding proteins: catalysts of protein folding and related processes in cells. *Cell* **1989**, *59*, 591–601.
44. Slepenkov, S. V.; Witt, S. N. The unfolding story of the *Escherichia coli* Hsp70 DnaK: is DnaK a holdase or an unfoldase? *Mol. Microbiol.* **2002**, *45*, 1197–1206.
45. Sharma, S. K.; Christen, P.; Goloubinoff, P. Disaggregating chaperones: an unfolding story. *Curr. Protein Pept. Sci.* **2009**, *10*, 432–446.
46. Rampelt, H.; Kirstein-Miles, J.; Nillegoda, N. B.; Chi, K.; Scholz, S. R.; Morimoto, R. I.; Bukau, B. Metazoan Hsp70 machines use Hsp110 to power protein disaggregation. *EMBO J.* **2012**, *21*, 4221–4235.
47. Lindquist, S. L.; Kelly, J. W. Chemical and biological approaches for adapting proteostasis to ameliorate protein misfolding and aggregation diseases—progress and prognosis. *Cold Spring Harbor Perspect. Biol.* **2011**, *3*, a004507.
48. Hershko, A.; Ciechanover, A. The ubiquitin system. *Annu. Rev. Biochem.* **1998**, *67*, 425–479.
49. He, C.; Klionsky, D. J. Regulation mechanisms and signaling pathways of autophagy. *Annu. Rev. Genet.* **2009**, *43*, 67–93.
50. Vijay-Kumar, S.; Bugg, C. E.; Cook, W. J. Structure of ubiquitin refined at 1.8 A resolution. *J. Mol. Biol.* **1987**, *194*, 531–544.
51. Bhaumik, S. R.; Malik, S. Diverse regulatory mechanisms of eukaryotic transcriptional activation by the proteasome complex. *Crit. Rev. Biochem. Mol. Biol.* **2008**, *43*, 419–433.
52. Choi, A. M. K.; Ryter, S. W.; Levine, B. Autophagy in human health and disease. *N. Engl. J. Med.* **2013**, *368*, 651–662.
53. Lamark, T.; Johansen, T. Aggrephagy: selective disposal of protein aggregates by macroautophagy. *Int. J. Cell Biol.* **2012**, 736905.
54. Eskelinen, E. L. Maturation of autophagic vacuoles in mammalian cells. *Autophagy* **2005**, *1*, 1–10.
55. Ramachandran, N.; Munteanu, I.; Wang, P.; Aubourg, P.; Rilstone, J. J.; Israelian, N., et al. VMA21 deficiency causes an autophagic myopathy by compromising V-ATPase activity and lysosomal acidification. *Cell* **2009**, *137*, 235–246.
56. Tyedmers, J.; Moegk, A.; Bukau, B. Cellular strategies for controlling protein aggregation. *Nat. Rev. Mol. Cell Biol.* **2010**, *11*, 777–788.
57. Hegde, A. N.; Upadhya, S. C. Role of ubiquitin-proteasome mediated proteolysis in nervous system disease. *Biochim. Biophys. Acta* **2011**, *1809*, 128–140.

58. Son, J. H.; Shim, J. H.; Kim, K. H.; Ha, J. Y.; Han, J. Y. Neuronal autophagy and neurodegenerative diseases. *Exp. Mol. Med.* **2012**, *44*, 89–98.
59. Sawkar, A. R.; D'Haeze, W.; Kelly, J. W. Therapeutic strategies to ameliorate lysosomal storage disorders—a focus on Gaucher disease. *Cell. Mol. Life Sci.* **2006**, *63*, 1179–1192.
60. Ross, C. A.; Poirier, M. A. Protein aggregation and neurodegenerative disease. *Nat. Med.* **2004**, *10*, S10–S17.
61. Gidalevitz, T.; Ben-Zvi, A.; Ho, K. H.; Brignull, H. R.; Morimoto, R. I. Progressive disruption of cellular protein folding in models of polyglutamine diseases. *Science* **2006**, *311*, 1471–1474.
62. Wong, E.; Cuervo, A. M. Integration of clearance mechanisms: the proteasome and autophagy. *Cold Spring Harbor Perspect. Biol.* **2010**, *2*, a006734.
63. Cohen, E.; Bieschke, J.; Perciavalle, R. M.; Kelly, J. W.; Dillin, A. Opposing activities protect against age-onset proteotoxicity. *Science* **2006**, *313*, 1604–1610.
64. Morimoto, R. I.; Cuervo, A. M. Protein homeostasis and aging: taking care of proteins from the cradle to the grave. *J. Gerontol. Ser. A* **2009**, *64A*, 167–170.
65. Akerfelt, M.; Morimoto, R. I.; Sistonen, L. Heat shock factors: integrators of cell stress, development and lifespan. *Nat. Rev. Mol. Cell. Biol.* **2010**, *11*, 545–555.
66. Balch, W. E.; Morimoto, R. I.; Dillin, A.; Kelly, J. W. Adapting proteostasis for disease intervention. *Science* **2008**, *319*, 916–919.
67. Hardy, J.; Selkoe, D. J. The amyloid hypothesis of Alzheimer's disease: progress and problems on the road to therapeutics. *Science* **2002**, *297*, 353–356.
68. Haass, C.; De Strooper, B. The presenilins in Alzheimer's disease—proteolysis holds the key. *Science* **1999**, *286*, 916–919.
69. Neugroschl, J.; Sano, M. Current treatment and recent clinical research in Alzheimer's disease. *Mount Sinai J. Med.* **2010**, *77*, 3–16.
70. Gerald, Z.; Ockert, W. Alzheimer's disease market: hope deferred. *Nat. Rev. Drug Discovery* **2013**, *12*, 19–20.
71. Goedert, M. α-Synuclein and neurodegenerative diseases. *Nat. Rev. Neurosci.* **2001**, *2*, 492–501.
72. Polymeropoulos, M. H.; Lavedan, C.; Leroy, E.; Ide, S. E.; Dehejia, A.; Dutra, A., et al. Mutation in the α-synuclein gene identified in families with Parkinson's disease. *Science* **1997**, *276*, 2045–2047.
73. Iwatsubo, T. Aggregation of α-synuclein in the pathogenesis of Parkinson's disease. *J. Neurol.* **2003**, *250*, III11–III14.
74. Conway, K. A.; Rochet, J. C.; Bieganski, R. M.; Lansbury, P. T., Jr. Kinetic stabilization of the α-synuclein protofibril by a dopamine-α-synuclein adduct. *Science* **2001**, *294*, 1346–1349.
75. Lee, E. N.; Cho, H. J.; Lee, C. H.; Lee, D.; Chung, K. C.; Paik, S. R. Phthalocyanine tetrasulfonates affect the amyloid formation and cytotoxicity of α-synuclein. *Biochemistry* **2004**, *43*, 3704–3715.
76. Zoghbi, H. Y.; Orr, H. T. Glutamine repeats and neurodegeneration. *Annu. Rev. Neurosci.* **2000**, *23*, 217–247.
77. Young, A. B. Huntingtin in health and disease. *J. Clin. Invest.* **2003**, *111*, 299–302.
78. Rosen, D. R.; Siddique, T.; Patterson, D.; Figlewicz, D. A.; Sapp, P.; Hentati, A., et al. Mutations in Cu/Zn superoxide dismutase gene are associated with familial amyotrophic lateral sclerosis. *Nature* **1993**, *362*, 59–62.
79. Neumann, M.; Sampathu, D. M.; Kwong, L. K.; Truax, A. C.; Micsenyi, M. C.; Chou, T. T., et al. Ubiquitinated TDP-43 in frontotemporal lobar degeneration and amyotrophic lateral sclerosis. *Science* **2006**, *314*, 130–133.
80. Kwiatkowski, T. J., Jr.; Bosco, D. A.; Leclerc, A. L.; Tamrazian, E.; Vanderburg, C. R.; Russ, C., et al. Mutations in the FUS/TLS gene on chromosome 16 cause familial amyotrophic lateral sclerosis. *Science* **2009**, *323*, 1205–1208.

81. Prusiner, S. B.; Scott, M. R.; DeArmond, S. J.; Cohen, F. E. Prion protein biology. *Cell* **1998**, *93*, 337–348.

82. Weingarten, M. D.; Lockwood, A. H.; Hwo, S. Y.; Kirschner, M. W. A protein factor essential for microtubule assembly. *Proc. Natl. Acad. Sci. U.S.A.* **1975**, *72*, 1858–1862.

83. Hirokawa, N. Microtubule organization and dynamics dependent on microtubule-associated proteins. *Curr. Opin. Cell Biol.* **1994**, *6*, 74–81.

84. Oldfield, C. J.; Cheng, Y.; Cortese, M. S.; Brown, C. J.; Uversky, V. N.; Dunker, A. K. Comparing and combining predictors of mostly disordered proteins. *Biochemistry* **2005**, *44*, 1989–2000.

85. Fink, A. L. Natively unfolded proteins. *Curr. Opin. Struct. Biol.* **2005**, *15*, 35–41.

86. Dunker, A. K.; Silman, I.; Uversky, V. N.; Sussman, J. L. Function and structure of inherently disordered proteins. *Curr. Opin. Struct. Biol.* **2008**, *18*, 756–764.

87. Dyson, H. J.; Wright, P. E. Intrinsically unstructured proteins and their functions. *Nat. Rev. Mol. Cell Biol.* **2005**, *6*, 197–208.

88. Mukrasch, M. D.; Bibow, S.; Korukottu, J.; Jeganathan, S.; Biernat, J.; Griesinger, C., et al. Structural polymorphism of 441-residue tau at single residue resolution. *PLoS Biol.* **2009**, *7*, 399–414.

89. Zheng, S.; Chen, Y.; Donahue, C. P.; Wolfe, M. S.; Varani, G. Structural basis for stabilization of the tau pre-mRNA splicing regulatory element by novantrone (mitoxantrone). *Chem. Biol.* **2009**, *16*, 557–566.

90. Neve, R. L.; Harris, P.; Kosik, K. S.; Kurnit, D. M.; Donlon, T. A. Identification of cDNA clones for the human microtubule-associated protein tau and chromosomal localization of the genes for tau and microtubule-associated protein 2. *Brain Res.* **1986**, *387*, 271–280.

91. Andreadis, A. Tau gene alternative splicing: expression pattern, regulation and modulation of function in normal brain and neurodegenerative disease. *Biochem. Biophys. Acta* **1739**, *2005*, 91–103.

92. Gustke, N.; Trinczek, B.; Biernat, J.; Mandelkow, E. M.; Mandelkow, E. Domains of tau protein and interactions with microtubules. *Biochemistry* **1994**, *33*, 9511–9522.

93. He, H. J.; Wang, X. S.; Pan, R.; Wang, D. L.; Liu, M. N.; He, R. Q. The proline-rich domain of tau plays a role in interactions with actin. *BMC Cell Biol.* **2009**, *10*, 81.

94. Pooler, A. M.; Hanger, D. P. Functional implications of the association of tau with the plasma membrane. *Biochem. Soc. Trans.* **2010**, *38*, 1012–1015.

95. Lee, G. Tau and src family tyrosine kinases. *Biochim. Biophys. Acta* **1739**, *2005*, 323–330.

96. Andreadis, A.; Brown, W. M.; Kosik, K. S. Structure and novel exons of the human T gene. *Biochemistry* **1992**, *31*, 10626–10633.

97. Hof, P. R.; Simic, G. Human fetal tau protein isoform: possibilities for Alzheimer's disease treatment. *Int. J. Biochem. Cell Biol.* **2012**, *44*, 1290–1294.

98. Goedert, M.; Spillantini, M. G. Pathogenesis of the tauopathies. *J. Mol. Neurosci.* **2011**, *45*, 425–431.

99. Crespo-Biel, N.; Theunis, C.; Van Leuven, F. Protein tau: prime cause of synaptic and neuronal degeneration in Alzheimer's disease. *Int. J. Alzheim. Dis.* **2012**, 251426.

100. Li, K.; Arikan, M. C.; Andreadis, A. Modulation of the membrane-binding domain of tau protein: splicing regulation of exon 2. *Mol. Brain Res.* **2003**, *116*, 94–105.

101. Magnani, E.; Fan, J.; Gasparini, L.; Golding, M.; Williams, M.; Schiavo, G., et al. Interaction of tau protein with the dynactin complex. *EMBO J.* **2007**, *26*, 4546–4554.

102. Andreadis, A. Tau splicing and the intricacies of dementia. *J. Cell. Physiol.* **2011**, *227*, 1220–1225.

103. Deshpande, A.; Win, K. M.; Busciglio, J. Tau isoform expression and regulation in human cortical neurons. *Faseb J.* **2008**, *22*, 2357–2367.

104. Goode, B. L.; Feinstein, S. C. Identification of a novel microtubule binding and assembly domain in the developmentally regulated inter-repeat region of tau. *J. Cell. Biol.* **1994**, *124*, 769–782.

105. Goedert, M.; Spillantini, M. G.; Potier, M. C.; Ulrich, J.; Crowther, R. A. Cloning and sequencing of the cDNA encoding an isoform of microtubule-associated protein tau containing four tandem repeats: differential expression of tau protein mRNAs in human brain. *EMBO J.* **1989**, *8*, 393–399.
106. Martin, L.; Latypova, X.; Terro, F. Post-translational modifications of tau protein: implications for Alzheimer's disease. *Neurochem. Int.* **2011**, *58*, 458–471.
107. Perry, G.; Zhu, X.; Smith, M. A.; Sorensen, A.; Avila, J.; Wang, J. Z.; Xia, Y. Y., et al. Abnormal hyperphosphorylation of tau: sites, regulation, and molecular mechanism of neurofibrillary degeneration. *J. Alzheimer's Dis.* **2013**, *33*, S123–S139.
108. Lee, V. M. Regulation of tau phosphorylation in Alzheimer's disease. *Ann. N. Y. Acad. Sci.* **1996**, *777*, 107–113.
109. Mazanetz, M. P.; Fischer, P. M. Untangling tau hyperphosphorylation in drug design for neurodegenerative diseases. *Nat. Rev. Drug Discov.* **2007**, *6*, 464–479.
110. Riederer, B. M.; Binder, L. I. Differential distribution of tau proteins in developing rat cerebellum. *Brain Res. Bull.* **1994**, *33*, 155–161.
111. Brion, J. P.; Octave, J. N.; Couck, A. M. Distribution of the phosphorylated microtubule-associated protein tau in developing cortical neurons. *Neuroscience* **1994**, *63*, 895–909.
112. Badiola, N.; Suarez-Calvet, M.; Lleo, A. Tau phosphorylation and aggregation as a therapeutic target in tauopathies. *CNS Neurol. Dis. Drug Targets* **2010**, *9*, 727–740.
113. Yu, Y.; Run, X.; Liang, Z.; Li, Y.; Liu, F.; Liu, Y., et al. Developmental regulation of tau phosphorylation, tau kinases, and tau phosphatases. *J. Neurochem.* **2009**, 1480–1494.
114. Su, B.; Wang, X.; Drew, K. L.; Perry, G.; Smith, M. A.; Zhu, X. Physiological regulation of tau phosphorylation during hibernation. *J. Neurochem.* **2008**, *105*, 2098–2108.
115. Dickey, C. A.; Koren, J.; Zhang, Y. J.; Xu, Y. F.; Jinwal, U. K.; Birnbaum, M. J., et al. Akt and CHIP co-regulate tau degradation through coordinated interactions. *Proc. Natl. Acad. Sci. U.S.A.* **2008**, *105*, 3622–3627.
116. Bertrand, J.; Plouffe, V.; Senechal, P.; Leclerc, N. The pattern of human tau phosphorylation is the result of priming and feedback events in primary hippocampal neurons. *Neuroscience* **2010**, *168*, 323–334.
117. Ploia, C.; Antoniou, X.; Sclip, A.; Grande, V.; Cardinetti, D.; Colombo, A., et al. JNK plays a key role in tau hyperphosphorylation in Alzheimer's disease models. *J. Alzheim. Dis.* **2011**, *26*, 315–329.
118. Wang, S.; Toth, M. E.; Bereczki, E.; Santha, M.; Guan, Z. Z.; Winblad, B.; Pei, J. J. Interplay between glycogen synthase kinase-3β and tau in the cerebellum of Hsp27 transgenic mouse. *J. Neurosci. Res.* **2011**, *89*, 1267–1275.
119. Bancher, C.; Brunner, C.; Lassmann, H.; Budka, H.; Jellinger, K.; Wiche, G., et al. Accumulation of abnormally phosphorylated x precedes the formation of neurofibrillary tangles in Alzheimer's disease. *Brain Res.* **1989**, *477*, 90–99.
120. Hanover, J. A.; Krause, M. W.; Love, D. C. Bittersweet memories: linking metabolism to epigenetics through O-GlcNacylation. *Nat. Rev. Mol. Cell Biol.* **2012**, *13*, 312–321.
121. Hart, G. W.; Slawson, C.; Ramirez-Correa, G.; Lagerlof, O. Cross talk between O-GlcNAcylation and phosphorylation: roles in signaling, transcription, and chronic disease. *Annu. Rev. Biochem.* **2011**, *80*, 825–858.
122. Lazarus, B. D.; Love, D. C.; Hanover, J. A. O-GlcNAc cycling: implications for neurodegenerative disorders. *Int. J. Biochem. Cell Biol.* **2009**, *41*, 2134–2146.
123. Liu, Y.; Liu, F.; Grundke-Iqbal, I.; Iqbal, K.; Gong, C. X. Brain glucose transporters, O-GlcNAcylation and phosphorylation of tau in diabetes and Alzheimer's disease. *J. Neurochem.* **2009**, *111*, 242–249.
124. Wedemeyer, W. J.; Welker, E.; Scheraga, H. A. Proline cis-trans isomerization and protein folding. *Biochemistry* **2002**, *41*, 14637–14644.
125. Zhou, X. Z.; Lu, P. J.; Wulf, G.; Lu, K. P. Phosphorylation-dependent prolyl isomerization. A novel signaling regulatory mechanism. *Cell. Mol. Life Sci.* **1999**, *56*, 788–806.

126. Shaw, P. E. Peptidyl-prolyl cis/trans isomerases and transcription: is there a twist in the tail? *EMBO Rep.* **2007**, *8*, 40–45.
127. Nakamura, K.; Greenwood, A.; Binder, L.; Bigio, E. H.; Denial, S.; Nicholson, L., et al. Proline isomer-specific antibodies reveal the early pathogenic tau conformation in Alzheimer's disease. *Cell* **2012**, *149*, 232–244.
128. Min, S. W.; Cho, S. H.; Zhou, Y.; Schroeder, S.; Haroutunian, V.; Seeley, W. W., et al. Acetylation of tau inhibits its degradation and contributes to tauopathy. *Neuron* **2010**, *67*, 953–966.
129. Cohen, T. J.; Guo, J. L.; Hurtado, D. E.; Kwong, L. K.; Mills, I. P.; Trojanowski, J. Q.; Lee, V. M. Y. The acetylation of tau inhibits its function and promotes pathological tau aggregation. *Nat. Commun.* **2011**, *2*, 252.
130. Cohen, T. J.; Friedmann, D.; Hwang, A. W.; Marmorstein, R.; Lee, V. M. Y. The microtubule-associated tau protein has intrinsic acetyltransferase activity. *Nat. Struct. Mol. Biol.* **2013**, *20*, 756–762.
131. Irwin, D. J.; Cohen, T. J.; Grossman, M.; Arnold, S. E.; Xie, S. X.; Lee, V. M.; Trojanowski, J. Q. Acetylated tau, a novel pathological signature in Alzheimer's disease and other tauopathies. *Brain* **2012**, *135*, 807–818.
132. Tenenholz Grinberg, L.; Wang, X.; Wang, C.; Dongmin Sohn, P.; Theofilas, P.; Sidhu, M., et al. Argyrophilic grain disease differs from other tauopathies by lacking tau acetylation. *Acta Neuropathol.* **2013**, *125*, 581–593.
133. Hanger, D. P.; Wray, S. Tau cleavage and tau aggregation in neurodegenerative disease. *Biochem. Soc. Trans.* **2010**, *38*, 1016–1020.
134. Proteases in the brain. *Proteases in biological diseases*, Volume 3, Lendeckel, U., Hooper, N. M., Eds.; Springer Science and Business Media, Inc: New York, 2005; 383 pages.
135. Johnson, G. V. W. Tau phosphorylation and proteolysis: Insights and perspectives. *J. Alzheim. Dis.* **2006**, *9*, 243–250.
136. Guillozet-Bongaards, A. L.; Cahill, M. E.; Cryns, V. L.; Reynolds, M. R.; Berry, R. W.; Binder, L. I. Pseudophosphorylation of tau at serine 422 inhibits caspase cleavage: in vitro evidence and implications for tangle formation. *J. Neurochem.* **2006**, *97*, 1005–1014.
137. Wang, I.; Garg, S.; Mandelkow, E. M.; Mandelkow, E. Proteolytic processing of tau. *Biochem. Soc. Trans.* **2010**, *38*, 955–961.
138. Rohn, T. T. The role of caspases in Alzheimer's disease; potential novel therapeutic opportunities. *Apoptosis* **2010**, *15*, 1403–1409.
139. Getz, G. S. Calpain inhibition as a potential treatment of Alzheimer's disease. *Am. J. Physiol.* **2012**, *181*, 388–391.
140. Basurto-Islas, G.; Luna-Munoz, J.; Guillozet-Bongaards, A. L.; Binder, L. I.; Mena, R.; Garcia-Sierra, F. Accumulation of aspartic acid 421- and glutamic acid 391-cleaved tau in neurofibrillary tangles correlates with progression in Alzheimer's disease. *J. Neuropathol. Exp. Neurol.* **2008**, *67*, 470–483.
141. Chung, C. W.; Song, Y. H.; Kim, I. K.; Yoon, W. J.; Ryu, B. R.; Jo, D. G., et al. Proapoptotic effects of tau cleavage product generated by caspase-3. *Neurobiol. Dis.* **2001**, *8*, 162–172.
142. Nacharaju, P.; Ko, L.; Yen, S. H. Characterization of in vitro glycation sites of tau. *J. Neurochem.* **1997**, *69*, 1709–1719.
143. Horiguchi, T.; Uryu, K.; Giasson, B. I.; Ischiropoulos, H.; LightFoot, R.; Bellmann, C., et al. Nitration of tau protein is linked to neurodegeneration in tauopathies. *Am. J. Pathol.* **2003**, *163*, 1021–1031.
144. Halverson, R. A.; Lewis, J.; Frausto, S.; Hutton, M.; Muma, N. A. Tau protein is cross-linked by transglutaminase in P301L tau transgenic mice. *J. Neurosci.* **2005**, *25*, 1226–1233.
145. Takahashi, K.; Ishida, M.; Komano, H.; Takahashi, H. SUMO-1 immunoreactivity colocalizes with phospho-Tau in APP transgenic mice but not in mutant Tau transgenic mice. *Neurosci. Lett.* **2008**, *441*, 90–93.

146. Schweers, O.; Mandelkow, E. M.; Biernat, J.; Mandelkow, E. Oxidation of cysteine-322 in the repeat domain of microtubule-associated protein tau controls the in vitro assembly of paired helical filaments. *Proc. Natl. Acad. Sci. U.S.A.* **1995**, *92*, 8463–8467.
147. Santa-Maria, I.; Varghese, M.; Ksie ak-Reding, H.; Dzhun, A.; Wang, J.; Pasinetti, G. M. Paired helical filaments from Alzheimer disease brain induce intracellular accumulation of tau protein in aggresomes. *J. Biol. Chem.* **2012**, *287*, 20522–20533.
148. Alonso, A.; Zaidi, T.; Novak, M.; Grundke-Iqbal, I.; Iqbal, K. Hyperphosphorylation induces self-assembly of tau into tangles of paired helical filaments/straight filaments. *Proc. Natl. Acad. Sci. U.S.A.* **2001**, *98*, 6923–6928.
149. Kato, S.; Nakamura, H.; Otomo, E. Reappraisal of neurofibrillary tangles. Immunohistochemical, ultrastructural, and immunoelectron microscopical studies. *Acta Neuropathol.* **1989**, *77*, 258–266.
150. Buee, L.; Bussie, T.; Buee-Scherrerb, V.; Delacourte, A.; Hof, P. R. Tau protein isoforms, phosphorylation and role in neurodegenerative disorders. *Brain Res. Rev.* **2000**, *33*, 95–130.
151. Lee, V. M. J.; Goedert, M.; Trojanowski, J. Q. Neurodegenerative tauopathies. *Annu. Rev. Neurosci.* **2001**, *24*, 1121–1159.
152. Hutton, M.; Lendon, C. L.; Rizzu, P.; Baker, M.; Froelich, S.; Houlden, H., et al. Association of missense and 50-splice-site mutations in tau with the inherited dementia FTDP-17. *Nature* **1998**, *393*, 702–705.
153. Spillantini, M. G.; Crowther, R. A.; Kamphorst, W.; Heutink, P.; Van Swieten, J. C. Tau pathology in two Dutch families with mutations in the microtubule-binding region of tau. *Am. J. Pathol.* **1998**, *153*, 1359–1363.
154. Sergeant, N.; Delacourte, A.; Buee, L. Tau protein as a differential biomarker of tauopathies. *Biochim. Biophys. Acta* 1739, *2005*, 179–197.
155. Zhukareva, V.; Vogelsberg-Ragaglia, V.; Van Deerlin, V. M.; Bruce, J.; Shuck, T.; Grossman, M., et al. Loss of brain tau defines novel sporadic and familial tauopathies with frontotemporal dementia. *Ann. Neurol.* **2001**, *49*, 165–175.
156. Gustafson, L. Clinical picture of frontal lobe degeneration of non-Alzheimer type. *Dementia* **1993**, *4*, 143–148.
157. Berchtold, N. C.; Cotman, C. W. Evolution in the conceptualization of dementia and Alzheimer's disease: Greco-Roman period to the 1960s. *Neurobiol. Aging* **1998**, *19*, 173–189.
158. Medeiros, R.; Chabrier, M. A.; LaFerla, F. M. Elucidating the triggers, progression, and effects of Alzheimer's disease. *J. Alzheimer's Dis.* **2013**, *33*, S195–S210.
159. Cras, P.; Kawai, M.; Lowery, D.; Gonzalez-DeWhitt, P.; Greenberg, B.; Perry, G. Senile plaque neurites in Alzheimer disease accumulate amyloid precursor protein. *Proc. Natl. Acad. Sci. U.S.A.* **1991**, *88*, 7552–7556.
160. Vermersch, P.; David, J. P.; Frigard, B.; Fallet-Bianco, C.; Wattez, A.; Petit, H.; Delacourte, A. Cortical mapping of Alzheimer pathology in brains of aged non-demented subjects. *Prog. Neuropsychopharmacol. Biol. Psychiatry* **1995**, *19*, 1035–1047.
161. Delacourte, A.; David, J. P.; Sergeant, N.; Buee, L.; Wattez, A.; Vermersch, P., et al. The biochemical pathway of neurofibrillary degeneration in aging and Alzheimer's disease. *Neurology* **1999**, *52*, 1158–1165.
162. Braak, H.; Braak, E. Neuropathological staging of Alzheimer-related changes. *Acta Neuropathol.* **1991**, *82*, 239–259.
163. Matsui, T.; Ingelsson, M.; Fukumoto, H.; Ramasamy, K.; Kowa, H.; Frosch, M. P., et al. Expression of APP pathway mRNAs and proteins in Alzheimer's disease. *Brain Res.* **2007**, *1161*, 116–123.
164. Cruts, M.; van Broeckhoven, C. Presenilin mutations in Alzheimer's disease. *Hum. Mutat.* **1998**, *11*, 183–190.
165. Waring, S. C.; Rosenberg, R. N. Genome-wide association studies in Alzheimer disease. *Arch Neurol.* **2008**, *65*, 329–334.

166. Strittmatter, W. J. Apolipoprotein E: high-avidity binding to beta-amyloid and increased frequency of type 4 allele in late-onset familial Alzheimer disease. *Proc. Natl. Acad. Sci. U.S.A.* **1993**, *90*, 1977–1981.

167. Mahley, R. W.; Weisgraber, K. H.; Huang, Y. Apolipoprotein E4: a causative factor and therapeutic target in neuropathology, including Alzheimer's disease. *Proc. Natl. Acad. Sci. U.S.A.* **2006**, *103*, 5644–5651.

168. Jonsson, T.; Stefansson, H.; Steinberg, S.; Jonsdottir, I.; Jonsson, P. V.; Snaedal, J., et al. Variant of TREM2 associated with the risk of Alzheimer's disease. *N. Engl. J. Med.* **2013**, *368*, 107–116.

169. Guerreiro, R.; Wojtas, A.; Bras, J.; Carrasquillo, M.; Rogaeva, E.; Majounie, E., et al. *TREM2* variants in Alzheimer's disease. *N. Engl. J. Med.* **2013**, *368*, 117–127.

170. Bancher, C.; Jellinger, K. A. Neurofibrillary tangle predominant form of senile dementia of Alzheimer type: a rare subtype in very old subjects. *Acta Neuropathol.* **1994**, *88*, 565–570.

171. Yamada, M.; Itoh, Y.; Sodeyama, N.; Suematsu, N.; Otomo, E.; Matsushita, M.; Mizusawa, H. Senile dementia of the neurofibrillary tangle type: a comparison with Alzheimer's disease. *Dement. Geriatr. Cogn. Disord.* **2001**, *12*, 117–126.

172. Jellinger, K. A.; Attems, J. Neurofibrillary tangle-predominant dementia: comparison with classical Alzheimer disease. *Acta Neuropathol.* **2007**, *113*, 107–117.

173. Vermersch, P.; Delacourte, A.; Javoy-Agid, F.; Hauw, J. J.; Agid, Y. Dementia in Parkinson's disease: biochemical evidence for cortical involvement using the immunodetection of abnormal Tau proteins. *Ann. Neurol.* **1993**, *33*, 445–450.

174. Tolosa, E.; Calandrella, D.; Gallardo, M. Caribbean parkinsonism and other atypical parkinsonian disorders. *Parkinsonism Related Disord.* **2004**, *10*, S19–S26.

175. Hirano, A.; Kurland, L. T.; Krooth, R. S.; Lessel, S. Parkinsonism-dementia complex, and endemic disease on the island of Guam. l. Clinical features. *Brain* **1961**, *84*, 642–661.

176. Buee-Scherrer, V.; Buee, L.; Hof, P. R.; Leveugle, B.; Gilles, C.; Loerzel, A. J., et al. Neurofibrillary degeneration in normal amyotrophic lateral sclerosis/parkinsonism–dementia complex of Guam. Immunochemical characterization of tau proteins. *Am. J. Pathol.* **1995**, *146*, 924–932.

177. Hasegawa, M.; Arai, T.; Akiyama, H.; Nonaka, T.; Mori, H.; Hashimoto, T., et al. TDP-43 is deposited in the Guam parkinsonism–dementia complex brains. *Brain* **2007**, *130*, 1386–1394.

178. DeJesus-Hernandez, M.; Mackenzie, I. R.; Boeve, B. F.; Boxer, A. L.; Baker, M.; Rutherford, N. J., et al. Expanded GGGGCC hexanucleotide repeat in noncoding region of C9ORF72 causes chromosome 9p-linked FTD and ALS. *Neuron* **2011**, *72*, 245–256.

179. Mahoney, C. J.; Beck, J.; Rohrer, J. D.; Lashley, T.; Mok, K.; Shakespeare, T., et al. Frontotemporal dementia with the C9ORF72 hexanucleotide repeat expansion: clinical, neuroanatomical and neuropathological features. *Brain* **2012**, *135*, 736–750.

180. Bieniek, K. F.; Murray, M. E.; Rutherford, N. J.; Castanedes-Casey, M.; DeJesus-Hernandez, M.; Liesinger, A. M., et al. Tau pathology in frontotemporal lobar degeneration with C9ORF72 hexanucleotide repeat expansion. *Acta Neuropathol.* **2013**, *125*, 289–302.

181. Lannuzel, A.; Ruberg, M.; Michel, P. P. Atypical parkinsonism in the Caribbean island of Guadeloupe: etiological role of the mitochondrial complex I inhibitor annonacin. *Movement Disorders* **2008**, *23*, 2122–2128.

182. Caparros-Lefebvre, D.; Steele, J. Atypical parkinsonism on Guadeloupe, comparison with the parkinsonism–dementia complex of Guam, and environmental toxic hypotheses. *Environ. Toxicol. Pharmacol.* **2005**, *19*, 407–413.

183. Lannuzel, A.; Hoglinger, G. U.; Verhaeghe, S.; Gire, L.; Belson, S.; Escobar-Khondiker, M., et al. Atypical parkinsonism in Guadeloupe: a common risk factor for two closely related phenotypes? *Brain* **2007**, *130*, 816–827.

184. Wenning, G. K.; Colosimo, C.; Geser, F.; Poewe, W. Multiple system atrophy. *Lancet Neurol.* **2004**, *3*, 93–103.

185. Arima, K.; Uéda, K.; Sunohara, N.; Arakawa, K.; Hirai, S.; Nakamura, M.; Tonozuka-Uehara, H.; Kawai, M. NACP/alpha-synuclein immunoreactivity in fibrillary components of neuronal and oligodendroglial cytoplasmic inclusions in the pontine nuclei in multiple system atrophy. *Acta Neuropathol.* **1998**, *96*, 439–444.

186. Cairns, N. J.; Atkinson, P. F.; Hanger, D. P.; Anderton, B. H.; Daniel, S. E.; Lantos, P. L. Tau protein in the glial cytoplasmic inclusions of multiple system atrophy can be distinguished from abnormal tau in Alzheimer's disease. *Neurosci. Lett.* **1997**, *230*, 49–52.

187. Nagaishi, M.; Yokoo, H.; Nakazato, Y. Tau-positive glial cytoplasmic granules in multiple system atrophy. *Neuropathology* **2011**, *31*, 299–305.

188. Vilensky, J. A.; Gilman, S.; McCall, S. A historical analysis of the relationship between encephalitis lethargica and postencephalitic parkinsonism: a complex rather than a direct relationship. *Movement Disorders* **2010**, *25*, 1116–1123.

189. Litvan, I. I.; Jankovic, J.; Goetz, C. G.; Wenning, G. K.; Sastry, N.; Jellinger, K., et al. Accuracy of the clinical diagnosis of postencephalitic parkinsonism: a clinicopathologic study. *Eur. J. Neurol.* **1998**, *5*, 451–457.

190. Jellinger, K. A. Absence of α-synuclein pathology in postencephalitic parkinsonism. *Acta Neuropathol.* **2009**, *118*, 371–379.

191. Buee-Scherrer, V.; Buee, L.; Leveugle, B.; Perl, D. P.; Vermersch, P.; Hof, P. R.; Delacourte, A. Pathological tau proteins in postencephalitic parkinsonism: comparison with Alzheimer's disease and other neurodegenerative disorders. *Ann. Neurol.* **1997**, *42*, 356–359.

192. Ikeda, K.; Akiyama, H.; Kondo, H.; Arai, T.; Arai, N.; Yagishita, S. Numerous glial fibrillary tangles in oligodendroglia in cases of subacute sclerosing panencephalitis with neurofibrillary tangles. *Neurosci. Lett.* **1995**, *194*, 133–135.

193. Bancher, C.; Leitner, H.; Jellinger, K.; Eder, H.; Setinek, U.; Fischer, P., et al. On the relationship between measles virus and Alzheimer neurofibrillary tangles in subacute sclerosing panencephalitis. *Neurobiol. Aging* **1996**, *17*, 527–533.

194. Yuksel, D.; Yilmaz, D.; Uyar, N. Y.; Senbil, N.; Gurer, Y.; Anlar, B. Tau proteins in the cerebrospinal fluid of patients with subacute sclerosing panencephalitis. *Brain Dev.* **2010**, *32*, 467–471.

195. Kosaka, K. Diffuse neurofibrillary tangles with calcification: a new presenile dementia. *J. Neurol. Neurosurg. Psychiatry* **1994**, *57*, 594–596.

196. Habuchi, C.; Iritani, S.; Sekiguchi, H.; Torii, Y.; Ishihara, R.; Arai, T., et al. Clinicopathological study of diffuse neurofibrillary tangles with calcification with special reference to TDP-43 proteinopathy and alpha-synucleinopathy. *J. Neurolog. Sci.* **2011**, *301*, 77–85.

197. Uchihara, T.; Tsuchiya, K.; Nakamura, A.; Akiyama, H. Argyrophilic grains are not always argyrophilic—distinction from neurofibrillary tangles of diffuse neurofibrillary tangles with calcification revealed by comparison between Gallyas and Campbell-Switzer methods. *Acta Neuropathol.* **2005**, *110*, 158–164.

198. Corsellis, J. A.; Bruton, C. J.; Freeman-Browne, D. The aftermath of boxing. *Psychol. Med.* **1973**, *3*, 270–303.

199. Blennow, K.; Hardy, J.; Zetterberg, H. The neuropathology and neurobiology of traumatic brain injury. *Neuron* **2012**, *76*, 886–899.

200. Roberts, G. W.; Allsop, D.; Bruton, C. The occult aftermath of boxing. *J. Neurol. Neurosurg. Psychiatry* **1990**, *53*, 373–378.

201. King, A.; Sweeney, F.; Bodi, I.; Troakes, C.; Maekawa, S.; Al-Sarraj, S. Abnormal TDP-43 expression is identified in the neocortex in cases of dementia pugilistica, but is mainly confined to the limbic system when identified in high and moderate stages of Alzheimer's disease. *Neuropathology* **2010**, *30*, 408–419.

202. Cardenas, A. M.; Ardiles, A. O.; Barraza, N.; Baez-Matus, X.; Caviedes, P. Role of tau protein in neuronal damage in Alzheimer's disease and Down syndrome. *Archiv. Med. Res.* **2012**, *43*, 645–654.

203. Hof, P. R.; Bouras, C.; Morrison, J. H. Cortical neuropathology in aging and dementing disorders: neuronal typology, connectivity, and selective vulnerability. *Cerebral Cortex*, Vol. 14, Peters, A., Morrison, J. H., Eds.; Kluwer Academic Plenum: New York, 1999; pp. 175–312.

204. Hof, P. R.; Bouras, C.; Perl, D. P.; Sparks, D. L.; Mehta, N.; Morrison, J. H. Age-related distribution of neuropathologic changes in the cerebral cortex of patients with Down's syndrome. Quantitative regional analysis and comparison with Alzheimer's disease. *Arch. Neurol.* **1995**, *52*, 379–391.

205. Vidal, R.; Frangione, B.; Rostagno, A.; Mead, S.; Revesz, T.; Plant, G.; Ghiso, J. A stop-codon mutation in the BRI gene associated with familial British dementia. *Nature* **1999**, *399*, 776–781.

206. Ghiso, J. A.; Holton, J.; Miravalle, L.; Calero, M.; Lashley, T.; Vidal, R., et al. Systemic amyloid deposits in familial British dementia. *J. Biol. Chem.* **2001**, *276*, 43909–43914.

207. Holton, J. L.; Ghiso, J.; Lashley, T.; Rostagno, A.; Guerin, C. J.; Gibb, G., et al. Regional distribution of amyloid-Bri deposition and its association with neurofibrillary degeneration in familial British dementia. *Am. J. Pathol.* **2001**, *158*, 515–526.

208. Schwab, C.; Arai, T.; Hasegawa, M.; Akiyama, H.; Yu, S.; McGeer, P. L. TDP-43 pathology in familial British dementia. *Acta Neuropathol.* **2009**, *118*, 303–311.

209. Stromgren, E.; Dalby, A.; Dalby, M. A.; Ranheim, B. Cataract, deafness, cerebellar ataxia, psychosis and dementia—a new syndrome. *Acta Neurol. Scand.* **1970**, *46* (S43), 261.

210. Gibson, G.; Gunasekera, N.; Lee, M.; Lelyveld, V.; El-Agnaf, O. M. A.; Wright, A.; Austen, B. Oligomerization and neurotoxicity of the amyloid ADan peptide implicated in familial Danish dementia. *J. Neurochem.* **2004**, *88*, 281–290.

211. Garringer, H. J.; Murrell, J.; Sammeta, N.; Gnezda, A.; Ghetti, B.; Vidal, R. Increased tau phosphorylation and tau truncation, and decreased synaptophysin levels in mutant BRI2/tau transgenic mice. *PLoS One* **2013**, *8*, e56426.

212. Gerstmann, J.; Straussler, E.; Scheinker, I. Uber eine eigernartige hereditar-familiare Erkrankung des Zentralnervensystems. *Z. Neurol.* **1936**, *154*, 736–762.

213. Collins, S.; McLean, C. A.; Masters, C. L. Gerstmann–Sträussler–Scheinker syndrome, fatal familial insomnia, and kuru: a review of these less common human transmissible spongiform encephalopathies. *J. Clin. Neurosci.* **2001**, *8*, 387–397.

214. Tranchant, C.; Sergeant, N.; Wattez, A.; Mohr, M.; Warter, J. M.; Delacourte, A. Neurofibrillary tangles in Gerstmann–Straussler–Scheinker syndrome with the A117V prion gene mutation. *J. Neurol. Neurosurg. Psychiatry* **1997**, *63*, 240–246.

215. Kitamoto, T.; Iizuka, R.; Tateishi, J. An amber mutation of prion protein in Gerstmann-Sträussler syndrome with mutant PrP plaques. *Biochem. Biophys. Res. Commun.* **1993**, *192*, 525–531.

216. Ghetti, B.; Piccardo, P.; Spillantini, M. G.; Ichimiya, Y.; Porro, M.; Perini, F., et al. Vascular variant of prion protein cerebral amyloidosis with t-positive neurofibrillary tangles: the phenotype of the stop codon 145 mutation in PRNP. *Proc. Natl. Acad. Sci. U.S.A.* **1996**, *93*, 744–748.

217. Schneider, S. A.; Bhatia, K. P. Excess iron harms the brain: the syndromes of neurodegeneration with brain iron accumulation (NBIA). *J. Neural Transm.* **2013**, *120*, 695–703.

218. Haraguchi, T.; Terada, S.; Ishizu, H.; Yokota, O.; Yoshida, H.; Takeda, N., et al. Coexistence of TDP-43 and tau pathology in neurodegeneration with brain iron accumulation type 1 (NBIA-1, formerly Hallervorden-Spatz syndrome). *Neuropathology* **2011**, *31*, 531–535.

219. Gregory, A.; Hayflick, S. J. Neurodegeneration with brain iron accumulation. *Folia Neuropathol.* **2005**, *43*, 286–296.

220. Love, S.; Bridges, L. R.; Case, C. P. Neurofibrillary tangles in Niemann–Pick disease type C. *Brain* **1995**, *118*, 119–129.

221. Rosenbaum, A. I.; Maxfield, F. R. Niemann-Pick type C disease: molecular mechanisms and potential therapeutic approaches. *J. Neurochem.* **2011**, *116*, 789–795.

222. Turpin, J. C.; Goas, J. Y.; Masson, M.; Zagnoli, F.; Mocquard, Y.; Baumann, N. Type C Niemann–Pick disease: supranuclear ophthalmoplegia associated with deficient biosynthesis of cholesterol esters. *Rev. Neurol.* **1991**, *147*, 28–34.

223. Auer, I. A.; Schmidt, M. L.; Lee, V. M. Y.; Curry, B.; Suzuki, K.; Shin, R. W., et al. Paired helical filament tau (PHF-tau) in Niemann–Pick type C disease is similar to PHF-tau in Alzheimer's disease. *Acta Neuropathol.* **1995**, *90*, 547–551.

224. Garbern, J. Y.; Neumann, M.; Trojanowski, J. Q.; Lee, V. M. Y.; Feldman, G., et al. A mutation affecting the sodium/proton exchanger, SLC9A6, causes mental retardation with tau deposition. *Brain* **2010**, *133*, 1391–1402.

225. Tolnay, M.; Schwietert, M.; Monsch, A. U.; Staehelin, H. B.; Langui, D.; Probst, A. Argyrophilic grain disease: distribution of grains in patients with and without dementia. *Acta Neuropathol.* **1997**, *94*, 353–358.

226. Togo, T.; Sahara, N.; Yen, S. H.; Cookson, N.; Ishizawa, T.; Hutton, M., et al. Argyrophilic grain disease is a sporadic 4-repeat tauopathy. *J. Neuropathol. Exp. Neurol.* **2002**, *61*, 547–556.

227. Braak, H.; Braak, E. Argyrophilic grain disease: frequency of occurrence in different age categories and neuropathological diagnostic criteria. *J. Neural Transm.* **1998**, *105*, 801–819.

228. Tenenholz Grinberg, L.; Wang, X.; Wang, C.; Dongmin Sohn, P.; Theofilas, P.; Sidhu, M., et al. Argyrophilic grain disease differs from other tauopathies by lacking tau acetylation. *Acta Neuropathol.* **2013**, *125*, 581–593.

229. Rebeiz, J. J.; Kolodny, E. H.; Richardson, E. P., Jr. Corticodentatonigral degeneration with neuronal achromasia. *Arch. Neurol.* **1968**, *18*, 20–33.

230. Feany, M. B.; Dickson, D. W. Widespread cytoskeletal pathology characterizes corticobasal degeneration. *Am. J. Pathol.* **1995**, *146*, 1388–1396.

231. Ksiezak-Reding, H.; Morgan, K.; Mattiace, L. A.; Davies, P.; Liu, W. K.; Yen, S. H., et al. Ultrastructure and biochemical composition of paired helical filaments in corticobasal degeneration. *Am. J. Pathol.* **1994**, *145*, 1496–1508.

232. Vandrovcova, J.; Anaya, F.; Kay, V.; Lees, A.; Hardy, J.; de Silva, R. Disentangling the role of the tau gene locus in sporadic tauopathies. *Curr. Alzheimer Res.* **2010**, *7*, 726–734.

233. Kouri, N.; Carlomagno, Y.; Baker, M.; Liesinger, A. M.; Caselli, R. J.; Wszolek, Z. K., et al. Novel mutation in *MAPT* exon 13 (p.N410H) causes corticobasal degeneration. *Acta Neuropathol.* **2014**, *127*, 271–282.

234. Ahmed, Z.; Bigio, E. H.; Budka, H.; Dickson, D. W.; Ferrer, I.; Ghetti, B., et al. Globular glial tauopathies (GGT): consensus recommendations. *Acta Neuropathol.* **2013**, *126*, 537–544.

235. Ahmed, Z.; Doherty, K. M.; Silveira-Moriyama, L.; Bandopadhyay, R.; Lashley, T.; Mamais, A., et al. Globular glial tauopathies (GGT) presenting with motor neuron disease or frontotemporal dementia: an emerging group of 4-repeat tauopathies. *Acta Neuropathol.* **2011**, *122*, 415–428.

236. Kovacs, G. G.; Majtenyi, K.; Spina, S.; Murrell, J. R.; Gelpi, E.; Hoftberger, R., et al. White matter tauopathy with globular glial inclusions: a distinct sporadic frontotemporal lobar degeneration. *J. Neuropathol. Exp. Neurol.* **2008**, *67*, 963–975.

237. Steele, J. C.; Richardson, J.; Olszewski, J. Progressive supranuclear palsy. A heterogeneous degeneration involving brain stem, basal ganglia and cerebellum with vertical gaze and pseudobulbar palsy, nuchal dystonia and dementia. *Arch. Neurol.* **1964**, 333–359.

238. Dickson, D. W. Parkinson's disease and parkinsonism: neuropathology. *Cold Spring Harb. Perspect. Med.* **2012**, *2*, a009258.

239. Williams, D. R.; Lees, A. J. Progressive supranuclear palsy: clinicopathological concepts and diagnostic challenges. *Lancet Neurol.* **2009**, *8*, 270–279.

240. Hauw, J. J.; Verny, M.; Delaere, P.; Cervera, P.; He, Y.; Duyckaerts, C. Constant neurofibrillary changes in the neocortex in progressive supranuclear palsy. Basic differences with Alzheimer's disease and aging. *Neurosci. Lett.* **1990**, *119*, 182–186.

241. Wenning, G. K.; Krismer, F.; Poewe, W. New insights into atypical parkinsonism. *Curr. Opin. Neurol.* **2011**, *24*, 331–338.

242. Constantinidis, J.; Richard, J.; Tissot, R. Pick's disease. Histological and clinical correlations. *Eur. Neurol.* **1974**, *11*, 208–217.

243. Brion, S.; Plas, J.; Jeanneau, A. Pick's disease. Anatomo-clinical point of view. *Rev. Neurol.* **1991**, *147*, 693–704.

244. Buee, L.; Delacourte, A. Comparative biochemistry of tau in progressive supranuclear palsy, corticobasal degeneration, FTDP-17 and Pick's disease. *Brain Pathol.* **1999**, *9*, 681–693.

245. Hogg, M.; Grujic, Z. M.; Baker, M.; Demirci, S.; Guillozet, A. L.; Sweet, A. P., et al. The L266V tau mutation is associated with frontotemporal dementia and Pick-like 3R and 4R tauopathy. *Acta Neuropathol.* **2003**, *106*, 323–336.

246. Probst, A.; Tolnay, M.; Langui, D.; Goedert, M.; Spillantini, M. G. Pick's disease: hyperphosphorylated tau protein segregates to the somatoaxonal compartment. *Acta Neuropathol.* **1996**, *92*, 588–596.

247. Harper, P. S. *Myotonic Dystrophy*, 2nd edition; Saunders: London, 1989.

248. Brook, J. D.; McCurrach, M. E.; Harley, H. G.; Buckler, A. J.; Church, D.; Aburatani, H., et al. Molecular basis of myotonic dystrophy: expansion of a trinucleotide (CTG) repeat at the 30 end of a transcript encoding a protein kinase family member. *Cell* **1992**, *68*, 799–808.

249. Dhaenens, C. M.; Tran, H.; Frandemiche, M. L.; Carpentier, C.; Schraen-Maschke, S.; Sistiaga, A., et al. Mis-splicing of Tau exon 10 in myotonic dystrophy type 1 is reproduced by overexpression of CELF2 but not by MBNL1 silencing. *Biochim. Biophys. Acta* **2011**, *1812*, 732–742.

250. Ghanem, D.; Tran, H.; Dhaenens, C. M.; Schraen-Maschke, S.; Sablonnière, B.; Buée, L., et al. Altered splicing of Tau in DM1 is different from the foetal splicing process. *FEBS Lett.* **2009**, *583*, 675–679.

251. Spillantini, M. G.; Goedert, M. Tau pathology and neurodegeneration. *Lancet Neurol.* **2013**, *12*, 609–622.

252. Goedert, M.; Ghetti, B.; Spillantini, M. G. Frontotemporal dementia: implications for understanding Alzheimer disease. *Cold Spring Harb. Perspect. Med.* **2012**, *2*, a006254.

253. Ghetti, B.; Wszolek, Z. W.; Boeve, B. F.; Spina, S.; Goedert, M. Frontotemporal dementia and parkinsonism linked to chromosome 17. In *Neurodegeneration: The molecular pathology of dementia and movement disorders*; Dickson, D., Weller, R. O., Eds.; 2nd ed.; Blackwell: Oxford, UK, 2011; pp. 110–134.

254. Hasegawa, M.; Smith, M. J.; Goedert, M. Tau proteins with FTDP-17 mutations have a reduced ability to promote microtubule assembly. *FEBS Lett.* **1998**, *437*, 207–210.

255. Pickering-Brown, S. M.; Baker, M.; Nonaka, T.; Ikeda, K.; Sharma, S.; Mackenzie, J., et al. Frontotemporal dementia with Pick-type histology associated with Q336R mutation in the tau gene. *Brain* **2004**, *127*, 1415–1426.

256. Rossi, G.; Bastone, A.; Piccoli, E.; Morbin, M.; Mazzoleni, G.; Fugnanesi, V., et al. Different mutations at V363 MAPT codon are associated with atypical clinical phenotypes and show unusual structural and functional features. *Neurobiol. Aging* **2014**, *35*, 408–417.

257. Hayashi, S.; Toyoshima, Y.; Hasegawa, M.; Umeda, Y.; Wakabayashi, K.; Tokiguchi, S., et al. Late-onset frontotemporal dementia with a novel exon 1 (Arg5His) tau gene mutation. *Ann. Neurol.* **2002**, *51*, 525–530.

258. Stanford, P. M.; Shepherd, C. E.; Halliday, G. M.; Brooks, W. S.; Schofield, P. W.; Brodaty, H., et al. Mutations in the tau gene that cause an increase in three repeat tau and frontotemporal dementia. *Brain* **2003**, *126*, 814–826.

259. van Swieten, J.; Spillantini, M. G. Hereditary frontotemporal dementia caused by *Tau* gene mutations. *Brain Pathol.* **2007**, *17*, 63–73.

260. Varani, L.; Hasegawa, M.; Spillantini, M. G.; Smith, M. J.; Murrell, J. R.; Ghetti, B., et al. Structure of tau exon 10 splicing regulatory element RNA and destabilization by mutations of frontotemporal dementia and parkinsonism linked to chromosome 17. *Proc. Natl. Acad. Sci. U.S.A.* **1999**, *96*, 8229–8234.

261. Ros, R.; Thobois, S.; Streichenberger, N.; Kopp, N.; Sanchez, M. P.; Perez, M., et al. A new mutation of the tau gene, G303V, in early-onset familial progressive supranuclear palsy. *Archiv. Neurol.* **2005**, *62*, 1444–1450.

262. D'Souza, I.; Schellenberg, G. D. Tau exon 10 expression involves a bipartite intron 10 regulatory sequence and weak 5' and 3' splice sites. *J. Biol. Chem.* **2002**, *277*, 26587–26599.

263. Rohrer, J. D.; Paviour, D.; Vandrovcova, J.; Hodges, J.; de Silva, R.; Rossor, M. N. Novel L284R MAPT mutation in a family with an autosomal dominant progressive supranuclear palsy syndrome. *Neurodeg. Dis.* **2011**, *8*, 149–152.

264. Kovacs, G. G.; Pittman, A.; Revesz, T.; Luk, C.; Lees, A.; Kiss, E., et al. MAPT S305I mutation: implications for argyrophilic grain disease. *Acta Neuropathol.* **2008**, *116*, 103–118.

265. Yoshida, H.; Crowther, R. A.; Goedert, M. Functional effects of tau gene mutations deltaN296 and N296H. *J. Neurochem.* **2002**, *80*, 548–551.

266. van Swieten, J. C.; Bronner, I. F.; Azmani, A.; Severijnen, L. A.; Kamphorst, W.; Ravid, R., et al. The ΔK280 mutation in MAP *tau* favours exon 10 skipping in vivo. *J. Neuropathol. Exp. Neurol.* **2007**, *66*, 17–25.

267. Hasegawa, M.; Smith, M. J.; Iijima, M.; Tabira, T.; Goedert, M. FTDP-17 mutations N279K and S305N in tau produce increased splicing of exon 10. *FEBS Lett.* **1999**, *443*, 93–96.

268. von Bergen, M.; Barghorn, S.; Li, L.; Marx, A.; Biernat, J.; Mandelkow, E. M.; Mandelkow, E. Mutations of tau protein in frontotemporal dementia promote aggregation of paired helical filaments by enhancing local β-structure. *J. Biol. Chem.* **2001**, *276*, 48165–48174.

269. Lippa, C. F.; Zhukareva, V.; Kawarai, T.; Uryu, K.; Shafiq, M.; Nee, L. E., et al. Frontotemporal dementia with novel tau pathology and a Glu342Val tau mutation. *Ann. Neurol.* **2000**, *48*, 850–858.

270. Iyer, A.; LaPointe, N. E.; Zielke, K.; Berdynski, M.; Guzman, E.; Barczak, A., et al. A novel MAPT mutation, G55R, in a frontotemporal dementia patient leads to altered tau function. *PLoS One* **2013**, *8*, e76409.

271. Combs, B.; Gamblin, T. C. FTDP-17 tau mutations induce distinct effects on aggregation and microtubule interactions. *Biochemistry* **2012**, *51*, 8597–8607.

272. Goedert, M.; Jakes, R.; Crowther, R. A. Effects of frontotemporal dementia FTDP-17 mutations on heparin-induced assembly of tau filaments. *FEBS Lett.* **1999**, *450*, 306–311.

273. Rossi, G.; Bastone, A.; Piccoli, E.; Mazzoleni, G.; Morbin, M.; Uggetti, A., et al. New mutations in *MAPT* gene causing frontotemporal lobar degeneration: biochemical and structural characterization. *Neurobiol. Aging* **2012**, *33*, 834e1–834e6.

274. Rizzini, C.; Goedert, M.; Hodges, J. R.; Smith, M. J.; Jakes, R.; Hills, R., et al. Tau gene mutation K257T causes a tauopathy similar to Pick's disease. *J. Neuropathol. Exp. Neurol.* **2000**, *59*, 990–1001.

275. Wszolek, Z.; Tsuboi, Y.; Ghetti, B.; Cheshire, W. Frontotemporal dementia and parkinsonism linked to chromosome 17 (FTDP-17). Orphanet Encyclopedia 2003, http://www.orpha.net/data/patho/GB/uk-FTDP.pdf.

276. Reed, L. A.; Wszolek, Z. K.; Hutton, M. Phenotypic correlations in FTDP-17. *Neurobiol. Aging* **2001**, *22*, 89–107.

277. Poorkaj, P.; Bird, T. D.; Wijsman, E.; Nemens, E.; Garruto, R. M.; Anderson, L., et al. Tau is a candidate gene for chromosome 17 frontotemporal dementia. *Ann. Neurol.* **1998**, *43*, 815–825.

278. Delisle, M. B.; Murrell, J. R.; Richardson, R.; Trofatter, J. A.; Rascol, O.; Soulages, X., et al. A mutation at codon 279 (N279K) in exon 10 of the Tau gene causes a tauopathy with dementia and supranuclear palsy. *Acta Neuropathol.* **1999**, *98*, 62–77.

279. D'Souza, I.; Poorkaj, P.; Hong, M.; Nochlin, D.; Lee, V. M. Y.; Bird, T. D.; Schellenberg, G. D. Missense and silent tau gene mutations cause frontotemporal dementia with parkinsonism-chromosome 17 type, by affecting multiple alternative RNA splicing regulatory elements. *Proc. Natl. Acad. Sci. U.S.A.* **1999**, *96*, 5598–5603.

280. Stanford, P. M.; Halliday, G. M.; Brooks, W. S.; Kwok, J. B.; Storey, C. E.; Creasey, H., et al. Progressive supranuclear palsy pathology caused by a novel silent mutation in exon 10 of the tau gene: expansion of the disease phenotype caused by tau gene mutations. *Brain* **2000,** *123,* 880–893.

281. Grover, A.; Houlden, H.; Baker, M.; Adamson, J.; Lewis, J.; Prihar, G., et al. 5' splice site mutations in tau associated with the inherited dementia FTDP-17 affect a stem loop structure that regulates alternative splicing of exon 10. *J. Biol. Chem.* **1999,** *274,* 15134–15143.

282. Pickering-Brown, S.; Baker, M.; Yen, S. H.; Liu, W. K.; Hasegawa, M.; Cairns, N., et al. Pick's disease is associated with mutations in the tau gene. *Ann. Neurol.* **2000,** *48,* 859–867.

283. Murrell, J. R.; Spillantini, M. G.; Zolo, P.; Guazzelli, M.; Smith, M. J.; Hasegawa, M., et al. Tau gene mutation G389R causes a tauopathy with Pick body-like inclusions and axonal deposits. *J. Neuropathol. Exp. Neurol.* **1999,** *58,* 1207–1226.

284. Spillantini, M. G.; Bird, T. D.; Ghetti, B. Frontotemporal dementia and parkinsonism linked to chromosome 17: a new group of tauopathies. *Brain Pathol.* **1998,** *8,* 387–402.

285. Bugiani, O.; Murrell, J. R.; Giaccone, G.; Hasegawa, M.; Ghigo, G.; Tabaton, M., et al. Frontotemporal dementia and corticobasal degeneration in a family with a P301S mutation in tau. *J. Neuropathol. Exp. Neurol.* **1999,** *58,* 667–677.

286. Zarranz, J. J.; Ferrer, I.; Lezcano, E.; Forcadas, M. I.; Eizaguirre, B.; Atares, B., et al. A novel mutation (K317M) in the MAPT gene causes FTDP and motor neuron disease. *Neurology* **2005,** *64,* 1578–1585.

287. Wszolek, Z. K.; Slowiński, J.; Golan, M.; Dickson, D. W. Frontotemporal dementia and parkinsonism linked to chromosome 17. *Folia Neuropathol.* **2005,** *43,* 258–270.

288. Grimm, S.; Hoehn, A.; Davies, K. J.; Grune, T. Protein oxidative modifications in the ageing brain: consequence for the onset of neurodegenerative disease. *Free Radical Res.* **2011,** *45,* 73–88.

289. Oueslati, A.; Fournier, M.; Lashuel, H. A. Role of post-translational modifications in modulating the structure, function and toxicity of α-synuclein: implications for Parkinson's disease pathogenesis and therapies. *Progr. Brain Res.* **2010,** *183,* 115–145.

2

Targeting the Protein Quality Control (PQC) Machinery
The Neuronal Salvation Army

2.1 MOLECULAR CHAPERONES, PQC, AND NEURODEGENERATION

Misfolded/misdecorated protein monomers in general [1], and tau monomers in particular [2], are processed through several paths in neurons. Such paths may be detrimental—leading to neurotoxic aggregates—or advantageous—either refolding abnormal protein copies, or disposing of them. The former must be antagonized, and the latter must be potentiated to prevent tau aggregation.

Abnormally decorated, misfolded protein copies are toxic to neurons [3,4]. The cellular quality control system takes care of misfolded proteins in normal cells. Protein chaperones, including—among others—the *heat shock protein* (*Hsp70* [5] and *Hsp90* [6]) systems, exert the control function participating to a network of protein quality control (PQC) activities briefly described in Chapter 1 (see Figure 1.1). Chaperones are intrinsically disordered proteins with broad substrate specificity, with high affinity for misfolded proteins and reduced affinity for their natively folded, physiologically useful counterparts [7].

Chaperones contribute to keep the needed level of functional proteins in the cell. Chaperone *holdases* hold aggregation-prone, partially misfolded substrates to prevent their aggregation and to stabilize them [8,9]. Chaperone *(un)foldases* bind partially or totally misfolded proteins, unfold and correctly refold them. The process requires ATP hydrolysis and leads to energy-minimized, physiologically competent client proteins through binding and localized reorganization with disordered chaperone regions [10,11]. Once refolding is complete, (un)foldases and correctly folded client proteins split, due to reduced binding affinity. *Disaggregating chaperones* bind many copies of aggregated, misfolded substrates in a

Molecular Targets in Protein Misfolding and Neurodegenerative Disease. DOI: 10.1016/B978-0-12-800186-8.00002-X concerted manner to unfold–refold them, causing their disaggregation and refolding–detoxification [11,12]. Overall, refolding misfolded proteins reduces the risk of aggregation and formation of toxic oligomers, and provides functionally competent protein copies [13].

The human chaperone network is complex, and its functions are only partially elucidated. A recent review [13] represents—in a simplified manner!—such a network, in an Hsp70 and Hsp90 chaperone family-centered manner. The network (graphically represented in Figure 2.1) entails, *inter alia*, 14 Hsp70 family members (pink circle); at least six homo- and heterodimeric Hsp90 isoforms (yellow); more than 40 Hsp40/J-domain co-chaperones (Hsp70 co-chaperones) belonging to three subfamilies (light blue); more than 15 nuclear exchange factors (NEFs, Hsp70 co-chaperones) belonging to four subfamilies (light green); an Hsp70 co-chaperone singleton (Hip, orange); seven Hsp90 co-chaperones (light brown); two dual Hsp70–Hsp90 co-chaperones (Hop and CHIP, gray); and 10 small Hsp chaperones (sHsps, white). The role of many co-chaperones in physiological

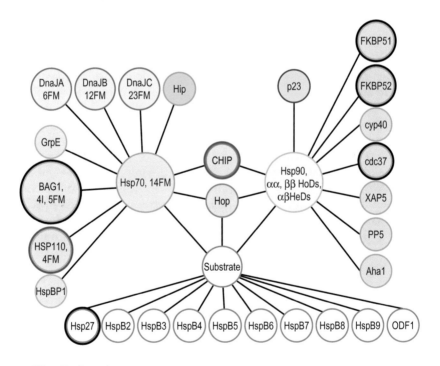

FM – Family members
I   – Isoforms
HoD – Homodimer
HeD – Heterodimer

FIGURE 2.1   The Hsp chaperone/co-chaperone network.

and pathological events related to neurodegeneration and tau will be mentioned; some of them will be covered in detail in this chapter (thick black) or in the following chapters (thick purple).

If the abnormal splicing/post-translational modification (PTM) pattern of tau is not rapidly reversed, or if pathological tau mutations are present (see Chapter 1, Figure 1.2), misdecorated tau copies adopt aberrant conformations. Abundant, misfolded tau copies tend to aggregate, recruiting misfolded tau copies but also non-hyper-phosphorylated (HP)-tau.

Chaperones are a prospective point of intervention against tauopathies, due to their interactions with tau [14,15]. sHsps counteract protein aggregation, and Hsp27 in particular shows strong interactions with tau and Aβ [16]. Hsp70 family members bind to tau [17], decrease its aggregation [18], and increase its microtubule (MT) binding as such [19], or through complexes with Hsp70 co-chaperones [20]. Hsp90 has a major impact on tau aggregation and folding *per se* [21], or through Hsp90 complexes with immunophilins FK-binding protein 51 (FKBP51) [22] and FK-binding protein 52 (FKBP52) [23]. Hsp70 and Hsp90 chaperone families show impaired functions in tauopathies [13], and are validated targets for the treatment of tauopathies. The same is true for Hsp70– and Hsp90–co-chaperone complexes, that also promote the degradation/elimination of misfolded tau copies, or aggregates [13].

2.2 MOLECULAR TARGETS

2.2.1 Hsp27

Hsp chaperones other than Hsp70 and Hsp90 have an impact on PQC [24,25], and specifically on tau misfolding, aggregation, and disaggregation. *Hsp27* [26] is a 205 amino acid-containing member of the small Hsp family (sHsp) [27], highly homologous with α-crystallin, the main eye lens protein. sHsps are ATP-independent chaperones capable of disrupting protein aggregation [28]. The core α-crystallin domain (ACD) is flanked by an N-terminal (WDPF) motif, and a short C-terminal domain [27]. Hsp27 exists as a mix of aggregated oligomers with size up to 800 kDa [29]. Oligomers are formed through the WDPF and ACD domains [30]. The phosphorylation level of Hsp27 is inversely proportional to its oligomerization state [31]. Oligomerization-regulating phosphorylation of Ser15, Ser78, and Ser82 is mostly due to mitogen-activated protein kinase-activated protein kinase 2 (MAPKAPK2, MK2) [32]. Highly phosphorylated, small Hsp27 dimers [30] and tetramers [33] increase their rescuing capacity of improperly folded/partially denatured proteins [16,34]. Protein rescuing in basal/physiological conditions correlates with phosphorylation-dependent translocation of Hsp27 from the cytosol to, *inter alia*, the nucleus and the cytoskeleton [35]. Hsp27 is then capable of handing over its misfolded client protein to ATP-dependent refolding chaperones [36]. Conversely, under stress conditions Hsp27 assists both the ubiquitylation-dependent [37] and ubiquitylation-independent [38] disposal of misfolded proteins through proteasomal degradation. Dynamic cycling of the oligomeric state of Hsp27 between phosphorylated dimers and tetramers (binding with aggregation-prone client proteins, inhibition of aggregation) and dephosphorylated larger oligomers (handover of client proteins to refolding chaperones and/or to the proteasome) is crucial for proper Hsp27 functioning in basal conditions [16].

Hsp27 is ubiquitously expressed in all human tissues, and in particular in the central nervous system (CNS) with a site- and cell-specific pattern [39,40]. Its expression is stimulated by stress triggers [41,42]. Around 20 Hsp27 mutations, equally partitioned among the three domains, correlate with the development of *distal hereditary motor neuropathy (DHMN)*, a severe motor neuron disease causing damage to the peripheral nervous system [43]. As to other neurodegenerative diseases (NDDs), Hsp27 levels are increased in patients affected by *PD* and *dementia with Lewy bodies* (DLB) [44], influencing the cytotoxic activity of α-synuclein aggregates [45]. Hsp27 seems to be neuroprotective in *Huntington disease (HD)* [46], possibly decreasing the level of reactive oxygen species (ROS) caused by huntingtin. Hsp27 is protective *in vitro* against β-amyloid toxicity [47], although its levels increase in diseased tissues from *Alzheimer's disease (AD)* [48,49]. One may presume that the PQC role of Hsp27 in physiological conditions could be reversed by NDD-inducing pathological events, and the tau–Hsp27 interactions illustrate this hypothesis.

A schematic representation of the Hsp27 network under basal/physiological (green arrows), stress/recoverable (blue arrows), and abnormal/pathological/NDD-promoting conditions (red arrows) is depicted in Figures 2.2 and 2.3.

Single Hsp27 copies aggregate into large Hsp27 oligomers (step 1, Figure 2.2), and most client proteins exist in their functional, properly folded conformation in basal conditions. Stress-induced client protein misfolding (step 2) elicits stress-induced phosphorylation of large Hsp27 oligomers by kinases such as MAPKAPK2 (step 3) [32], leading to large Hsp27 oligomer dismantling, small Hsp27 oligomer formation (dimer/tetramer, depicted as tetramers in Figure 2.2) and misfolded client protein capture by small Hsp27 oligomers (steps 4 and 4a). Then, client protein-charged Hsp27 oligomers either hand over their cargos to Hsp70-based chaperones [50] (step 5; physiological conditions, leading to proper refolding and Hsp70 release, respectively steps 6 and 7), or they pass it over to Hsp90-based chaperones [37] (step 8; stressed but recoverable conditions, leading to proteasome elimination of misfolded client proteins, step 9). In either case, small phosphorylated Hsp27 oligomers are dephosphorylated

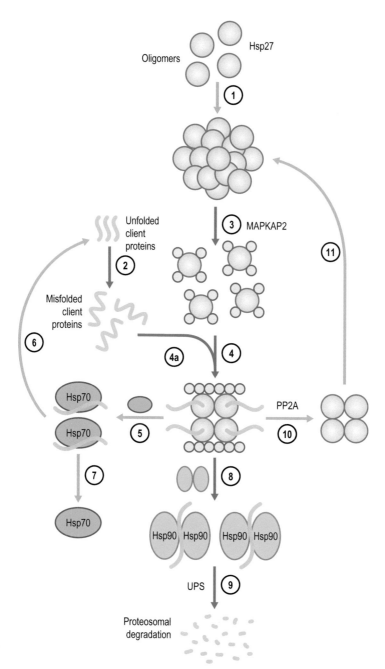

FIGURE 2.2 The Hsp27 chaperone cycle: physiological and stress-recoverable aspects.

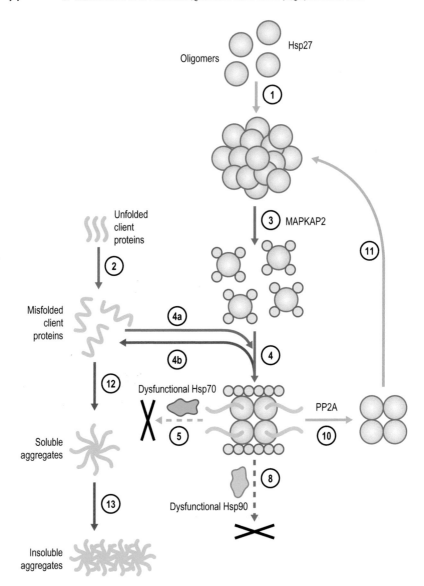

FIGURE 2.3   The Hsp27 chaperone cycle: pathological aspects.

by phosphatases such as protein phosphatase 2A (step 10, [51]), and then reassemble in large, dephosphorylated Hsp27 oligomers (step 11, Figure 2.2). In either case, misfolded client proteins are disposed of before their pathological aggregation.

Conversely, if the PQC machinery downstream of Hsp27 is pathologically impaired, another scenario takes place (Figure 2.3).

Steps 1 to 4a are identical, and lead to the formation of client protein-charged Hsp27 oligomers. Step 5/Hsp70-driven refolding and step 8/ Hsp90- and UPS-driven degradation, though, are impaired and cannot relieve the burden of misfolded client proteins. Thus, the misfolded protein copies are released from client protein-charged Hsp27 oligomers (step 4b, Figure 2.3), and aggregate into neurotoxic species (soluble oligomers, step 12, and eventually insoluble protein aggregates, step 13, Figure 2.3).

Hsp27 acts as a "holdase" on HP-tau, but does not bind to non-HP-tau. It hinders tau aggregation *in vitro*, and prevents neuronal apoptosis [52]. Gene delivery of Hsp27 to a transgenic (TG) tauopathy model reduces tau neuronal levels [16], due to the handover of aggregation-prone HP-tau by Hsp27 to other Hsp chaperones. A dynamic PTM–phosphorylation equilibrium of Hsp27 is needed to clear aggregation-prone tau from neurons. In fact, a permanent pseudo-phosphorylated Hsp27 mutant with three Ser-Asp substitutions still binds HP-tau, but causes an increase of tau levels [16]. A large excess of highly HP, non-refoldable tau—i.e., a common situation in tauopathies—may turn the basal PQC function of Hsp27 into a neurotoxic stabilization of tauopathy-prone HP-tau. Thus, inhibition of Hsp27 and/or of some of its roles should have a therapeutic outcome against tauopathies, and in animal models of tauopathies.

In fact, an Hsp27 TG mouse shows increased activation of glycogen synthase kinase (GSK-3β) and, consequently, HP-tau at pathological Ser262 and Ser396/404 residues [53]. The interplay between GSK-3β-mediated tau HP, the Hsp27 "holdase"-handover role and neuronal rescuing needs further elucidation, but its relevance to tauopathies is evident. In fact, Hsp27 levels increase in individuals affected by two tauopathies, progressive supranuclear palsy (PSP) and corticobasal degradation (CBD) [54].

2.2.2 Hsp70

The Hsp70 protein family is composed of 14 members in humans [55]. *Hsc70* [56] and *Hsp70* [57]—respectively a 73 kDa constitutively expressed and a 72kDa stress-induced heat shock protein—are the most important family members. They share an ≈92% sequence homology and a similar, but not identical, substrate specificity [55].

Hsp70 proteins contain two conserved binding domains for nucleotides (N-terminal, NBD) and for peptide substrates (SBD), and a less conserved C-terminal lid domain [58,59]. When the NBD binds ATP, the lid domain is in an "open" state, and the SBD assumes a conformation that promotes dynamic cycling between binding and releasing a given peptide substrate with moderate affinity and fast kinetics [60]. When the ATPase function in the NBD hydrolyzes ATP to ADP, the interaction between SBD and substrate is tightened, the lid assumes a closed conformation, and the kinetics for substrate release are slowed down [61]. The energy produced by ATP hydrolysis is used to unfold misfolded regions of the bound substrates, and to promote their native-physiological refolding [13].

A small number of Hsp70 family members bind a large number of protein substrates. The loose sequence specificity of Hsp70 proteins targets extended, accessible ≈seven-member-long peptide sequences of non-bulky, hydrophobic amino acids neighboring positively charged residues [62]. A conservative estimation sets the Hsp70 client protein pool at hundreds, if not thousands, of members. The number exponentially increases when one considers the possible folding conformations of Hsp70-binding sequences for any given client protein. Hsp70 family members need optimized substrate specificity, (un)folding efficiency, and dynamic turnover to direct the fate of nascent proteins—mostly by constitutive Hsc70—and to fix misfolded protein copies in stress conditions—mostly by stress-induced Hsp70 [63]. Hsp70 co-chaperones [64] make the whole machinery selective, efficient, and robust.

Hsp40/J-domain co-chaperones [65,66] bind onto the Hsp70 NBD through a conserved, ≈70 amino acidic sequence (J-domain). Their binding stimulates ATP hydrolysis [67] and provides substrate specificity [68], as more than 40 Hsp40 co-chaperones are known in humans. The stability of ADP-bound Hsp70 is increased by parallel binding of Hsp70 NBD with the *Hsp interacting protein (Hip)* chaperone [69], preventing premature substrate release [70]. *Nuclear exchange factor (NEF)* co-chaperones [71,72], then, stimulate release of ADP and terminate the ATP cycling-substrate processing circle [73]. Four main NEF families bind the Hsp70 NBD with different mechanisms/without interfamily conserved binding domains [74,75]. More than 15 NEFs are known in humans [13]. Finally, *carboxy-terminus of Hsp70-interacting protein (CHIP)* [76] and *Hsp70/Hsp90 organizing protein (Hop)* [77] bind simultaneously to Hsp70 and Hsp90. The former orients the Hsp70 chaperone towards client degradation, while the latter drives it towards client folding [14]. They both regulate the Hsp70 and Hsp90 chaperone systems [78], their interactions [79], and the fate of released peptide substrates—for example, Hop-promoted folding [80] and client protein degradation *via* CHIP-mediated ubiquitination [81]. If Hsp70 family members bind to one representative of each co-chaperone family at a given time, their combinations into multi protein complexes amount to tens of thousands [63]. Such a complex network, and its disease implications, needs further clarification. Drug discovery efforts should target a specific, disease-validated Hsp70–co-chaperone complex, rather than the whole Hsp70 family, to focus on desired biological effects and to minimize unwanted side effects [14].

Hsp70 family members, and their co-chaperone complexes, perform various control functions on their protein substrates. An ATP-independent "holdase" role is usually coupled with an ATP-dependent "(un)foldase" activity [82,83]. Often the Hsp70 machinery can disaggregate soluble oligomers by binding to multiple, misfolded client copies in the oligomer and pulling them apart through an ATPase-dependent mechanism [84]. When the ATPase activity of Hsp70 is inhibited by small molecules [82] or by biological means [83] the "holdase" role may lead to a toxic "gain of function" stabilization of misfolded proteins which cannot be rescued [85]. Alternatively, inhibition of Hsp70 ATPase activity in stress conditions may switch the Hsp70 quality control machinery towards the needed elimination of damaged proteins, through the Hsp70 complex with co-chaperones such as CHIP [81].

A sketched rendition of the Hsp70 refolding chaperone cycle is reported in Figure 2.4 [64].

The first interaction with a misfolded protein copy may happen through an Hsp40 family member (step 1a, top, Figure 2.4), or directly through Hsp70 (step 1b, bottom, Figure 2.4). In the former case ATP-bound Hsp70 binds to the Hsp40–client protein complex (step 2a), while in the latter an Hsp40 family member binds the ATP-bound Hsp70–client protein complex (step 2b). In both cases, the result is the ternary Hsp40–ATP-bound Hsp70–client protein complex, where the client protein is bound with low affinity in the misfolded state. A non-productive Hsp70 chaperone cycle may result from client protein dissociation (step 3c), causing ATP-bound Hsp70–Hsp40 dissociation (respectively steps 3a, 3b). Such a cycle, nevertheless, leaves the misfolded client protein ready for a new Hsp40- or Hsp70-driven interaction.

A productive Hsp70 chaperone cycle (Figure 2.5) may stem from the ternary Hsp40–ATP-bound Hsp70–client protein complex.

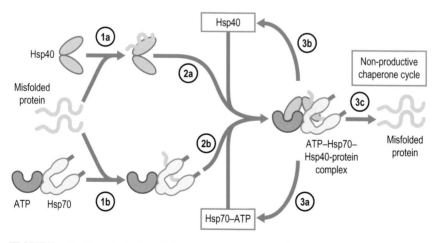

FIGURE 2.4 The Hsp70-based chaperone cycle: non-productive interactions with client proteins and co-chaperones.

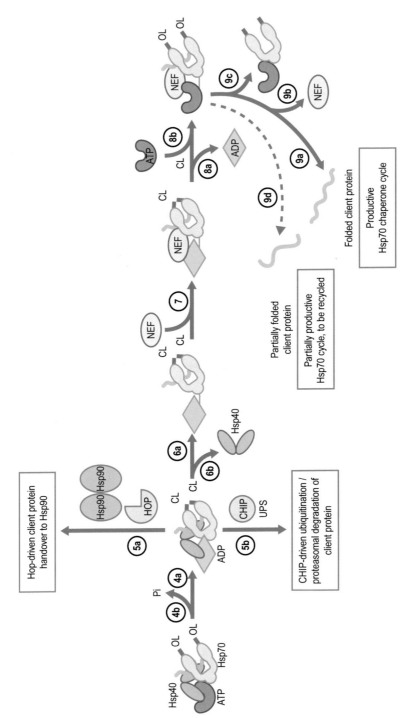

FIGURE 2.5　The Hsp70-based chaperone cycle: productive and partially productive interactions with client proteins and co-chaperones.

The ternary complex may undergo Hsp40 J domain-promoted ATP hydrolysis (step 4a) with inorganic phosphate elimination (step 4b). The conformational change of Hsp70 following ATP hydrolysis causes the previously mentioned lid closure, and the strengthening of the ADP-bound Hsp70–client protein complex. The energy from ATP hydrolysis is used to unfold and properly refold the misfolded regions of the client protein. The ternary Hsp40–ADP-bound Hsp70–client protein complex may then interact with either one of the dual Hsp70–Hsp90 chaperones. Hop induces the Hsp70–Hsp90 interaction, followed by handover of the client protein to Hsp90 (step 5a, see next paragraph) [86]. CHIP interacts with Hsp70 and induces the ubiquitylation of client proteins, directing them to degradation *via* the proteasome (step 5b, see Chapter 3) [87]. The ratio of Hsp70–CHIP/Hsp70–Hop bound complexes is influenced by the degree of C-terminus phosphorylation of Hsp70. Higher phosphorylation levels favor Hsp70–Hop interactions and refolding (basal conditions, physiological growth), while lower phosphorylation levels favor Hsp70–CHIP interactions (stress conditions, disposal of unrecoverable protein copies) [88].

Alternatively, if the ternary complex continues its progress through the Hsp70 refolding cycle, at first a binary ADP-bound Hsp70–client protein complex is formed (step 6a) by release of Hsp40 (step 6b). Then, an NEF family member binds the binary protein complex (step 7), leading to sequential release of ADP (step 8a, due to the distortion in the NBD domain by NEF binding) and binding to ATP (step 8b, due to much higher concentration of ATP than ADP in cells). Finally, ATP binding weakens the refolded client protein–Hsp70 interaction and causes the overall dismantling of the ternary NEF–ATP-bound Hsp70–client protein complex into its three individual partners (steps 9a–d, Figure 2.5). If complete refolding has taken place, the native protein is now ready to perform its functions (step 9a). Both the NEF family member and ATP-bound Hsp70 re-enter the productive Hsp70 refolding chaperone cycle (steps 9b, c). If refolding is not complete, the partially folded protein (step 9d) may re-enter another Hsp70 chaperone cycle to complete its refolding.

Protein misfolding is a major factor in aging and in neurological disturbances [89,90]. PQC and proteostasis are important in neurons, to avoid aggregation diseases. Consequently, the Hsp70 machinery is essential for healthy neurons, while its impairment shows a strong connection with NDDs [13,56].

Polyglutamine (polyQ) diseases [91], and HD in particular, are caused by the expansion of tri-nucleotide CAG repeats, leading to the inclusion of long poly-Gln sequences in abnormal, aggregation-prone proteins. Hsp70 is co-located with polyQ aggregates in HD [92]; it inhibits aggregation and cytotoxicity of polyQ proteins in cell-free and cellular models, together with J-domain co-chaperones [93,94]; and it has a modest anti-aggregation efficacy, when overexpressed, in transgenic HD models [95]. Hsp70 is also related to *PD* [96]. Lewy bodies, the typical protein aggregates in PD, contain Hsp70 [97]; Hsp70 overexpression reduces the aggregation of α-synuclein *in vitro* [98]; and it prevents neuronal loss in transgenic PD models [99]. Beneficial effects caused by reducing protein aggregation are observed in animal models of *amyotrophic lateral sclerosis (ALS)* [100], *spinocerebellar ataxia* [101], and *muscular atrophy* [102].

Increased levels of Hsp70 proteins are observed in hippocampal sections from *AD* patients [103]. Association studies show that genetic variations in some Hsp70 co-chaperones are more frequent in AD patients [104]. The complex between Hsp70 and J-domain co-chaperones blocks Aβ self-assembly *in vitro* [105], and prevents its cytotoxicity [106]. Aβ-TG mice crossed with Hsp70 overexpression show lower levels of Aβ, of Aβ plaque deposition, and of neuronal and synaptic loss than Aβ-TG mice [107].

Two binding motifs in the MT-binding region (MTBR) of *tau* promote a strong binding with the C-terminal region of Hsp70 family members— Hsc70 in particular [17]. They mostly contain bulky, hydrophobic residues (275-VQII-278 and 306-VQIV-309). Tau-Hsc/Hsp70 binding inhibits tau aggregation, as the same binding motifs promote tau–tau interactions with ≈100 times lower affinity. Tau-Hsc/Hsp70 binding is not mediated by co-chaperones, and is ATPase activity-independent [17,83]. Hsp70 family members prevent tau aggregation by binding soluble tau [85].

Human Hsp70 binds to preformed soluble tau oligomers but not to insoluble tau aggregates [84]. Anterograde fast axonal transport (FAT), inhibited by tau mixtures containing soluble and insoluble aggregates [108], is restored by Hsp70. Thus, soluble tau oligomers are likely to be the true neurotoxic species [84]. Tau–Hsp70 binding is isoform-dependent, with a 4R > 3R binding strength, and does not affect tau–MT interactions [83]. Thus, its role is to prevent pathological tau dysfunctions while preserving tau binding to MTs.

In basal conditions, Hsp70-bound tau is recycled into newly polymerized MTs in an Hsp70-promoted process [85]. In stress-pathological conditions, the excess of misfolded HP-tau is preserved from elimination by binding to Hsp70 family members, and awaits to be unfolded–refolded in an ATPase-dependent process. If the ATPase activity is overloaded, HP soluble tau becomes neurotoxic [85]. Hsp70 ATPase inhibition results in tau release from Hsp70 and subsequent degradation [82,83], and should have therapeutic usefulness against tauopathies.

The Hsp70 machinery has a broad client specificity. Although ATPase inhibition seems to cause tau-specific effects [82], one cannot exclude side effects on other client proteins. Specific Hsp70 family member–co-chaperone complexes are strongly validated as tauopathy-related targets for a therapeutic strategy. The J-domain co-chaperone DnaJA1 is a major tau regulator, switching it towards refolding or ubiquitin-mediated degradation [109]. The complex between Hsp70 and the NEF co-chaperone BAG-2 may drive HP, misfolded prone tau—even when bound to MTs—towards ubiquitin-mediated degradation [110].

More is known about the *Hsp70–BAG-1* (Bcl-2-associated athanogene 1) [111] complex. BAG-1 is a neuroprotective, anti-apoptotic factor with differentiating activity in neurons. Neuroprotection depends on Hsp70-mediated folding, while differentiation is Raf-1 kinase-dependent [112]. BAG-1 associates with the Hsp70–tau complex, but not with tau alone, in a tau isoform-independent manner. The ternary complex is found in the soluble cytosolic fraction while BAG-1 and Hsp70 cannot destabilize MT-bound tau [113]. BAG-1 overexpression does not alter tau expression, but increases tau levels through Hsp70-dependent inhibition of tau degradation *via* ubiquitin-independent 20S proteasome degradation [113].

BAG-1 is found in ≈30% of neurons from triple TG mice (amyloid, tau, and presenilin mutations). The percentage escalates to 95% when HP-tau-containing neurons are considered [113]. A comparison between hippocampal tissue from rapidly autopsied AD patients and non-diseased individuals shows similar abundance and distribution for small (BAG-1S, 33 kDa) and long (BAG-1L, 50 kDa) BAG isoforms. The BAG-1M–46 kDa isoform, conversely, is largely increased in the cytosolic fraction of tissues from AD patients [114]. BAG-1 co-localizes with tau tangles and with the amyloid precursor protein APP in the cytosol of neurons. BAG-1M overexpression increases tau levels by reducing its proteasome degradation, and indirectly increases Aβ levels by reducing APP degradation [114]. A genetic association between BAG-1 and frontotemporal lobar degeneration, a tauopathy, is established [115]. A similar genetic connection with AD is currently not established.

HP-tau is increased in cells overexpressing BAG-1—more tau, more access to tau for HP-promoting kinases—and in cells where BAG-1 is knocked out by small interfering RNA (siRNA)—decreased association of tau with Hsp70 and BAG-1, more soluble-unhindered tau, more access to tau for HP-promoting kinases [113]. Observations with TG mice and hippocampal tissues from AD patients indicate that BAG-1 contributes to stabilize aggregation-prone, misfolded, HP-tau in aging, tauopathy-prone brains. Thus, selective inhibitors of the BAG-1–Hsp70–tau complex should have a therapeutic role against tauopathies.

2.2.3 Hsp90

The Hsp90 protein family consists of two major cytoplasmic isoforms, constitutively expressed *Hsp90β* and stress-induced *Hsp90α* [116]. The two isoforms have a >85% sequence homology [117]. They are not completely redundant in terms of client proteins [118]. Three other isoforms—truncated Hsp90N [118], mitochondrial Hsp75, and endoplasmic reticulum-located Grp94 [119]—are much less characterized.

Hsp90 contains three major domains. An N-terminal, ATP-binding domain—NBD, missing in Hsp90N—precedes a middle segment containing a client protein domain and a catalytic loop to promote ATP hydrolysis—SBD. Finally, the C-terminal domain CTD is responsible for Hsp90 dimerization [6]. Hsp90 proteins mostly exist as $\alpha\alpha$ or $\beta\beta$ homodimers, but monomers, heterodimers, and higher oligomers are known [118].

Binding to ATP forces two C-terminally associated Hsp90 monomers to fully dimerize through swapping of their most N-terminal strand and rotation of a "lid" region in the N-terminal domain [120]. The lid closes over the ATP binding region, and related conformational changes bring the middle segment residue Arg380 close to bound ATP, assisting ATP hydrolysis. Both ADP-bound and nucleotide-free Hsp90 are open, less defined structures [120].

Hsp90 functions are largely ATP-dependent, but their nature is less clear than Hsp70. Rather than (un)folding and refolding of abnormal clients, Hsp90 seems to promote subtler conformational changes on partially or even largely folded proteins [6]. Hsp90 may stabilize and even activate bound client proteins. It invariably shows higher binding affinity for mutant/misfolded proteins than for their wild-type (WT)/correctly folded counterparts [121]. Hsp90 often determines the fate of a protein, switching between refolding and degradation in a client-dependent manner [122].

The client list for Hsp90 is constantly updated on the Web [123], and is likely to contain up to 10% of the total protein content of a given organism [124]. Clients are structurally unrelated, as is the detailed mechanism of their interaction with Hsp90 family members [120]. Kinases [125] and transcription factors [126] are the most represented client families. The cross-talk between Hsp70 and Hsp90 systems is extensive, although examples of the Hsp90 machinery as a single chaperone system for a few client proteins are known [13].

Hsp90 binds to the heat shock factor 1 (HSF1) and prevents HSF1-dependent heat shock response. Inhibition of Hsp90 leads to HSF1 release, and to stress-activated Hsp70 and Hsp40/J-domain co-chaperone induction [127]. The Hsp90 machinery often picks up an Hsp70 client protein by binding to the Hop–Hsp70 complex [14]. The ternary Hsp90–Hop–Hsp70 complex then loses Hsp70, and proceeds towards either binding to Hsp90 co-chaperones, refolding-release or degradation [128]. Client processing also depends on CHIP [128], the other shared Hsp70–Hsp90 co-chaperone. While Hop binds Hsp70 and Hsp90 on different binding sites, the Hsp-binding site on CHIP is shared by Hsp70 and Hsp90. Their switch in a CHIP-containing complex drives CHIP-ubiquitinated client proteins towards degradation [129].

A sketched depiction of the Hsp90 refolding chaperone cycle is reported in Figure 2.6. In addition to Hsp70–Hsp90 co-chaperones Hop and CHIP, Hsp90-specific co-chaperones described later in this paragraph may

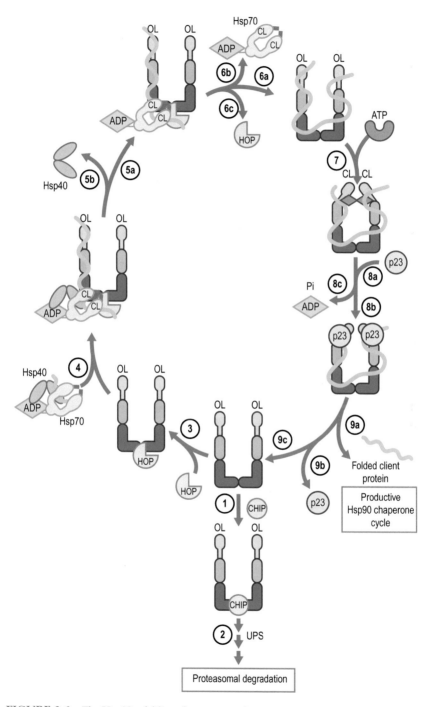

FIGURE 2.6 The Hsp90 refolding chaperone cycle.

influence this cycle and target it towards client protein subfamilies (e.g., cdc37 and protein kinases).

Initially, open, non-ATP-bound Hsp90 dimers may bind either to CHIP or to Hop. In the former case (step 1, Figure 2.6), any client protein bound to the CHIP–Hsp90 complex will be directed towards proteasomal degradation (step 2, described in detail in Chapter 3). In the latter case, Hop binds through one of its three tetratricopeptide (TPR) domains to a TPR acceptor sequence in the CTD domain of ADP–Hsp90 open dimers (step 3) [130]. The Hop–Hsp90 complex then binds through another Hop TPR domain to a different TPR acceptor sequence on Hsp70 in the ternary Hsp40–ADP-bound Hsp70–client protein complex, forming a five protein-containing complex (step 4, see also Section 2.2.2) [6]. While remaining Hsp70-bound, the client protein also interacts with the SBD of Hsp90. The loss of Hsp40 (step 5b) leads to the quaternary Hop–Hsp90–ADP-bound Hsp70–client protein complex (step 5a). The client protein transfer from Hsp70 to Hsp90 (step 6a) is completed by release of ADP-bound Hsp70 (step 6b) and Hop (step 6c). The NBDs of the resulting Hsp90–client protein complex (which were unavailable earlier to ATP due to Hop–Hsp90 binding) bind to two ATP molecules (step 7). ATP binding causes a major conformational change, leading to earlier mentioned full Hsp90 dimerization through NBD-binding/swapping, and to lid closure on the ATP binding domain [6]. Binding with Hsp90 co-chaperone progesterone receptor complex 3 (p23, step 8a) provides stabilization to the closed conformation of the ATP–Hsp90–client protein complex, does not significantly hinder the hydrolysis of ATP promoted by the closed conformation (step 8b), and provides needed time to Hsp90 to use the energy released by ATP hydrolysis to refold the bound client protein [6]. ADP and inorganic phosphate are then released (8c). The Hsp90 refolding cycle ends by release of the functional, refolded client protein (9a), of p23 (9b), and of non-ATP-bound Hsp90 (9c, Figure 2.6), ready to restart the cycle by binding with the ternary Hsp40–ADP-bound Hsp70–client protein complex and/or CHIP.

Hsp90 family proteins are mostly targeted in oncology [131,132] but their relevance against NDDs is clear [133]. Hsp90 complexes the polyQ expansion-containing mutant androgen receptor (mAR), typical of *spinal and bulbar muscular atrophy* (*SBMA*), a motor neuron disease [134]. An Hsp90 ATPase inhibitor shows motor improvements and biochemical effects in an *in vivo* SBMA model [135]. Hsp90 is linked to *PD* through aggregation—binding with α-synuclein [136]—and neurotoxicity—binding with mutant leucine repeat-rich kinase 2 (LRRK2) [137]. Finally, Hsp90 prevents the development of a form of familial PD through binding and stabilization of the phosphatase and tensin homolog (PTEN)-inducible kinase 1 (PINK1) [138].

AD and Hsp90 are connected [15]. Hsp90 proteins prevent the early stages of Aβ aggregation *in vitro* in an ATPase activity-dependent manner [105]. As to *tau*, Hsp90 binds to mutated P301L tau and positively regulates its stability, while failing to bind to WT human tau [21]. A brain-permeable Hsp90 inhibitor causes mutant tau destabilization and reduction of the mutated protein levels in a tau TG mouse mode [21]. WT tau–Hsp90 binding is reported in another paper [139], where such binding induces an HP- and aggregation-inducing conformation in tau. The tau–Hsp90 and tau–Hsp70 binding sites overlap, and a transfer of tau from Hsp70 to Hsp90 seems to take place during tau degradation [140]. A decrease of Hsp90 levels in the brains of AD patients is observed, leading to retention/stabilization of non-functional tau [140].

Hsp90 co-chaperones provide optimized substrate specificity, client-targeted efficiency, and dynamic turnover to direct the fate of Hsp90 substrate proteins [15]. Hsp90 co-chaperones may activate or inhibit the ATPase activity of Hsp90, and may modulate the Hsp90–client interaction. Some Hsp90 co-chaperones have peptidyl–prolyl isomerase (PPIase), phosphatase, or ubiquitylating activities [6]. Targeting specific Hsp90–co-chaperone complexes relevant for tau pathologies should represent a safer therapeutic approach than targeting Hsp90 proteins.

ATPase activators include the stress-regulated activator of Hsp90 ATPase (Aha1) [141] and the cyclophilin 40/cytoplasmic ribosomal protein-6 (Cyp40/Cpr6) PPIase [142]. Aha1 is involved in repositioning of the activation loop followed by Arg380-mediated ATP hydrolysis [142], and plays a role in v-Src kinase activation [143]. A positive regulatory role on the Hsp90 chaperone system is suggested for Aha1 [144], as its binding to Hsp90 is compatible with binding to most other co-chaperones. Cyp40 is a PPIase co-chaperone, involved in the late stages of client folding [145], and shares its Hsp90 binding site with FKBP51, FKBP52, and protein phosphatase 5 (PP5) [146]. Neither Aha1 nor Cyp40 have a clear connection to tau.

p23 [147] is an ATPase-inhibiting co-chaperone involved in late, client-releasing steps of the Hsp90 chaperone machinery [6]. It has a 1:2 binding stoichiometry with Hsp90, and consequently it may inhibit only up to 50% of its ATPase activity [148]. siRNA-mediated knockdown of p23 in cells overexpressing tau leads to a significant reduction of tau levels [22]. The wide range of p23-mediated functions, and the embryonic lethality of p23-null mice [149], does not encourage its consideration as a therapeutic target.

PP5 [150] is a negative modulator of Hsp90, possessing PPIase and phosphatase activities. Its enzymatic activity as a tau phosphatase is activated by Hsp90 binding [151]. Its action leads to increased tau–MT binding through dephosphorylation of protein kinase A (PKA)- and GSK-3-phosphorylated residues *in vitro* [152], including several phospho-epitopes linked to AD [153].

An opposite role is performed in tauopathies by two closely related immunophilins, *FKBP51* and *FKBP52* [154,155]. FKBP51 binds tau in cell cultures, and in *postmortem* brain lysates from AD patients and controls

[149]. siRNA depletion of FKBP51 in cells leads to a significant reduction of tau levels. Its overexpression leads to a decrease of ubiquitinated tau, to an increase of Hsp90–tau interactions, and to a decrease of Hsp90-mediated interactions between tau and the co-chaperones Aha1 and p23 [149]. FKBP51 appears in the brains of WT mice at around 5.5 months of age, and stabilizes tau through direct binding. Tau stabilization is independent from the PPIase activity of FKBP51, but tau rescuing (measured as increase of tubulin polymerization caused by rescuing of functional tau) is PPIase-dependent (FKBP51 mutants with no PPIase activity lead to accumulation of HP, aggregation-prone tau [149]).

FKBP52 binds tubulin *in vitro* and *ex vivo*, while FKBP51 does not [156]. Tubulin binding takes place either with each FKBP co-chaperone as such, or in the presence of a mixture of MT-associated proteins (MAPs), which assist MT assembly. Tubulin polymerization is inhibited by FKBP52, and tubulin disassembly is promoted in a MAP-dependent manner. FKBP52–tubulin binding is PPIase-independent, and involves multiple interactions on the C-terminus of FKBP52 [156]. Tau has a key role in tubulin–FKBP52 interactions, as tau–FKBP52 complexes—but not complexes with other MAPs—are immunoprecipitated from rat brain homogenates [23]. FKBP52 has a higher affinity for HP-tau than for normally phosphorylated tau, and prevents tau accumulation in cells [23].

The opposite roles played by >75% homologs FKBP51 and FKBP52 can be rationalized [155]. FKBP51 has a higher affinity for HP-tau, and isomerizes tau through its PPIase activity to a PP5 dephosphorylation-susceptible conformation. Dephosphorylation leads to tau recycling and MT stabilization–polymerization. FKBP52 preferentially binds to HP-tau too, but it also interacts with tubulin [155]. Its PPIase domain has enough difference from its FKBP51 counterpart to prevent isomerization-dependent tau rescuing. FKBP51 and FKBP52 compete for the same Hsp90 binding site, and a dynamic equilibrium—involving also handover from FKBP51 to PP5—is established in an Hsp90-dependent manner. Tau rescuing–recycling is promoted when microtubules are stabilized (e.g., FKBP51 prevalence). MT destabilization is promoted when neurite outgrowth is needed by removing tau (e.g., FKBP52 prevalence) [154]. The physiological equilibrium is pathologically exploited in tauopathies. FKBP51-based, PPIase-dependent tau rescuing is impaired, and HP-tau is protected by PP5 dephosphorylation in the development of tauopathies. A positive role of FKBP52 in tauopathies, possibly through the switch of the Hsp90 machinery towards tau degradation, is indirectly supported by observed lower FKBP52 levels (up to 75% in the brains of AD and FTDP-17 patients) [157]. Thus, FKBP51-negative and FKBP52-positive modulators may have a therapeutic and diagnostic role in tauopathies.

An Hsp90 co-chaperone involved in tau processing is *cell division cycle 37 homolog (cdc37)* [158]. Cdc37 interacts with kinases, the largest client family for Hsp90. Namely, cdc37 is an ATPase-inhibiting Hsp90 chaperone that first promotes and then stabilizes the tertiary Hsp90 (ADP-bound)–cdc37–client kinase complex [125]. Cdc37 siRNA silencing abolishes most Hsp90–kinase interactions, causing instability-proteasome degradation of client kinases [159]. The effect of cdc37 siRNA silencing varies depending on kinases, hinting to a complex regulation role for cdc37 [160].

Cdc37 is being evaluated as an oncology target, as many of its client kinases are involved in malignancy [161]. Lack of HSF1-mediated induction of heat shock proteins by inhibiting the cdc37–Hsp90 complex may lead to safer and more selective treatments than broad-spectrum, HSF1-activating Hsp90 ATPase inhibitors [159].

More than 60% of human kinases are true Hsp90 interactors, and the large majority of these interactions are cdc37-dependent [162]. Validated, cdc37–Hsp90 client kinases include protein kinase B (Akt), cyclin-dependent kinase 5 (Cdk5), death-associated protein kinase (DAPK), Fyn and other physiological and pathological tau kinases (see also Chapter 1). The tau kinase GSK-3 is also connected to Hsp90 [163], and may be affected by cdc37 targeting. Overall, a cdc37 inhibition-mediated reduction of HP-tau through destabilization of pathological tau kinases in diseased cells and tissues is a validated hypothesis [164]. siRNA silencing of cdc37 causes the reduction of tau levels, while its overexpression stabilizes both WT and P301L-mutated tau. The effect is protein-specific, as aggregation-prone α-synuclein is not affected by cdc37 modulation [164]. Cdc37 co-immuno-precipitates with tau in *postmortem* brain lysates from AD patients and controls, and its levels are similar. Cdc37 levels increase with age in the brains of WT mice and induce a higher stabilization of HP, compared to "normal" tau [164]. The activity of cdc37 on tau HP is kinase-specific, as the levels of several phosphoepitopes are increased when cdc37 is silenced. Among four main tau kinases two—Akt and Cdk5—are significantly reduced, while two—GSK-3 and mitogen-activated protein kinase 2 (MAPK2)—are largely unchanged by cdc37 siRNA silencing [164]. Thus, cdc37 operates as a phosphorylation switch for tau, preventing its transition to misfolded/tauopathy-prone tau. In stress conditions, though, cdc37 may contribute to stabilize HP-misfolded tau, and may justify the use of inhibitors of the cdc37–Hsp90 interaction as neuroprotective agents.

Cdc37 may be indirectly inhibited, by preventing its activation. Cdc37 needs to be phosphorylated on Ser13 to efficiently bind its client kinases and to form a ternary complex with Hsp90 [165]. *Casein kinase 2 (CK2)* is the enzyme responsible for cdc37 phosphorylation on Ser13 [166]. CK2 is a constitutively active kinase with a tetrameric complex structure comprising two catalytic (CK2α and CK2α') and two regulatory (CK2β) subunits [167]. CK2 phosphorylates a large panel—probably beyond 1000 [168]—of substrates, whose regulation impacts on most cellular functions. Its activity is constantly elevated in tumors, and makes it a popular oncology target [169,170]. Its implication in other therapeutic areas is also substantiated [171].

CK2 is more abundant in the brain than in any other organ, and its role (according to the presence of a myriad of CK2 substrates in CNS) must be essential [172]. Its association with misfolding proteins includes α-synuclein (hyperphosphorylation-dependent progression of PD and other α-synucleinopathies [173], localization of regulatory CK2 subunits in Lewy bodies [174]), synphilin-1 (phosphorylation-mediated interaction with α-synuclein in Lewy bodies [175]), and the prion protein (molecular interaction between catalytic CK2 subunits and prions *in vivo*) [176]. CK2 and the amyloid pathway are linked by the activating effect of oligomeric, neurotoxic Aβ peptides on CK2, resulting in the inhibition of bidirectional axonal transport [177]. Aβ-dependent CK2 activation has an effect on GSK-3 and other tau-related kinases [178], and is prevented by CK2 inhibitors [179].

CK2 is among the first identified, pathology-related tau kinases [180]. It shows Aβ-induced phosphorylation of tau [181], co-localizes with NFTs [182], and has an altered–deregulated distribution in AD patients [183]. While direct CK2 phosphorylation of tau has not been recently pursued [184], its impact on cdc37-mediated stabilization of client kinases is actively studied. CK2 inhibitors show effects compatible with the dissociation of cdc37–Hsp90–client kinase complexes [166,185]. Thus, HP-tau could be reduced in tauopathy-modulating models by treatment with CK2 inhibitors. The presence of tau itself, Hsp90 [186] and FKBP52 [187] among CK2 substrates; the activating role played by Hsp90 on CK2 [188]; and the presence of CK2 among the client proteins of cdc37–Hsp90 in a "feedback loop" [189] make the CK2–cdc37–Hsp90–tau interaction network extremely complex. CK2 may be a crucial node in progressing tau towards rescuing pathways, but its inhibition could also lead to a variety of effects on tau.

The complexity of cdc37 functions in neurodegeneration and its impairment in pathological conditions are substantiated by a recent paper [190]. Trans-active response DNA-binding protein 43 (TDP-43) [191] is an aggregation-prone nuclear protein that suffers mislocalization, truncation, and aggregation in tauopathies such as FTDP-17, ALS, and—although much less than Aβ and tau—AD [192]. Its interaction with Hsp90 [193] leads to the reduction of TDP-43 levels by inhibition of the ATPase activity of Hsp90. An even stronger effect on TDP-43 is obtained through cdc37 silencing [190]. Cdc37 knockdown-dependent reduction of TDP-43 is executed by caspases. They truncate TDP-43, so that the aggregation-prone, truncated form moves from nucleus to cytoplasm and becomes accessible for autophagy-dependent degradation [190]. Adding tau to TDP-43-overexpressing cells—i.e., artificially recreating a tauopathy environment—causes a gradual switch of the cdc37–Hsp90 chaperone complex from TDP-43 binding to tau binding, leading to TDP-43 accumulation and aggregation [190]. A concurrent increase of two or more misfolded/unstable client proteins may, thus, saturate-impair a chaperone system, leading to the neurotoxic accumulation of pathological species.

2.3 DISEASE-MODIFYING COMPOUNDS

This chapter deals with early/mid steps leading to neuropathological alterations related to protein misfolding and aggregation in general, and to tau and/or tau-connected events in particular. Three potential therapeutic mechanisms were examined in detail, and at least one target was chosen for each mechanism. Tens of other targets—some of which are validated and actively pursued by various labs—were neglected here, mostly for reasons of space. One hundred compounds/scaffolds acting on the three selected targets are diffusely covered in a chemistry-oriented companion book [194] devoted to disease-modifying compounds, and are briefly summarized in Table 2.1. Each compound class is numbered as in the chemistry-oriented companion book, and its chemical core is structurally defined; its mechanism of action and molecular target are mentioned; the public or private laboratory that develops the compound is listed; and the development status—according to publicly available information—is provided.

TABLE 2.1   Compounds 2.1–2.83: Chemical Class, Target, Developing Organization, Development Status

Number	Chemical cpd./class	Target	Organization	Dev. status
2.1	KRIBB3	Hsp27 inhibition	Korea Research Institute of Bioscience and Biotechnology	LO
2.2	Zerumbone	Hsp27 inhibition	Korea Institute of Radiological and Medical Science	LO
2.4, 2.5	Clerodanes, hardwickiic acid	Hsp27 inhibition	Salerno University, Italy	DD
2.6	Brivudine, RP-101	Hsp27 inhibition	RESprotect, Dresden, Germany	Ph II
2.7	Arylsulfonamides, JCC76	Hsp27 inhibition	Cleveland State University	LO
2.9	Arylethynyltriazolyl ribonucleoside	Hsp27 down-regulation	Wuhan University, China	LO
2.10	Xanthones, TDP	Hsp27 down-regulation	The Chinese University of Hong Kong	LO

(Continued)

TABLE 2.1   Compounds 2.1–2.83: Chemical Class, Target, Developing Organization, Development Status (cont.)

Number	Chemical cpd./class	Target	Organization	Dev. status
2.11	Quercetin	CK2, CamKII	Chung San Medical University, Taiwan	PE
2.12	LY303511	mTOR, CK2	National University of Singapore	DD
2.13	Benzopyranopyridines	MK2	Pfizer	LO
2.14	Aminopyrazinthioureas	MK2	Merck	LO
2.15	Indazolepyrimidines	MK2	Abbott	DD
2.16	Indazole-pyrrolo[3,2-d] pyrimidines	MK2	Abbott	LO
2.17	2-Pyridyl-tetrahydro-4H-pyrrolo[3,2-c] pyridine-4-ones	MK2	Pfizer	LO
2.18a,b	2-Pyrimidinyl-tetrahydro-4H-pyrrolo[3,2-c]pyridine-4-one spiro compounds	MK2	Merck	LO
2.19	Benzamide-substituted tetrahydro-4H-pyrrolo[3,2-c]pyridine-4-one spiro compounds	MK2	Merck	DD
2.20	Pyrrole-based tetracycles	MK2	Novartis	LO
2.21	2-Pyridyl-dihydropyrro-lopyrimidinones	MK2	Novartis	DD
2.22	1,4-Bisaryl-3-aminopyrazoles	MK2	Novartis	LO
2.23a,b	Carbolines	MK2	Pfizer	LO
2.24	Thiazolamide-substituted carbolines	MK2	Boehringer Ingelheim	DD
2.25	Piperidine-containing spiroderivatives	MK2	Boehringer Ingelheim	LO
2.26, 2.27	Benzothiophene lactams	MK2	Pfizer	DD
2.28	Bicyclol	Hsp27 inducer	Chinese Academy of Medical Sciences	MKTD
2.29	Squamosamide analogs, FLZ	HSF1 activator, Akt up-regulation	Chinese Academy of Medical Sciences	DD
2.30	Apigenin	CK2, cdc37 inhibitor	Cognitive Res. Center, Bejing, China	LO

TABLE 2.1 Compounds 2.1–2.83: Chemical Class, Target, Developing Organization, Development Status (cont.)

Number	Chemical cpd./class	Target	Organization	Dev. status
2.31	Resveratrol	Hsp27 inducer/down-regulator	Several	PE
2.32	Gusperimus	Hsp70	Nordic	Ph III
2.33	Tresperimus	Hsp70	Laboratoires Fournier	PE
2.34	DHP-spergualin	Hsp70, ATPase inhib.	University of Pittsburgh	DD
2.35a,d,e	DHPs	Hsp70, ATPase stimul.	Alzheimer's Institute, Florida	LO
2.35b,c	DHPs	Hsp70, ATPase inhib.	Alzheimer's Institute, Florida	LO
2.36	MKT-077	Hsp70	Harvard Med. School	Ph I
2.37	Sulfogalactosyl ceramides	Hsp70	University of Toronto	Ph I
2.38	Acylbenzamide fatty acids	Hsp70	Max Planck, Halle, D	DD
2.39a,b	Gallates—ECG, EGCG	Hsp70, plus others	Several	PE
2.40a,b	Luteolin, myricetin—flavonoids	Hsp70, plus others	Several	PE
2.41	Apoptozole	Hsp70, ATPase inhib.	Yonsei University, South Korea	LO
2.42	Heterocyclic carboxamides	Hsp70, ATPase inhib.	Burnham Institute, Calif.	DD
2.43	Phenoxy-N-arylacetamides	Hsp70, ATPase inhib.	ALS Biopharma	DD
2.44	Pifithrin	Hsp70, plus others	University of Pennsylvania	LO
2.45a,b	Thioflavin S	Hsp70-BAG-1	Cancer Research, UK	DD
2.46	VER-155008	Hsp70-BAG-1	Vernalis	LO
2.47	Sildenafil	Hsp70-BAG-1	University of Genova, I	DD
2.48	KM11060	Hsp70-BAG-1	University of Genova, I	DD

(Continued)

TABLE 2.1   Compounds 2.1–2.83: Chemical Class, Target, Developing Organization, Development Status *(cont.)*

Number	Chemical cpd./class	Target	Organization	Dev. status
2.49a–d	Geldanamycin derivatives	Hsp90	BMS, Infinity	Ph III
2.50	Retaspimycin	Hsp90	Infinity	Ph III
2.51	BIIB021	Hsp90	Biogen Idec	Ph II
2.52a–c	Purine-like compounds	Hsp90	Sloan-Kettering NY, Myrexis, DebioPharm	Ph I
2.53	Radicicol	Hsp90	–	NP
2.54	Ganetespib	Hsp90	Synta	Ph III
2.55	AUY922	Hsp90	Vernalis, Novartis	Ph II
2.56	KW-2478	Hsp90	Kiowa Hakko Kirin	Ph II
2.57	AT13387	Hsp90	Astex	Ph II
2.58	HSP990	Hsp90	Novartis	Ph I
2.59	SNX-5422	Hsp90	Esanex	Ph I
2.60	XL888	Hsp90	Exelixis	Ph I
2.61	FK506	Hsp90 FKBPs	Astellas	MKTD
2.62	GPI-1046	Hsp90 FKBPs	Guilford, Amgen	Ph I
2.63a,b	V1-0367, VX-710	Hsp90 FKBPs	Vertex	Ph II
2.64	Pipecolyl ketoamide	Hsp90 FKBPs	Max Planck, Munich, D	DD
2.65	Pipecolyl sulfonamide	Hsp90 FKBPs	Max Planck, Munich, D	DD
2.66	Celastrol	Hsp90 cdc37, plus others	–	NP
2.68	DRB	CK2	University of Pennsylvania	LO
2.69a–c	Tetrahalo benzimidazoles	CK2	University of Padova	LO
2.70	Tetrabromo benzotriazole	CK2	University of Padova	LO
2.71a–c	Quinones	CK2	University of Padova	DD
2.72	Ellagic acid	CK2	–	NP
2.73	Resorufin	CK2	Southern Denmark University	DD
2.74	Pyrazolotriazines	CK2	Polaris	DD

TABLE 2.1   Compounds 2.1–2.83: Chemical Class, Target, Developing Organization, Development Status (cont.)

Number	Chemical cpd./class	Target	Organization	Dev. status
2.75	Macrocyclic pyrazolotriazines	CK2	Polaris	DD
2.76	IQA	CK2	University of Padova	LO
2.77	Xanthenyl benzoic acids	CK2	INSERM, Grenoble	DD
2.78	CC04820	CK2	Kyoto University, Jp	DD
2.79	Thiazolylbenzoic acids	CK2	Kyoto University, Jp	DD
2.80	Benzo[g]indazole carboxylate	CK2	Kyoto University, Jp	DD
2.81a,b	Tricyclic carboxylates	CK2	Cylene Pharmaceuticals	Ph I
2.82	Benzothiazolium sulfonates	CK2	INSERM, Grenoble	DD
2.83	W16	CK2	INSERM, Grenoble	DD

Not progressed, NP; early discovery, DD; lead optimization, LO; preclinical evaluation, PE; clinical Phase I-II-III, Ph I–Ph III; marketed, MKTD.

References

1. Friedman, R. Aggregation of amyloids in a cellular context: modeling and experiment. Biochem. J. 2011, 438, 415–426.
2. Badiola, N.; Suarez-Calvet, M.; Lleo, A. Tau phosphorylation and aggregation as a therapeutic target in tauopathies. CNS Neurol. Dis. Drug Targets 2010, 9, 727–740.
3. Warren, J. D.; Rohrer, J. D.; Schott, J. M.; Fox, N. C.; Hardy, J.; Rossor, M. N. Molecular nexopathies: a new paradigm of neurodegenerative disease. Tr. Neurosci. 2013, 36, 561–569.
4. Martin, L.; Latypova, X.; Terro, F. Post-translational modifications of tau protein: implications for Alzheimer's disease. Neurochem. Int. 2011, 58, 458–471.
5. Mayer, M. P.; Bukau, B. Hsp70 chaperones: cellular functions and molecular mechanisms. Cell. Mol. Life Sci. 2005, 62, 670–684.
6. Pearl, L. H.; Prodromou, C. Structure and mechanism of the Hsp90 molecular chaperone machinery. Annu. Rev. Biochem. 2006, 75, 271–294.
7. Bukau, B.; Weissman, J.; Horwich, A. Molecular chaperones and protein quality control. Cell 2006, 125, 443–451.
8. Kim, R.; Lai, L.; Lee, H. H.; Cheong, G. W.; Kim, K. K.; Wu, Z., et al. On the mechanism of chaperone activity of the small heat-shock protein of Methanococcus jannaschii. Proc. Natl. Acad. Sci. U.S.A. 2003, 100, 8151–8155.
9. Haslbeck, M.; Miess, A.; Stromer, T.; Walter, S.; Buchner, J. Disassembling protein aggregates in the yeast cytosol: the cooperation of HSP26 with Ssa1 and Hsp104. J. Biol. Chem. 2005, 280, 23861–23868.
10. Slepenkov, S. V.; Witt, S. N. The unfolding story of the Escherichia coli Hsp70 DnaK: is DnaK a holdase or an unfoldase? Mol. Microbiol. 2002, 45, 1197–1206.

11. Sharma, S. K.; Christen, P.; Goloubinoff, P. Disaggregating chaperones: an unfolding story. *Curr. Protein Pept. Sci.* **2009**, *10*, 432–446.
12. De Los Rios, P.; Ben-Zvi, A.; Slutsky, O.; Azem, A.; Goloubinoff, P. Hsp70 chaperones accelerate protein translocation and the unfolding of stable protein aggregates by entropic pulling. *Proc. Natl. Acad. Sci. U.S.A.* **2006**, *103*, 6166–6171.
13. Uversky, V. N. Flexible nets of malleable guardians: intrinsically disordered chaperones in neurodegenerative diseases. *Chem. Rev.* **2011**, *111*, 1134–1166.
14. Evans, C. G.; Chang, L.; Gestwicki, J. E. Heat shock protein 70 (Hsp70) as an emerging drug target. *J. Med. Chem.* **2010**, *53*, 4585–4602.
15. Salminen, A.; Ojala, J.; Kaarniranta, K.; Hiltunen, M.; Soininen, H. Hsp90 regulates tau pathology through co-chaperone complexes in Alzheimer's disease. *Progr. Neurobiol.* **2011**, *93*, 99–110.
16. Abisambra, J. F.; Blair, L. J.; Hill, S. E.; Jones, J.; Kraft, C.; Rogers, J., et al. Phosphorylation dynamics regulate Hsp27-mediated rescue of neuronal plasticity deficits in tau transgenic mice. *J. Neurosci.* **2010**, *30*, 15374–15382.
17. Sarkar, M.; Kuret, J.; Lee, G. Two motifs within the tau-microtubule-binding domain mediate its association with the hsc70 molecular chaperone. *J. Neurosci. Res.* **2008**, *86*, 2763–2773.
18. Petrucelli, L.; Dickson, D.; Kehoe, K.; Taylor, J.; Snyder, H.; Grover, A., et al. CHIP and Hsp70 regulate tau ubiquitination, degradation and aggregation. *Hum. Mol. Genet.* **2004**, *13*, 703–714.
19. Dou, F.; Netzer, W. J.; Tanemura, K.; Li, F.; Hartl, F. U.; Takashima, A., et al. Chaperones increase association of tau protein with microtubules. *Proc. Natl. Acad. Sci. U.S.A.* **2003**, *100*, 721–726.
20. Elliott, E.; Tsvetkov, P.; Ginzburg, I. BAG-1 associates with Hsc70-tau complex and regulates the proteasomal degradation of tau protein. *J. Biol. Chem.* **2007**, *282*, 37276–37284.
21. Luo, W.; Dou, F.; Rodina, A.; Chip, S.; Kim, J.; Zhao, Q., et al. Roles of heat-shock protein 90 in maintaining and facilitating the neurodegenerative phenotype in tauopathies. *Proc. Natl. Acad. Sci. U.S.A.* **2007**, *104*, 9511–9516.
22. Jinwai, U. K.; Koren, J., III.; Borysov, S. I.; Schmid, A. B.; Abisambra, J. F.; Blair, L. J., et al. The Hsp90 cochaperone, FKBP51, increases tau stability and polymerizes microtubules. *J. Neurosci.* **2010**, *30*, 591–599.
23. Chambraud, B.; Sardin, E.; Giustiniani, J.; Dounane, O.; Schumacher, M.; Goedert, M.; Baulieu, E. E. A role for FKBP52 in tau protein function. *Proc. Natl. Acad. Sci. U.S.A.* **2010**, *107*, 2658–2663.
24. Carra, S.; Crippa, V.; Rusmini, P.; Boncoraglio, A.; Minoia, M.; Giorgetti, E., et al. Alteration of protein folding and degradation in motor neuron diseases: implications and protective functions of small heat shock proteins. *Prog. Neurobiol.* **2012**, *97*, 83–100.
25. Carra, S.; Rusmini, P.; Crippa, V.; Giorgetti, E.; Boncoraglio, A.; Cristofani, R., et al. Different anti-aggregation and prodegradative functions of the members of the mammalian sHSP family in neurological disorders. *Phil. Trans. R. Soc. B* **2013**, *368*, 20110409.
26. Hickey, E.; Brandon, S. E.; Potter, R.; Stein, G.; Stein, J.; Weber, L. A. Sequence and organization of genes encoding the human 27 kDa heat shock protein. *Nucleic Acids Res.* **1986**, *14*, 4127–4145.
27. Mymrikov, E. V.; Seit-Nebi, A. S.; Gusev, N. B. Large potentials of small heat shock proteins. *Physiol. Rev.* **2011**, *91*, 1123–1159.
28. Stromer, T.; Ehrnsperger, M.; Gaestel, M.; Buchner, J. Analysis of the interaction of small heat shock proteins with unfolding proteins. *J. Biol. Chem.* **2003**, *278*, 18015–18021.
29. Stamler, R.; Kappe, G.; Boelens, W.; Slingsby, C. Wrapping the α-crystallin domain fold in a chaperone assembly. *J. Mol. Biol.* **2005**, *353*, 68–79.
30. Lambert, H.; Charette, S. J.; Bernier, A. F.; Guimond, A.; Landry, J. HSP27 multimerization mediated by phosphorylation-sensitive intermolecular interactions at the amino terminus. *J. Biol. Chem.* **1999**, *274*, 9378–9385.

31. Hayes, D.; Napoli, V.; Mazurkie, A.; Stafford, W. F.; Graceffa, P. Phosphorylation dependence of hsp27 multimeric size and molecular chaperone function. *J. Biol. Chem.* **2009**, *284*, 18801–18807.

32. Stokoe, D.; Engel, K.; Campbell, D. G.; Cohen, P.; Gaestel, M. Identification of MAPKAP kinase 2 as a major enzyme responsible for the phosphorylation of the small mammalian heat shock proteins. *FEBS Lett.* **1992**, *313*, 307–313.

33. Rogalla, T.; Ehrnsperger, M.; Preville, X.; Kotlyarov, A.; Lutsch, G.; Ducasse, C., et al. Regulation of Hsp27 oligomerization chaperone function, and protective activity against oxidative stress/tumor necrosis factor alpha by phosphorylation. *J. Biol. Chem.* **1999**, *274*, 18947–18956.

34. Haley, D. A.; Bova, M. P.; Huang, Q. L.; Mchaourab, H. S.; Stewart, P. L. Small heat-shock protein structures reveal a continuum from symmetric to variable assemblies. *J. Mol. Biol.* **2000**, *298*, 261–272.

35. Bryantsev, A. L.; Loktionova, S. A.; Ilyinskaya, O. P.; Tararak, E. M.; Kampinga, H. H.; Kabakov, A. E. Distribution, phosphorylation, and activities of Hsp25 in heat-stressed H9c2 myoblasts: a functional link to cytoprotection. *Cell Stress Chaperones* **2002**, *7*, 146–155.

36. Haslbeck, M. shsps and their role in the chaperone network. *Cell. Mol. Life Sci.* **2002**, *59*, 1649–1657.

37. Parcellier, A.; Brunet, M.; Schmitt, E.; Col, E.; Didelot, C.; Hammann, A., et al. HSP27 favors ubiquitination and proteasomal degradation of p27Kip1 and helps S-phase re-entry in stressed cells. *FASEB J.* **2006**, *20*, 1179–1181.

38. Mathew, S. S.; Della Selva, M. P.; Burch, A. D. Modification and reorganization of the cytoprotective cellular chaperone Hsp27 during herpes simplex virus type 1 infection. *J. Virol.* **2009**, *83*, 9304–9312.

39. Mehlen, P.; Coronas, V.; Ljubic-Thibal, V.; Ducasse, C.; Granger, L.; Jourdan, F.; Arrigo, A. P. Small stress protein Hsp27 accumulation during dopamine-mediated differentiation of rat olfactory neurons counteracts apoptosis. *Cell Death Differ.* **1999**, *6*, 227–233.

40. Chen, S.; Brown, I. R. Neuronal expression of constitutive heat shock proteins: implications for neurodegenerative diseases. *Cell Stress Chaperones* **2007**, *12*, 51–58.

41. Kato, H.; Liu, Y.; Kogure, K.; Kato, K. Induction of 27-kDa heat shock protein following cerebral ischemia in a rat model of ischemic tolerance. *Brain Res.* **1994**, *634*, 235–244.

42. Bechtold, D. A.; Brown, I. R. Heat shock proteins Hsp27 and Hsp32 localize to synaptic sites in the rat cerebellum following hyperthermia. *Brain Res. Mol. Brain Res.* **2000**, *75*, 309–320.

43. Datskevich, P. N.; Nefedova, V. V.; Sudnitsyna, M. V.; Gusev, N. B. Mutations of small heat shock proteins and human congenital diseases. *Biochemistry (Moscow)* **2012**, *77*, 1500–1514.

44. McLean, P. J.; Kawamata, H.; Shariff, S.; Hewett, J.; Sharma, N.; Ueda, K., et al. TorsinA and heat shock proteins act as molecular chaperones: suppression of alpha-synuclein aggregation. *J. Neurochem.* **2002**, *83*, 846–854.

45. Outeiro, T. F.; Klucken, J.; Strathearn, K. E.; Liu, F.; Nguyen, P.; Rochet, J. C., et al. Small heat shock proteins protect against alpha-synuclein-induced toxicity and aggregation. *Biochem. Biophys. Res. Commun.* **2006**, *351*, 631–638.

46. Wyttenbach, A.; Sauvageot, O.; Carmichael, J.; Diaz-Latoud, C.; Arrigo, A. P.; Rubinsztein, D. C. Heat shock protein 27 prevents cellular polyglutamine toxicity and suppresses the increase of reactive oxygen species caused by huntingtin. *Hum. Mol. Genet.* **2002**, *11*, 1137–1151.

47. Wilhelmus, M. M.; Boelens, W. C.; Otte-Holler, I.; Kamps, B.; de Waal, R. M.; Verbeek, M. M. Small heat shock proteins inhibit amyloid-beta protein aggregation and cerebrovascular amyloid-beta protein toxicity. *Brain Res.* **2006**, *1089*, 67–78.

48. Bjorkdahl, C.; Sjogren, M. J.; Zhou, X.; Concha, H.; Avila, J.; Winblad, B.; Pei, J. J. Small heat shock proteins Hsp27 or alphaB-crystallin and the protein components of neurofibrillary tangles: tau and neurofilaments. *J. Neurosci. Res.* **2008**, *86*, 1343–1352.

49. Di Domenico, F.; Sultana, R.; Tiu, G. F.; Scheff, N. N.; Perluigi, M.; Cini, C.; Butterfield, D. A. Protein levels of heat shock proteins 27, 32, 60, 70, 90 and thioredoxin-1 in amnestic mild cognitive impairment: an investigation on the role of cellular stress response in the progression of Alzheimer disease. *Brain Res.* **2010,** *1333,* 72–81.

50. Ehrnsperger, M.; Graber, S.; Gaestel, M.; Buchner, J. Binding of non-native protein to Hsp25 during heat shock creates a reservoir of folding intermediates for reactivation. *EMBO J.* **1997,** *16,* 221–229.

51. Cairns, J.; Qin, S.; Philp, R.; Tan, Y. H.; Guy, G. R. Dephosphorylation of the small heat shock protein Hsp27 *in vivo* by protein phosphatase 2A. *J. Biol. Chem.* **1994,** *269,* 9176–9183.

52. Shimura, H.; Miura-Shimura, Y.; Kosik, K. S. Binding of tau to heat shock protein 27 leads to decreased concentration of hyperphosphorylated tau and enhanced cell survival. *J. Biol. Chem.* **2004,** *279,* 17957–17962.

53. Wang, S.; Toth, M. E.; Bereczki, E.; Santha, M.; Guan, Z. Z.; Winblad, B.; Pei, J. J. Interplay between glycogen synthase kinase-3β and tau in the cerebellum of Hsp27 transgenic mouse. *J. Neurosci.* **2011,** *89,* 1267–1275.

54. Schwarz, L.; Vollmer, G.; Richter-Landsberg, C. The small heat shock protein HSP25/27 (HspB1) is abundant in cultured astrocytes and associated with astrocytic pathology in progressive supranuclear palsy and corticobasal degeneration. *Int. J. Cell Biol.* **2010,** 717520.

55. Daugaard, M.; Rohde, M.; Jaattela, M. The heat shock protein 70 family: Highly homologous proteins with overlapping and distinct functions. *FEBS Lett.* **2007,** *581,* 3702–3710.

56. Liu, T.; Daniels, C. K.; Cao, S. Comprehensive review on the HSC70 functions, interactions with related molecules and involvement in clinical diseases and therapeutic potential. *Pharmacol. Therap.* **2012,** *136,* 354–374.

57. Abravaya, K.; Myers, M. P.; Murphy, S. P.; Morimoto, R. I. The human heat shock protein Hsp70 interacts with HSF, the transcription factor that regulates heat shock gene expression. *Genes Devel.* **1992,** *6,* 1153–1164.

58. McCarty, J. S.; Buchberger, A.; Reinstein, J.; Bukau, B. The role of ATP in the functional cycle of the DnaK chaperone system. *J. Mol. Biol.* **1995,** *249,* 126–137.

59. Bertelsen, E. B.; Chang, L.; Gestwicki, J. E.; Zuiderweg, E. R. Solution conformation of wild-type *E. coli* Hsp70 (DnaK) chaperone complexed with ADP and substrate. *Proc. Natl. Acad. Sci. U.S.A.* **2009,** *106,* 8471–8476.

60. Russell, R.; Jordan, R.; McMacken, R. Kinetic characterization of the ATPase cycle of the DnaK molecular chaperone. *Biochemistry* **1998,** *37,* 596–607.

61. Mayer, M. P.; Schroder, H.; Rudiger, S.; Paal, K.; Laufen, T.; Bukau, B. Multistep mechanism of substrate binding determines chaperone activity of Hsp70. *Nat. Struct. Biol.* **2000,** *7,* 586–593.

62. Frydman, J. Folding of newly translated proteins in vivo: the role of molecular chaperones. *Annu. Rev. Biochem.* **2001,** *70,* 603–647.

63. Patury, S.; Miyata, Y.; Gestwicki, J. E. Pharmacological targeting of the Hsp70 chaperone. *Curr. Top. Med. Chem.* **2009,** *9,* 1337–1351.

64. Kampinga, H. H.; Craig, E. A. The HSP70 chaperone machinery: J proteins as drivers of functional specificity. *Nat. Rev. Mol. Cell. Biol.* **2010,** *11,* 579–592.

65. Qiu, X. B.; Shao, Y. M.; Miao, S.; Wang, L. The diversity of the DnaJ/Hsp40 family, the crucial partners for Hsp70 chaperones. *Cell. Mol. Life Sci.* **2006,** *63,* 2560–2570.

66. Jiang, J.; Maes, E. G.; Taylor, A. B.; Wang, L.; Hinck, A. P.; Lafer, E. M.; Sousa, R. Structural basis of J cochaperone binding and regulation of Hsp70. *Mol. Cell* **2007,** *28,* 422–433.

67. Wall, D.; Zylicz, M.; Georgopoulos, C. The NH_2-terminal 108 amino acids of the *Escherichia coli* DnaJ protein stimulate the ATPase activity of DnaK and are sufficient for lambda replication. *J. Biol. Chem.* **1994,** *269,* 5446–5451.

68. Gibbs, S. J.; Braun, J. E. Emerging roles of J proteins in neurodegenerative diseases. *Neurobiol. Dis.* **2008,** *32,* 196–199.

69. Velten, M.; Villoutreix, B. O.; Ladjimi, M. M. Quaternary structure of the HSC co-chaperone HIP. *Biochemistry* **2000**, *39*, 307–315.
70. Hohfeld, J.; Minami, Y.; Hartl, F. U. Hip, a novel cochaperone involved in the eukaryotic Hsc70/Hsp40 reaction cycle. *Cell* **1995**, *83*, 589–598.
71. Kabbage, M.; Dickman, M. B. The BAG proteins: a ubiquitous family of chaperone regulators. *Cell. Mol. Life Sci.* **2008**, *65*, 1390–1402.
72. Vos, M. J.; Hageman, J.; Carra, S.; Kampinga, H. H. Structural and functional diversities between members of the human HSPB, HSPH, HSPA, and DNAJ chaperone families. *Biochemistry* **2008**, *47*, 7001–7011.
73. Shaner, L.; Morano, K. A. All in the family: atypical Hsp70 chaperones are conserved modulators of Hsp70 activity. *Cell Stress Chaper.* **2007**, *12*, 1–8.
74. Liu, Y.; Gierasch, L. M.; Bahar, I. Role of Hsp70 ATPase domain intrinsic dynamics and sequence evolution in enabling its functional interactions with NEFs. *PLoS Comput. Biol.* **2010**, *6*, e1000931.
75. Bukau, B.; Weissman, J.; Horwich, A. Molecular chaperones and protein quality control. *Cell* **2006**, *125*, 443–451.
76. Ballinger, C. A.; Connell, P.; Wu, Y.; Hu, Z.; Thompson, L. J.; Yin, L. Y.; Patterson, C. Identification of CHIP, a novel tetratricopeptide repeat-containing protein that interacts with heat shock proteins and negatively regulates chaperone functions. *Mol. Cell. Biol.* **1999**, *19*, 4535–4545.
77. Scheufler, C.; Brinker, A.; Bourenkov, G.; Pegoraro, S.; Moroder, L.; Bartunik, H.; Hartl, F. U.; Moarefi, I. Structure of TPR domain-peptide complexes: critical elements in the assembly of the Hsp70-Hsp90 multichaperone machine. *Cell* **2000**, *101*, 199–210.
78. Connell, P.; Ballinger, C. A.; Jiang, J.; Wu, Y.; Thompson, L. J.; Hohfeld, J.; Patterson, C. The co-chaperone CHIP regulates protein triage decisions mediated by heat-shock proteins. *Nat. Cell Biol.* **2001**, *3*, 93–96.
79. Hernandez, M. P.; Sullivan, W. P.; Toft, D. O. The assembly and intermolecular properties of the hsp70-Hop-hsp90 molecular chaperone complex. *J. Biol. Chem.* **2002**, *277*, 38294–38304.
80. Onuoha, S. C.; Coulstock, E. T.; Grossmann, J. G.; Jackson, S. E. Structural studies on the co-chaperone Hop and its complexes with Hsp90. *J. Mol. Biol.* **2008**, *379*, 732–744.
81. Murata, S.; Minami, Y.; Minami, M.; Chiba, T.; Tanaka, K. CHIP is a chaperone-dependent E3 ligase that ubiquitylates unfolded protein. *EMBO Rep.* **2001**, *2*, 1133–1138.
82. Jinwal, U. K.; Miyata, Y.; Koren, J. III.; Jones, J. R.; Trotter, J. H.; Chang, L., et al. Chemical manipulation of Hsp70 ATPase activity regulates Tau stability. *J. Neurosci.* **2009**, *29*, 12079–12088.
83. Voss, K.; Combs, B.; Patterson, K. R.; Binder, L. I.; Gamblin, T. C. Hsp70 alters tau function and aggregation in an isoform specific manner. *Biochemistry* **2012**, *51*, 888–898.
84. Patterson, K. R.; Ward, S. M.; Combs, B.; Voss, K.; Kanaan, N. M.; Morfini, G., et al. Heat shock protein 70 prevents both tau aggregation and the inhibitory effects of preexisting tau aggregates on fast axonal transport. *Biochemistry* **2011**, *50*, 10300–10310.
85. Jinwal, U. K.; O'Leary, J. C. III.; Borysov, S. I.; Jones, J. R.; Li, Q., et al. Hsc70 rapidly engages Tau after microtubule destabilization. *J. Biol. Chem.* **2010**, *285*, 16798–16805.
86. http://openi.nlm.nih.gov/detailedresult.php?img=2816227_11693_2009_9046_Fig1_ HTML&req=4
87. Luo, W.; Zhong, J.; Chang, R.; Hu, H.; Pandey, A.; Semenza, G. L. Hsp70 and CHIP selectively mediate ubiquitination and degradation of hypoxia-inducible factor (HIF)-1α but not HIF-2α. *J. Biol. Chem.* **2010**, *285*, 3651–3663.
88. Muller, P.; Ruckova, E.; Halada, P.; Coates, P. J.; Hrstka, R.; Lane, D. P.; Vojtesek, B. C-terminal phosphorylation of Hsp70 and Hsp90 regulates alternate binding to co-chaperones CHIP and HOP to determine cellular protein folding/degradation balances. *Oncogene* **2013**, *32*, 3101–3110.

89. Soskic, V.; Groebe, K.; Schrattenholz, A. Nonenzymatic posttranslational protein modifications in ageing. *Exp. Gerontol.* **2008**, *43*, 247–257.
90. Shpund, S.; Gershon, D. Alterations in the chaperone activity of HSP70 in aging organisms. *Arch. Gerontol. Geriatr.* **1997**, *24*, 125–131.
91. Bauer, P. O.; Nukina, N. The pathogenic mechanisms of polyglutamine diseases and current therapeutic strategies. *J. Neurochem.* **2009**, *110*, 1737–1765.
92. Jana, N. R.; Tanaka, M.; Wang, G.; Nukina, N. Polyglutamine length-dependent interaction of Hsp40 and Hsp70 family chaperones with truncated N-terminal huntingtin: their role in suppression of aggregation and cellular toxicity. *Hum. Mol. Genet.* **2000**, *9*, 2009–2018.
93. Muchowski, P. J.; Schaffar, G.; Sittler, A.; Wanker, E. E.; Hayer-Hartl, M. K.; Hartl, F. U. Hsp70 and hsp40 chaperones can inhibit self-assembly of polyglutamine proteins into amyloid-like fibrils. *Proc. Natl. Acad. Sci. U.S.A.* **2000**, *97*, 7841–7846.
94. Rujano, M. A.; Kampinga, H. H.; Salomons, F. A. Modulation of polyglutamine inclusion formation by the Hsp70 chaperone machine. *Exp. Cell Res.* **2007**, *313*, 3568–3578.
95. Hansson, O.; Nylandsted, J.; Castilho, R. F.; Leist, M.; Jaattela, M.; Brundin, P. Overexpression of heat shock protein 70 in R6/2 Huntington's disease mice has only modest effects on disease progression. *Brain Res.* **2003**, *970*, 47–57.
96. Auluck, P. K.; Chan, H. Y.; Trojanowski, J. Q.; Lee, V. M.; Bonini, N. M. Chaperone suppression of alpha-synuclein toxicity in a Drosophila model for Parkinson's disease. *Science* **2002**, *295*, 865–868.
97. Witt, S. N. Hsp molecular chaperones and Parkinson's disease. *Biopolymers* **2010**, *93*, 218–228.
98. Klucken, J.; Shin, Y.; Hyman, B. T.; McLean, P. J. A single amino acid substitution differentiates Hsp70-dependent effects on alpha-synuclein degradation and toxicity. *Biochem. Biophys. Res. Commun.* **2004**, *325*, 367–373.
99. Dong, Z.; Wolfer, D. P.; Lipp, H. P.; Bueler, H. Hsp70 gene transfer by adeno-associated virus inhibits MPTP-induced nigrostriatal degeneration in the mouse model of Parkinson disease. *Mol. Ther.* **2005**, *11*, 80–88.
100. Watanabe, M.; Dykes-Hoberg, M.; Culotta, V. C.; Price, D. L.; Wong, P. C.; Rothstein, J. D. Histological evidence of protein aggregation in mutant SOD1 transgenic mice and in amyotrophic lateral sclerosis neural tissues. *Neurobiol. Dis.* **2001**, *8*, 933–941.
101. Cummings, C. J.; Sun, Y.; Opal, P.; Antalffy, B.; Mestril, R.; Orr, H. T., et al. Over-expression of inducible HSP70 chaperone suppresses neuropathology and improves motor function in SCA1 mice. *Hum. Mol. Genet.* **2001**, *10*, 1511–1518.
102. Adachi, H.; Katsuno, M.; Minamiyama, M.; Sang, C.; Pagoulatos, G.; Angelidis, C., et al. Heat shock protein 70 chaperone overexpression ameliorates phenotypes of the spinal and bulbar muscular atrophy transgenic mouse model by reducing nuclear-localized mutant androgen receptor protein. *J. Neurosci.* **2003**, *23*, 2203–2211.
103. Sahara, N.; Maeda, S.; Yoshiike, Y.; Mizoroki, T.; Yamashita, S.; Murayama, M., et al. Molecular chaperone-mediated tau protein metabolism counteracts the formation of granular tau oligomers in human brain. *J. Neurosci. Res.* **2007**, *85*, 3098–3108.
104. Broer, L.; Arfan Ikram, M.; Schuur, M.; DeStefano, A. L.; Bish, J. C.; Liu, F., et al. Association of HSP70 and its co-chaperones with Alzheimer's disease. *J. Alzheimer Dis.* **2011**, *25*, 93–102.
105. Evans, C. G.; Wisen, S.; Gestwicki, J. E. Heat shock proteins 70 and 90 inhibit early stages of amyloid beta-(1-42) aggregation in vitro. *J. Biol. Chem.* **2006**, *281*, 33182–33191.
106. Kumar, P.; Ambasta, R. K.; Veereshwarayya, V.; Rosen, K. M.; Kosik, K. S.; Band, H., et al. CHIP and HSPs interact with beta-APP in a proteasome dependent manner and influence Abeta metabolism. *Hum. Mol. Genet.* **2007**, *16*, 848–864.
107. Hoshino, T.; Murao, N.; Namba, T.; Takehara, M.; Adachi, H.; Katsuno, M., et al. Suppression of Alzheimer's disease-related phenotypes by expression of heat shock protein 70 in mice. *J. Neurosci.* **2011**, *31*, 5225–5234.

108. LaPointe, N. E.; Morfini, G.; Pigino, G.; Gaisina, I. N.; Kozikowski, A. P.; Binder, L. I.; Brady, S. T. The amino terminus of tau inhibits kinesin-dependent axonal transport: Implications for filament toxicity. *J. Neurosci. Res.* **2009**, *87*, 440–451.

109. Abisambra, J. F.; Jinwal, U. K.; Suntharalingam, A.; Arulselvam, K.; Brady, S.; Cockman, M., et al. DnaJA1 antagonizes constitutive Hsp70-mediated stabilization of tau. *J. Mol. Biol.* **2012**, *421*, 653–661.

110. Carrettiero, D. C.; Hernandez, I.; Neveu, P.; Papagiannakopoulos, T.; Kosik, K. S. The cochaperone BAG2 sweeps paired helical filament-insoluble tau from the microtubule. *J. Neurosci.* **2009**, *29*, 2151–2161.

111. Takayama, S.; Sato, T.; Krajewski, S.; Kochel, K.; Irie, S.; Millan, J. A.; Reed, J. C. Cloning and functional analysis of BAG-1: a novel Bcl-2-binding protein with anti-cell death activity. *Cell* **1995**, *80*, 279–284.

112. Liman, J.; Ganesan, S.; Dohm, C. P.; Krajewski, S.; Reed, J. C.; Bahr, M., et al. Interaction of BAG1 and Hsp70 mediates neuroprotectivity and increases chaperone activity. *Mol. Cell. Biol.* **2005**, *25*, 3715–3725.

113. Elliott, E.; Tsvetkov, P.; Ginzburg, I. BAG-1 associates with Hsc70.Tau complex and regulates the proteasomal degradation of Tau protein. *J. Biol. Chem.* **2007**, *282*, 37276–37284.

114. Elliott, E.; Laufer, O.; Ginzburg, I. BAG-1M is up-regulated in hippocampus of Alzheimer's disease patients and associates with tau and APP patients. *J. Neurochem.* **2009**, *109*, 1168–1178.

115. Venturelli, E.; Villa, C.; Fenoglio, C.; Clerici, F.; Marcone, A.; Benussi, L., et al. *BAG1* is a protective factor for sporadic frontotemporal lobar degeneration but not for Alzheimer's disease. *J. Alzheimer's Dis.* **2011**, *23*, 701–707.

116. Csermely, P.; Schnaider, T.; Soti, C.; Prohaszka, Z.; Nardai, G. The 90 kDa molecular chaperone family: structure, function and clinical applications. A comprehensive review. *Pharmacol. Ther.* **1998**, *79*, 129–168.

117. Hickey, E.; Brandon, S. E.; Smale, G.; Lloyd, D.; Weber, L. A. Sequence and regulation of a gene encoding a human 89-kilodalton heat shock protein. *Mol. Cell. Biol.* **1989**, *9*, 2615–2626.

118. Sreedhar, A. S.; Kalmar, E.; Csermely, P.; Shen, Y. F. Hsp90 isoforms: functions, expression and clinical importance. *FEBS Lett.* **2004**, *562*, 11–15.

119. Grammatikakis, N.; Vultur, A.; Ramana, C. V.; Siganou, A.; Schweinfest, C. W.; Watson, D. K.; Raptis, L. The role of Hsp90N, a new member of the Hsp90 family, in signal transduction and neoplastic transformation. *J. Biol. Chem.* **2002**, *277*, 8312–8320.

120. Pearl, L. H.; Prodromou, C.; Workman, P. The Hsp90 molecular chaperone: an open and shut case for treatment. *Biochem. J.* **2008**, *410*, 439–453.

121. Koren, J., III.; Jinwal, U. K.; Lee, D. C.; Jones, J. R.; Shults, C. L.; Johnson, A. G., et al. Chaperone signalling complexes in Alzheimer's disease. *J. Cell. Mol. Med.* **2009**, *13*, 619–630.

122. Schneider, C.; Sepp-Lorenzino, L.; Nimmesgern, E.; Ouerfelli, O.; Danishefsky, S.; Rosen, N.; Hartl, F. U. Pharmacologic shifting of a balance between protein refolding and degradation mediated by Hsp90. *Proc. Natl. Acad. Sci. U.S.A.* **1996**, *93*, 14536–14541.

123. http://www.picard.ch/downloads/Hsp90interactors.pdf.

124. Echeverria, P. C.; Bernthaler, A.; Dupuis, P.; Mayer, B.; Picard, D. An interaction network predicted from public data as a discovery tool: Application to the Hsp90 molecular chaperone machine. *PLoS One* **2011**, *6*, e26044.

125. Roe, S. M.; Ali, M. M. U.; Meyer, P.; Vaughan, C. K.; Panaretou, B.; Piper, P. W., et al. The mechanism of Hsp90 regulation by the protein kinase-specific cochaperone p50^{cdc37}. *Cell* **2004**, *116*, 87–98.

126. Picard, D. Heat-shock protein 90, a chaperone for folding and regulation. *Cell. Mol. Life Sci.* **2002**, *59*, 1640–1648.

127. Zou, J.; Guo, Y.; Guettouche, T.; Smith, D. F.; Voellmy, R. Repression of heat shock transcription factor HSF1 activation by HSP90 (HSP90 complex) that forms a stress-sensitive complex with HSF1. *Cell* **1998**, *94*, 471–480.
128. Wegele, H.; Muller, L.; Buchner, J. Hsp70 and Hsp90—a relay team for protein folding. *Rev. Physiol. Biochem. Pharmacol.* **2004**, *151*, 1–44.
129. Theodoraki, M. A.; Caplan, A. J. Quality control and fate determination of Hsp90 client proteins. *Biochim. Biophys. Acta* **1823**, *2012*, 683–688.
130. Young, J. C.; Moarefi, I.; Hartl, F. U. Hsp90: a specialized but essential protein-folding tool. *J. Cell Biol.* **2001**, *154*, 267–273.
131. Nahleh, Z.; Tfayli, A.; Najm, A.; El Sayed, A.; Nahle, Z. Heat shock proteins in cancer: targeting the chaperones. *Fut. Med. Chem.* **2012**, *4*, 927–935.
132. Travers, J.; Sharp, S.; Workman, P. HSP90 inhibition: two-pronged exploitation of cancer dependencies. *Drug Discov. Today* **2012**, *17*, 242–252.
133. Luo, W.; Sun, W.; Taldone, T.; Rodina, A.; Chiosis, G. Heat shock protein 90 in neurodegenerative diseases. *Mol. Neurodeg.* **2010**, *5*, 24.
134. Thomas, M.; Harrell, J. M.; Morishima, Y.; Peng, H. M.; Pratt, W. B.; Lieberman, A. P. Pharmacologic and genetic inhibition of hsp90-dependent trafficking reduces aggregation and promotes degradation of the expanded glutamine androgen receptor without stress protein induction. *Hum. Mol. Genet.* **2006**, *15*, 1876–1883.
135. Adachi, H.; Waza, M.; Katsuno, M.; Tanaka, F.; Doyu, M.; Sobue, G. Pathogenesis and molecular targeted therapy of spinal and bulbar muscular atrophy. *Neuropathol. Applied Neurobiol.* **2007**, *33*, 135–151.
136. Falsone, S. F.; Kungl, A. J.; Rek, A.; Cappai, R.; Zangger, K. The molecular chaperone Hsp90 modulates intermediate steps of amyloid assembly of the Parkinson-related protein alpha-synuclein. *J. Biol. Chem.* **2009**, *284*, 31190–31199.
137. Wang, L.; Xie, C.; Greggio, E.; Parisiadou, L.; Shim, H.; Sun, L., et al. The chaperone activity of heat shock protein 90 is critical for maintaining the stability of leucine rich repeat kinase 2. *J. Neurosci.* **2008**, *28*, 3384–3891.
138. Moriwaki, Y.; Kim, Y. J.; Ido, Y.; Misawa, H.; Kawashima, K.; Endo, S.; Takahashi, R. L347P PINK1 mutant that fails to bind to Hsp90/Cdc37 chaperones is rapidly degraded in a proteasome-dependent manner. *Neurosci Res.* **2008**, *61*, 43–48.
139. Tortosa, E.; Santa-Maria, I.; Moreno, F.; Lima, F.; Perez, M.; Avila, J. Binding of Hsp90 to tau promotes a conformational change and aggregation of tau protein. *J. Alzheimer's Dis.* **2009**, *17*, 319–325.
140. Thompson, A. D.; Scaglione, K. M.; Prensner, J.; Gillies, A. T.; Chinnaiyan, A.; Paulson, H. L., et al. Analysis of the tau-associated proteome reveals that exchange of Hsp70 for Hsp90 is involved in tau degradation. *ACS Chem. Biol.* **2012**, *7*, 1677–1686.
141. Lotz, G. P.; Lin, H.; Harst, A.; Obermann, W. M. Aha1 binds to the middle domain of Hsp90, contributes to client protein activation, and stimulates the ATPase activity of the molecular chaperone. *J. Biol. Chem.* **2003**, *278*, 17228–17235.
142. Mok, D.; Allan, R. K.; Carrello, A.; Wangoo, K.; Walkinshaw, M. D.; Ratajczak, T. The chaperone function of cyclophilin 40 maps to a cleft between the prolyl isomerase and tetratricopeptide repeat domains. *FEBS Lett.* **2006**, *580*, 2761–2768.
143. Siligardi, G.; Hu, B.; Panaretou, B.; Piper, P. W.; Pearl, L. H.; Prodromou, C. Co-chaperone regulation of conformational switching in the Hsp90 ATPase cycle. *J. Biol. Chem.* **2004**, *279*, 51989–51998.
144. Panaretou, B.; Siligardi, G.; Meyer, P.; Maloney, A.; Sullivan, J. K.; Singh, S., et al. Activation of the ATPase activity of Hsp90 by the stress regulated cochaperone Aha1. *Mol. Cell* **2002**, *10*, 1307–1318.
145. http://discovery.ucl.ac.uk/1348542/1/1348542.pdf
146. Davies, T. H.; Ning, Y. M.; Sanchez, E. R. A new first step in activation of steroid receptors: hormone-induced switching of FKBP51 and FKBP52 immunophilins. *J. Biol. Chem.* **2002**, *277*, 4597–4600.

147. Bose, S.; Weikl, T.; Bugl, H.; Buchner, J. Chaperone function of Hsp90-associated proteins. *Science* **1996**, *274*, 1715–1717.

148. Richter, K.; Walter, S.; Buchner, J. The co-chaperone Sba1 connects the ATPase reaction of Hsp90 to the progression of the chaperone cycle. *J. Mol. Biol.* **2004**, *342*, 1403–1413.

149. Grad, I.; McKee, T. A.; Ludwig, S. M.; Hoyle, G. W.; Ruiz, P.; Wurst, W., et al. The Hsp90 cochaperone p23 is essential for perinatal survival. *Mol. Cell. Biol.* **2006**, *26*, 8976–8983.

150. Hinds, T. D., Jr.; Sanchez, E. R. Protein phosphatase 5. *Int. J. Biochem. Cell Biol.* **2008**, *40*, 2358–2362.

151. Conde, R.; Xavier, J.; McLoughlin, C.; Chinkers, M.; Ovsenek, N. Protein phosphatase 5 is a negative modulator of heat shock factor 1. *J. Biol. Chem.* **2005**, *280*, 28989–28996.

152. Gong, C. X.; Liu, F.; Wu, G.; Rossie, S.; Wegiel, J.; Li, L., et al. Dephosphorylation of microtubule-associated protein tau by protein phosphatase 5. *J. Neurochem.* **2004**, *88*, 298–310.

153. Liu, F.; Iqbal, K.; Grundke-Iqbal, I.; Rossie, S.; Gong, C. X. Dephosphorylation of tau by protein phosphatase 5. Impairment in Alzheimer's disease. *J. Biol. Chem.* **2005**, *280*, 1790–1796.

154. Cau, W.; Konsolaki, M. FKBP immunophilins and Alzheimer's disease: a chaperoned affair. *J. Biosci.* **2011**, *36*, 493–498.

155. Cioffi, D. L.; Hubler, T. R.; Scammell, J. G. Organization and function of the FKBP52 and FKBP51 genes. *Curr. Opin. Pharmacol.* **2011**, *11*, 308–313.

156. Chambraud, B.; Belabes, H.; Fontaine-Lenoir, V.; Fellous, A.; Baulieu, E. E. The immunophilin FKBP52 specifically binds to tubulin and prevents microtubule formation. *FASEB J.* **2007**, *21*, 2787–2797.

157. Giustiniani, J.; Sineus, M.; Sardin, E.; Dounane, O.; Panchal, M.; Sazdovitch, V., et al. Decrease of the immunophilin FKBP52 accumulation in human brains of Alzheimer's disease and FTDP-17. *J. Alzheimer's Dis.* **2012**, *29*, 471–483.

158. Pearl, L. H. Hsp90 and cdc37—a chaperone cancer conspiracy. *Curr. Opin. Genet. Dev.* **2005**, *15*, 55–61.

159. Smith, J. R.; Clarke, P. A.; de Billy, E.; Workman, P. Silencing the cochaperone CDC37 destabilises kinase clients and sensitises cancer cells to HSP90 inhibitors. *Oncogene* **2009**, *28*, 157–169.

160. Smith, J. R.; Workman, P. Targeting CDC37—an alternative, kinase-directed strategy for disruption of oncogenic chaperoning. *Cell Cycle* **2009**, *8*, 362–372.

161. Gray, P. J., Jr.; Prince, T.; Cheng, J.; Stevenson, M. A.; Calderwood, S. K. Targeting the oncogene and kinome chaperone CDC37. *Nat. Rev. Cancer* **2008**, *8*, 491–495.

162. Taipale, M.; Krykbaeva, I.; Koeva, M.; Kayatekin, C.; Westover, K. D.; Karras, G. I.; Lindquist, S. Quantitative analysis of Hsp90–client interactions reveals principles of substrate recognition. *Cell* **2012**, *150*, 987–1001.

163. Dou, F.; Chang, X.; Ma, D. Hsp90 maintains the stability and function of the tau phosphorylating kinase GSK3β. *Int. J. Mol. Sci.* **2007**, *8*, 51–60.

164. Jinwal, U. K.; Trotter, J. H.; Abisambra, J. F.; Koren, J., III.; Lawson, L. Y.; Vestal, G. D., et al. The Hsp90 kinase co-chaperone Cdc37 regulates tau stability and phosphorylation dynamics. *J. Biol. Chem.* **2011**, *286*, 16976–16983.

165. Shao, J.; Prince, T.; Hartson, S. D.; Matts, R. L. Phosphorylation of serine 13 is required for the proper function of the Hsp90 co-chaperone, Cdc37. *J. Biol. Chem.* **2003**, *278*, 38117–38120.

166. Miyata, Y.; Nishida, E. CK2 controls multiple protein kinases by phosphorylating a kinase targeting molecular chaperone, Cdc37. *Mol. Cell. Biol.* **2004**, *24*, 4065–4074.

167. Graham, K. C.; Litchfield, D. W. The regulatory beta subunit of protein kinase CK2 mediates formation of tetrameric CK2 complexes. *J. Biol. Chem.* **2000**, *275*, 5003–5010.

168. Pagano, M. A.; Cesaro, L.; Meggio, F.; Pinna, L. A. Protein kinase CK2: a newcomer in the "druggable kinome". *Biochem. Soc. Trans.* **2006**, *34*, 1303–1306.

169. Trembley, J. H.; Wang, G.; Unger, G.; Slaton, J.; Ahmed, K. CK2: a key player in cancer biology. *Cell. Mol. Life Sci.* **2009**, *66*, 1858–1867.
170. Duncan, J. S.; Litchfield, D. W. Too much of a good thing: the role of protein kinase CK2 in tumorigenesis and prospects for therapeutic inhibition of CK2. *Biochim. Biophys. Acta* **1784**, *2008*, 33–47.
171. Cozza, G.; Bortolato, A.; Moro, S. How druggable is protein kinase CK2? *Med. Res. Rev.* **2010**, *30*, 419–462.
172. Blanquet, P. R. Casein kinase 2 as a potentially important enzyme in the nervous system. *Prog. Neurobiol.* **2000**, *60*, 211–246.
173. Ishii, A.; Nonaka, T.; Taniguchi, S.; Saito, T.; Arai, T.; Mann, D., et al. Casein kinase 2 is the major enzyme in brain that phosphorylates Ser129 of human alpha-synuclein: Implication for alpha-synucleinopathies. *FEBS Lett.* **2007**, *581*, 4711–4717.
174. Ryu, M. Y.; Kim, D. W.; Arima, K.; Mouradian, M. M.; Kim, S. U.; Lee, G. Localization of CKII beta subunits in Lewy bodies of Parkinson's disease. *J. Neurol. Sci.* **2008**, *266*, 9–12.
175. Lee, G.; Tanaka, M.; Park, K.; Lee, S. S.; Kim, Y. M.; Junn, E.; Lee, S. H.; Mouradian, M. M. Casein kinase II-mediated phosphorylation regulates alpha-synuclein/synphilin-1 interaction and inclusion body formation. *J. Biol. Chem.* **2004**, *279*, 6834–6839.
176. Chen, J.; Gao, C.; Shi, Q.; Wang, G.; Lei, Y.; Shan, B., et al. Casein kinase II interacts with prion protein in vitro and forms complex with native prion protein in vivo. *Acta Biochim. Biophys. Sin.* **2008**, *40*, 1039–1047.
177. Pigino, G.; Morfini, G.; Atagi, Y.; Deshpande, A.; Yu, C.; Jungbauer, L., et al. Disruption of fast axonal transport is a pathogenic mechanism for intraneuronal amyloid beta. *Proc. Natl. Acad. Sci. U.S.A.* **2009**, *106*, 5907–5912.
178. Moreno, F. J.; Munoz-Montano, J. R.; Avila, J. Glycogen synthase kinase 3 phosphorylation of different residues in the presence of different factors: analysis on tau protein. *Mol. Cell. Biochem.* **1996**, *165*, 47–54.
179. Moreno, H.; Yu, E.; Pigino, G.; Hernandez, A. I.; Kim, N.; Moreira, J. E., et al. Synaptic transmission block by presynaptic injection of oligomeric amyloid beta. *Proc. Natl. Acad. Sci. U.S.A.* **2009**, *106*, 5901–5906.
180. Iimoto, D. S.; Masliah, E.; DeTeresa, R.; Terry, R. D.; Saitoh, T. Aberrant casein kinase II in Alzheimer's disease. *Brain Res.* **1990**, *507*, 273–280.
181. Chauhan, A.; Chauhan, V. P.; Murakami, N.; Brockerhoff, H.; Wisniewski, H. M. Amyloid beta-protein stimulates casein kinase I and casein kinase II activities. *Brain Res.* **1993**, *629*, 47–52.
182. Baum, L.; Masliah, E.; Iimoto, D. S.; Hansen, L. A.; Halliday, W. C.; Saitoh, T. Casein kinase II is associated with neurofibrillary tangles but is not an intrinsic component of paired helical filaments. *Brain Res.* **1992**, *573*, 126–132.
183. Masliah, E.; Iimoto, D. S.; Mallory, M.; Albright, T.; Hansen, L.; Saitoh, T. Casein kinase II alteration precedes tau accumulation in tangle formation. *Am. J. Pathol.* **1992**, *140*, 263–268.
184. Jin, L. W.; Saitoh, T. Changes in protein kinases in brain aging and Alzheimer's disease. Implications for drug therapy. *Drugs Aging* **1995**, *6*, 136–149.
185. Zhao, M.; Ma, J.; Zhu, H.-Y.; Zhang, X.-H.; Du, Z.-Y.; Xu, Y.-J.; Yu, X.-D. Apigenin inhibits proliferation and induces apoptosis in human multiple myeloma cells through targeting the trinity of CK2, Cdc37 and Hsp90. *Mol. Cancer* **2011**, *10*, 104.
186. Lees-Miller, S. P.; Anderson, C. W. Two human 90-kDa heat shock proteins are phosphorylated in vivo at conserved serines that are phosphorylated in vitro by casein kinase II. *J. Biol. Chem.* **1989**, *264*, 2431–2437.
187. Miyata, Y.; Chambraud, B.; Radanyi, C.; Leclerc, J.; Lebeau, M.-C.; Renoir, J.-M., et al. Phosphorylation of the immunosuppressant FK506-binding protein FKBP52 by casein kinase II (CK2): regulation of HSP90-binding activity of FKBP52. *Proc. Natl. Acad. Sci. U.S.A.* **1997**, *94*, 14500–14505.

188. Miyata, Y.; Yahara, I. The 90-kDa heat shock protein, HSP90, binds and protects casein kinase II from self aggregation and enhances its kinase activity. *J. Biol. Chem.* **1992**, *267*, 7042–7047.

189. Bandhakavi, S.; McCann, R. O.; Hanna, D. E.; Glover, C. V. C. A positive feedback loop between protein kinase CKII and Cdc37 promotes the activity of multiple protein kinases. *J. Biol. Chem.* **2003**, *278*, 2829–2836.

190. Jinwal, U. K.; Abisambra, J. F.; Zhang, J.; Dharia, S.; O'Leary, J. C.; Patel, T., et al. Cdc37/Hsp90 protein complex disruption triggers an autophagic clearance cascade for TDP-43 protein. *J. Biol. Chem.* **2012**, *287*, 24814–24820.

191. Hanson, K. A.; Kim, S. H.; Tibbetts, R. S. RNA-binding proteins in neurodegenerative disease: TDP-43 and beyond. *Wiley Interdiscip. Rev.: RNA* **2012**, *3*, 265–285.

192. Neumann, M.; Sampathu, D. M.; Kwong, L. K.; Truax, A. C.; Micsenyi, M. C.; Chou, T. T., et al. Ubiquitinated TDP-43 in frontotemporal lobar degeneration and amyotrophic lateral sclerosis. *Science* **2006**, *314*, 130–133.

193. Lee, E. B.; Lee, V. M.; Trojanowski, J. Q. Gains or losses: molecular mechanisms of TDP43-mediated neurodegeneration. *Nat. Rev. Neurosci.* **2012**, *13*, 38–50.

194. Seneci, P. Chemical modulators of protein misfolding and neurodegenerative disease. Elsevier, accepted for publication, **2015**.

3

Proteasomal Degradation of Soluble, Misfolded Proteins
Throwing out the Bath Water, but Where's the Baby?

3.1 UPS-MEDIATED DEGRADATION OF MISFOLDED PROTEINS

Hsp27-, Hsp70-, and Hsp90-mediated rescuing of misfolded tau copies and/or oligomers is thoroughly described in Chapter 2. In parallel, the elimination of misfolded protein copies—especially when tau soluble oligomers/aggregates are involved—is another appealing option to tackle neurodegenerative diseases (NDDs) and tauopathies. The *ubiquitin-proteasome system (UPS)* [1,2] is the most important cellular pathway to dispose of soluble cytosolic proteins.

UPS embraces most regulated proteolytic events, and depends on the 76-mer protein ubiquitin (UBQ) [3,4]. A first UBQ molecule is anchored onto substrate proteins through an isopeptide bond involving the C-terminus of UBQ and the ε-NH$_2$ group of a Lys residue in the substrate proteins. Mono-ubiquitination may target a specific Lys residue [5], or a domain [6] on the substrate protein. Multiple monoubiquitination on different Lys residues of the substrate protein (multi-monoubiquitination) is also observed [7]. Monoubiquitination modulates the properties of a protein substrate such as increasing its activity [8], or increasing its transcription [9]. Monoubiquitination may promote [10] or prevent [11] the interaction of a protein with its binding partners. Monoubiquitination influences the internalization of plasma membrane proteins [12], cell signaling [13], and endocytosis [14]. Multi-monoubiquitination is connected to proteasomal processing [15] and nuclear export of multi-monoubiquitinated proteins [16].

Ubiquitinated proteins usually are tagged with polyUBQ chains. UBQ chain elongation implies the formation of an isopeptide bond between one

Molecular Targets in Protein Misfolding and Neurodegenerative Disease. DOI: 10.1016/B978-0-12-800186-8.00003-1

of seven Lys residues (K6, K11, K27, K29, K33, K48, and K63), or the Met1 residue of a substrate-anchored proximal UBQ molecule with the C-terminus of a free, distal UBQ protein—see below. UBQ is a stable, compact protein with the six C-terminus amino acids (AAs) arranged as a flexible tail [4]. Met1 and the Lys residues of UBQ are distributed throughout its structure, with K6 and K11 being in the most dynamic/flexible environment, and K27 requiring conformational UBQ changes to be exposed from its buried locus [17]. Different UBQ chain elongation enzymes bind to interaction surfaces on UBQ with diverse specificities, and promote UBQ elongation on a specific anchoring point. Three main hydrophobic patches are centered around Phe4 (including Gln2 and Thr12) [18], Ile36 (including Leu71 and Leu73) [19], and Ile44 (including Leu8 and Leu70) [20]. The hydrophilic TEK-box region includes Thr12, Thr14, and Glu34 residues together with K6 and K11 [21]. A hydrophilic patch is centered around Asp58, and includes the Arg54, Thr55, and Ser57 residues [22].

The Ile44 patch is most widely recognized by the UBQ-binding domains (UBDs) of different protein partners, and is used to establish binding interactions with the proteasome and most UBDs [17,18]. The Ile44 patch interacts differently with the UBDs of different proteins, sometimes even without involving the Ile44 residue [23]. The same is true for other, previously mentioned, interaction surfaces, and possibly for yet-undiscovered binding regions on UBQ. As a result, a complex UBQ code that is relevant for each and every cellular process is established [17].

Once a di-UBQ chain is formed, the two UBQ molecules assume an anchoring residue-dependent conformation. Some of their interaction surfaces stabilize the conformation of the di-UBQ chain, while others are exposed to recruit binding partners (Figure 3.1). Five out of the eight UBQ connections are structurally characterized by X-ray crystallography and/ or nuclear magnetic resonance (NMR) [17].

K48-linked di-UBQ chains adopt a "closed," compact conformation, where the two UBQ proteins interact with each other. A preferred conformation (A1, Figure 3.1) entails the strong interaction between two Ile44 patches in K48-linked di-UBQ chains [24]. Another conformation where the Ile36 patch on the distal UBQ interacts with the Ile44 patch on the proximal UBQ is observed [25] (A2). The dynamic equilibrium among them may be shifted towards each conformation by specific binding partners that interact at best with either conformation. Larger K48-linked UBQ oligomers, such as tetrameric A3, are also known [26].

Similarly to K48, *K6-* (structures B1/di-UBQ and B2/tetra-UBQ) and *K11-linked* di-UBQ chains (structures C1–C2/di-UBQ and C3/octa-UBQ) adopt closed/UBQ-interacting conformations. The former chains are compacted by interactions involving the Ile36 patch on the distal UBQ and the Ile44 patch on the proximal UBQ (different from K48-linked di-UBQ chains involving the same patches, compare A1 with B1) [27].

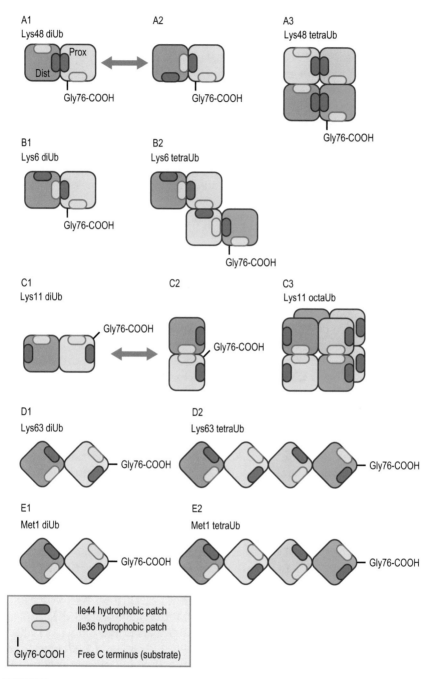

FIGURE 3.1 Molecular architecture of UBQ dimers and oligomers.

K11-linked di-UBQ chains may interact asymmetrically (C1, different α-helix regions on the two UBQ proteins [28]) or symmetrically (C2, through the Ile36 patch on both UBQs [29]) in two highly populated conformations sharing the exposure of the interaction-prone Ile44 patch. Each preferred conformation for K48-, K6-, and K11-linked di-UBQ chains is observed in polyUBQ chains, either as single conformation-containing polyUBQ (e.g., K48 tetramers/Ile44-Ile44, A3 [26]) and as conformational hybrids (e.g., K11 octamers/α-helix-α-helix and Ile36-Ile36, C3 [30]) [17].

K63-linked di-UBQ chains adopt an "open" and more flexible conformation, where the two UBQ molecules only interact through their isopeptide bond connection [31]. In addition to the interaction surfaces of each UBQ molecule, binding partners of K63-linked UBQ chains (di-, D1, or higher order oligomers, D2) recognize/interact also with the linkage between the two UBQs [32]. Similarly, *Met1-linked* di- and polyUBQ chains adopt a linear, open conformation [33] (respectively E1 and E2, Figure 3.1). The higher flexibility of open conformations of di-UBQ chains provides even more alternative binding modes with their protein partners. Such flexibility allows conformational rearrangements leading to closed conformers of K63- [34] and Met1-linked di-UBQ chains [35] in particular conditions.

The structure of *K27-, K29-, and K33-linked* di-UBQ chains is yet to be reported. A molecular modeling study [36] predicts a closed/compact conformation for K27-linked di-UBQ chains, and an open/flexible conformation for K29- and K33-linked di-UBQ chains.

Although UBQ is a key component of the UPS, protein ubiquitination encompasses additional functions in cells [37]. Thus, many UBQ labels are needed. PolyUBQ homotypic chains, containing varying numbers of UBQ residues connected through the same Lys or Met residue *via* an isopeptide bond with the C-terminus of UBQ, are known for each of the eight mentioned UBQ anchor points [17]. PolyUBQ heterotypic chains, including mixed K11-K63 [38], and K6-K27-K48 polyUBQ sequences [39], are also known, but their frequence and relevance appear to be limited.

K48-linked polyUBQ chains are the most abundant homotypic polyUBQ species [40], acting as labels on proteins to be processed for degradation through the UPS [41]. Proteasome inhibition causes a fast increase of K48-polyUBQ proteins [42], and mutation of Lys residues in yeast UBQ shows that K48 is the only essential residue among them [43].

Ubiquitination of proteins on Lys residues other than K48 leads to a number of functions, including proteasomal degradation [44]. K11-linked polyUBQ chains [45] promote proteasomal degradation of K11-linked proteins during cell cycle progression [46], and are enriched when the proteasome is inhibited [47]. They are involved in endoplasmic reticulum-associated degradation (ERAD) [48] and in membrane trafficking [49], and regulate nuclear factor kappa-light-chain-enhancer of activated B cells (NF-κB) essential modulator (NEMO)-dependent NF-κB activation [50].

PolyUBQ chains connected to substrate proteins through K6, K27, K29, and K33 are less characterized and more difficult to identify through mass spectrometry [51], as linkage-specific antibodies targeting them are still unknown [52]. K6-linked polyUBQ chains are observed on proteins involved in the DNA repair pathways, such as the breast cancer type 1 susceptibility (BRCA1)-associated RING domain protein 1 (BARD1) complex [53], nucleophosmin (NPM) [54], and the RNA polymerase subunit RPB8 [55]. Although little is known about K6-linked polyUBQ chains [52], they appear to be unrelated to proteasome degradation [42]. K27-linked poly-UBQ chains are connected to mitochondrial biology, and more specifically to the response to mitochondrial damage through polyubiquitination of, *inter alia*, voltage-dependent anion-selective channel protein 1 (VDAC1) [56]. They also inhibit nuclear translocation of the transcription factor TIEG1 (transforming growth factor-β (TGFβ)-inducible early growth response protein 1) [57]. K29-linked polyUBQ chains contribute to cellular proteostasis [58], and are enriched when the proteasome is inhibited [25]. They may also contribute to target ubiquitinated proteins to lysosomes [59]. K33-linked polyUBQ chains reduce phosphorylation and receptor association of T-cell receptor ζ (TCR) [60], and—as is the case for K29-linked polyUBQ chains—regulate the activity of AMP-activated protein kinase (AMPK) kinases [61].

K63-linked polyUBQ chains confer UPS-independent functions to their tagged substrates [62]. They exert a scaffolding/stabilizing/protein–protein interaction (PPI)-promoting role on spliceosome- [63] and polysome-driven [64] processes. Promoted PPIs may lead to substrate protein activation (i.e., inhibitor of NF-κB kinase (IKK) activation by binding to the transforming growth factor-β-activated kinase 1 (TAK1) complex [65]). K63-linked polyUBQ chains promote the recruitment of UBQ E3 ligases—see below—to sites of DNA damage [66]. Surprisingly, even unanchored K63-linked polyUBQ chains act as molecular scaffolds to promote PPIs. [67]. In particular, though, K63-linked polyUBQ chains have an essential role in directing K63-labeled polyUBQ protein towards selective autophagy/aggrephagy [68]. This key function of K63-linked polyUBQ chains is extensively covered in Chapter 5.

UBQ is initially translated as a linear polymer [69] that links the two terminal AAs—Met1 and Gly76—that are post-translationally processed/cleaved to give monoUBQ. The linear UBQ chain assembly complex (LUBAC) is responsible for linear polyUBQ chain assembly, which is relevant for other functions. For example, LUBAC and Met1-linked polyUBQ chains are essential components of the NF-κB activation process [70,71].

Eight Lys/Met anchoring points on UBQ; UBQ chain lengths varying between 1 (monoUBQ) and >10 polyUBQs on each anchoring point; >700 enzymes involved in a multistep process (see below), including UBQ activation, conjugation, transfer to a protein substrate and chain

trimming [17]. The combinations of UBQ codes easily match the experimental observation of thousands of UBQ-labeled protein substrates on
multiple sites, and ensure an exquisitely specific UBQ/UPS-dependent
regulation of the functions ultimately reconducible to ubiquitinated proteins. How can the UBQ machinery select the anchoring point, the UBQ
chain length and nature, and the specific substrate to be ubiquitinated
or de-ubiquitinated in a dynamic equilibrium? The UBQ activation–
conjugation–ligation–trimming cycle, and its connection with proteasomal degradation are briefly summarized in Figures 3.2 and 3.3 [72].

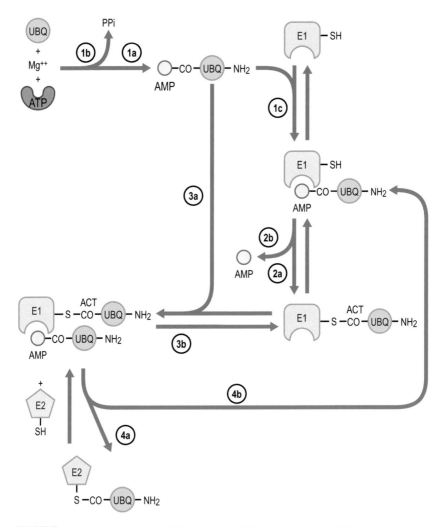

FIGURE 3.2   The UBQ cycle: UBQ-activating (E1) and conjugating (E2) enzymes.

UBQ is activated by two *UBQ-activating (E1) enzymes*, UBA1 [73] and UBA6 [74]. Both E1 enzymes work by first activating and binding UBQ. Namely, ATP- and Mg^{2+}-dependent acyladenylation of its C-terminus (step 1a, Figure 3.2) and release of pyrophosphate (step 1b) produce UBQ-AMP, which binds to the E1 ATP binding domain (step 1c) [75]. Then, intramolecular *trans*-esterification of $E1_{ATP}$-AMP-CO-UBQ-NH$_2$ with a Cys residue in the E1 active site provides the high-energy thioester H_2N-UBQ-CO-S-$_{ACT}$E1 (step 2a) with release of AMP (step 2b). A second UBQ-AMP molecule binds then to the E1 ATP binding domain (step 3a), forming the key di-UBQ intermediate H_2N-UBQ-CO-S-$_{ACT}$E1$_{ATP}$-AMP-CO-UBQ-NH$_2$ (step 3b) [72,76].

UBA1 catalyzes the vast majority of UBQ-activating reactions, while UBA6 is used in a limited set of UBQ cycles [77] in conjunction with a single E2-conjugating enzyme [78]—see below. A few other E1 enzymes work similarly to activate UBQ-like proteins for protein tagging [79].

Around 40 *UBQ-conjugating (E2) enzymes* relieve the E1 enzymes from Cys-bound UBQ [80]. Namely, a Cys residue in their active site receives UBQ through a *trans*-thiolation reaction with the key intermediate H_2N-UBQ-CO-S-$_{ACT}$E1$_{ATP}$-AMP-CO-UBQ-NH$_2$ to yield the conjugated E2-S-CO-UBQ-NH$_2$ intermediate (step 4a) [72]. The reaction is driven by the S-CO bond breakage in the high-energy-containing intermediate H_2N-UBQ-CO-S-$_{ACT}$E1$_{ATP}$-AMP-CO-UBQ-NH$_2$, and produces an E1$_{ATP}$-AMP-CO-UBQ-NH$_2$ molecule that re-enters the UBQ cycle (step 4b, Figure 3.2).

E2 enzymes contain a highly conserved 150–200 AA UBQ-conjugating catalytic (UBC) fold, which acts as a scaffold for E1 enzymes, E3 ligases (see below), and activated UBQ [80]. An E2 enzyme may contain only the UBC fold (class I E2s), an N-extension (class II) or a C-extension (class III), or both (class IV) [81]. The extensions may regulate the enzyme activity and the length of polyUBQ chains [82], may influence the localization of the enzyme [83], or may act as true protein domains mediating PPIs [84,85].

Around 10 E2 enzymes either lack a Cys residue in their active site [86], or use UBQ-like proteins as tags [87], or discharge UBQ to Cys, rather than Lys residues of the substrate protein [88]. The majority of E2 enzymes acts as in Figure 3.2, controlling the length of UBQ tags (monoUBQ, or poly-UBQs with varying lengths) [89], and the sites of UBQ tagging (which Lys residue on the substrate protein is tagged by UBQ, and which Lys/Met residue of UBQ connects polyUBQ chains) [86]. UBQ-conjugating enzymes E2T (UBE2T, HSPC150) [90] and E2W (UBE2W) [91] are *monoubiquitinating E2 enzymes* that transfer a single UBQ molecule onto their substrate proteins, and prevent UBQ chain elongation through an unknown mechanism [17]. UBQ-conjugating enzymes E2C (UBE2C, UbcH10) [92] and E2D (UBE2D1–2D4, UbcH5A–5C, HBUCE1) [93] are *chain priming E2 enzymes* that functionalize the substrate proteins with one or few UBQ molecules before chain elongation. UBE2C is highly specific and works with a single E3 ligase complex [82]. UBE2D family members interact with many E3 enzymes, priming a large population of substrate proteins on various positions and with various UBQ Lys/Met linkages [94]. *Chain elongating E2 enzymes* build polyUBQ chains with linkage selectivity [52]. K48-specific elongating E2 enzymes include UBQ-conjugating enzymes E2K (UBE2K, E2-25K, Hip2) [95], E2R1 (UBE2R1, Cdc34, Ubc3) [96], E2R2 (UBE2R2, Cdc34B) [97], E2G1 (UBE2G1) [98], and E2G2 (UBE2G2, Ubc7) [99]. K11 specificity is provided by the UBQ-conjugating enzyme E2S (UBE2S, E2-EPF), through an interaction with the K11-containing TEK box of UBQ [100]. K63 specificity is ensured by the activity of complexes containing the UBQ-conjugating enzyme E2N (UBE2N, Ubc13) [101] and either one among the UBQ-binding cofactors UBE2V1 [102] and UBE2V2 [103]. The UBE2V cofactors accommodate UBQ in such a way that K63 is close to the active site Cys of UBE2N [104]. K6-, K27-, K29-, and K33-specific chain elongation E2 enzymes are currently unknown.

As already mentioned, E2-S-CO-UBQ-NH$_2$ complexes then bind to *E3 ligases* [105], and transfer UBQ to substrates through three main mechanisms (Figure 3.3). Namely, Zn^{2+}-dependent binding of an E2-S-CO-UBQ-NH$_2$ complex and a protein substrate to the family of *really interesting new gene (RING) E3 ligases* (step 1a, Figure 3.3) [106] leads to the direct transfer of UBQ from E2-S-CO-UBQ-NH$_2$ complexes to RING E3-bound protein substrates (step 1b), and to the release of ubiquitinated substrate (step 1c). The release of E2s (step 1d) and of RING E3s enzymes (step 1f) recycles them back to the UBQ cycle. Binding of an E2-S-CO-UBQ-NH$_2$ complex with the family of *homologous to the E6AP carboxyl terminus (HECT)* E3 ligases (step 2a) [106] and *trans*-thiolation of UBQ from E2-S-CO-UBQ-NH$_2$ complexes to a Cys residue of HECT E3s (step 2b) is followed by the release of E2s (step 2c), binding with a protein substrate and transferring UBQ to the protein (step 2e), release of the ubiquitinated substrate (step 2f) and of HECT E3 enzymes (step 2g). Finally, *RING-in-between-RING (RBR)* E3 ligases [108] act through a RING/HECT hybrid mechanism. They bind E2-S-CO-UBQ-NH$_2$ complexes on a RING domain (step 3a), catalyze the *trans*-thiolation of UBQ from E2-S-CO-UBQ-NH$_2$ complexes to a Cys residue on a RING-like domain and bind a substrate protein (step 3b), release E2s (step 3c), transfer UBQ to the substrate protein (step 3e), and release ubiquitinated substrates (step 3f) and RBR E3 enzymes (step 3g, Figure 3.3).

The RING domain is a rigid, globular scaffold protein with the general structure   Cys-X$_2$-Cys-X$_{9-39}$-Cys-X$_{1-3}$-His-X$_{2-3}$-Cys/His-X$_2$-Cys-X$_{4-48}$-Cys-X$_2$-Cys [109]. The spaced Cys and His residues form a Zn^{2+}-coordinating frame buried deep into the RING E3 structure [110]. Bioinformatics predict that more than 600 RING domain-containing proteins exist [105], and

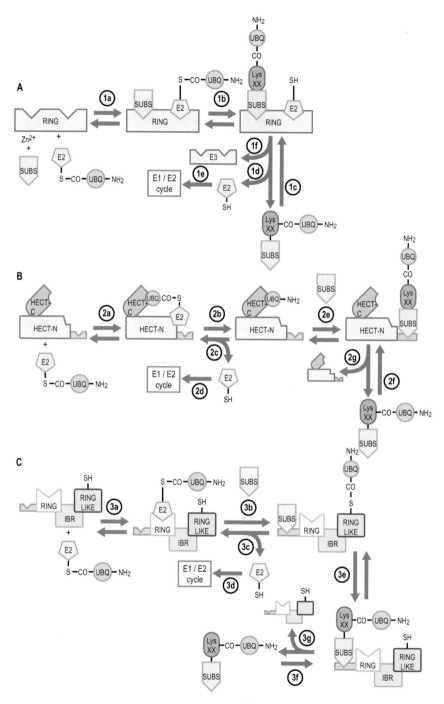

FIGURE 3.3   The UBQ cycle: UBQ ligases (E3).

an E3 ligase activity is confirmed for many proteins encoded by them [109]. Around 300 RING E3 ligases are centered around the RING domain, which binds both to UBQ-CO-S-E2 complexes and substrate proteins [107]. More than 230 RING E3s contain one of eight cullins, a family of scaffold proteins that promotes the assembly of large, multimeric *cullin RING E3 ligases* (*CRLs*) [109,111]. Additionally, nine proteins act as RING-like E3 ligases through a *U-box domain*, where polar and charged active site residues maintain the U-box shape and promote a Zn^{2+}-independent, RING-like mechanism of substrate protein ubiquitination [112]. RING domains, either in monomeric and in multimeric RING E3 ligases, often require dimerization to be functional [113]. RING E3 homodimers are well known [114,115] but heterodimers ensure even more structural complexity and, consequently, create supramolecular architectures with sophisticated—and often unclarified—mechanisms of action [116,117].

Surprisingly, UBQ-CO-S-E2 complexes and substrate proteins bind either to separate domains (CRLs) or to elongated locations on the RING domain (monomeric RING E3s) [118,119]. In both cases, the reaction partners lie at least 50–60 Å apart, and a major conformational change, mostly regarding the UBQ-CO-S-E2 complex [120], is needed to catalyze the *trans*-thiolation reaction [121]. As observed with E2 conjugating enzymes, RING E3s may be involved in UBQ transfer to a few or many protein substrates. Each E2-RING E3 combination shows different properties, and the E2 enzyme (where the *trans*-thiolation/UBQ transfer to the substrate takes place) is usually determining the substrate specificity [109]. E2-RING E3 complexes catalyse both UBQ chain initiation and chain elongation (even if their requirements are unalike [122]), either through the same complex or by varying one of its components. Chain initiation reactions are usually slow, with the accessibility of Lys substrate residues determining any specificity in UBQ attachment [109]. Chain elongation reactions are faster and more specific, due to either UBQ Lys/Met accessibility [123] or sequence element recognition [21]. E2-RING E3 complexes specific for K6- [53], K48- [124], and K63-linked polyUBQ chains [125] are known. Chain elongation may either entail a stepwise process, or RING E3-catalyzed *en bloc* transfer to the substrate protein [126]. The last process requires the binding of two (not necessarily identical) E2 enzymes onto the RING E3 domain, the pre-formation of a polyUBQ chain on one of the E2s by stepwise UBQ transfer from the other E2 enzyme, followed by polyUBQ transfer to the substrate protein in a single step [109].

The activity of E2-RING E3 complexes is subjected to a sophisticated regulation network. Phosphorylation of the substrate protein often activates it towards ubiquitination [127,128], and the same happens with glycosylation [129] and proline hydroxylation [130]. Phosphorylation of E2 enzymes triggers their activity [97], while phosphorylation of RING

E3 domains may stimulate [131] or inhibit [132] their activity. Negative regulation is reported for protein partners of multimeric RING E3s [133], and for non-ubiquitinated pseudo-substrates competing with substrate proteins [134]. Small molecules act either as positive [135] or negative [136] RING E3 regulators. Auto-ubiquitination is the most relevant regulation mechanism, with a RING E3-dependent outcome. It may decrease the RING E3 ligase activity by increasing its proteasomal degradation [137], which may be prevented by binding with the substrate protein (i.e., increased degradation of the RING E3 ligase only when there is shortage of the substrate protein [138]). It may also lead to increased RING E3 ligase activity [39], or to activation of signaling pathways [139]. Finally, conjugation of multimeric CRLs with the UBQ-like protein neural precursor cell expressed, developmentally down-regulated 8 (NEDD8), activates CRLs. Their activation implies increased E2 enzyme recruitment [140] and conformational changes [121].

The HECT domain is a ≈350 AA bilobal domain, usually found at the C-terminus of HECT E3 ligases [106]. Twenty-eight human proteins contain a HECT domain and have E3 ligase activity. Among them is the *neural precursor cell expressed, developmentally down-regulated 4 (NEDD4)* [141] protein family, composed of nine members and characterized by an N-terminal phospholipid-binding C2 domain and by two to four WW domains [106]. The C2 domain acts as a binding domain for specific proteins [142], including other HECT E3 ligases [143]. It promotes the intracellular targeting of NEDD4 E3 ligase activity to, *inter alia*, endosomal cargo [144] and Wnt signaling proteins [145]. The Trp-containing WW domains are responsible for substrate protein binding [146]. The *HECT and regulator of chromosome condensation 1 (RCC1)-like domains* (overall, *HERC*) [147] protein family, composed of six members, is characterized by one or more RCC1-like domains (RLDs) [106]. RLDs have a β-propeller fold, whose sides respectively bind to chromatin and act as a guanine nucleotide exchange factor [148]. HERC family members act as modulators of centrosome architecture, and contribute to maintain its integrity [149]. The other 13 HETC E3 ligases do not show common structural features apart from the HECT domain [106]. They contain additional domains, such as UBQ-associated/UBA domains and WWE domains (HECT, UBA, and WWE domain-containing protein 1, HUWE1 [150]), ankyrin repeats (HECT domain and ankyrin repeat-containing E3 ubiquitin-protein ligase 1, HACE1 [151]), and plant homeo domains (PHD) and RING domains (G2/M-phase-specific E3 UBQ ligase, G2E3 [152]).

The lobe towards the N-terminus of HECT domains (N-lobe) binds to the UBQ-CO-S-E2 complexes, while the C-lobe contains the conserved Cys residue that receives the UBQ in the *trans*-thiolation reaction. The substrate protein binds to other domains in the HECT E3 ligase structure [106]. The distance between the Cys residues involved in the *trans*-thiolation varies between ≈15 Å [153] and ≈50 Å [154]. Thus, large conformational changes allowed by the flexible hinge loop between the N- and C-lobes are needed for the HECT domains in order to be functional [155]. Mutation of a highly conserved Phe residue close to the C-terminus of HECT E3 ligases does not impair the *trans*-thiolation/UBQ transfer from E2s to the HECT domain, but prevents UBQ transfer to the Lys residue of substrate proteins [156]. Its absence, or mutation, may misorient E3-bound UBQ and prevent the approach/block the access to E3-bound UBQ for substrate proteins [156]. As to linkage specificity in E2-HECT E3 couples, the HECT E3 partner (i.e., where the *trans*-thiolation/ UBQ transfer to the substrate takes place) is determining linkage specificity. Namely, specificity stems from the 60 C-terminal AAs in the HECT domain [157]. HECT E3 ligases show K48- (E6-associated protein, E6AP [158]), K63- (NEDD4 [159]), or mixed K29/K48 linkage specificity (UBE3C [160]). The mechanism of the K63-selective *trans*-thiolation reaction for NEDD4 is structurally clarified [161].

HECT E3 ligases can be negatively regulated through the recruitment of partner proteins, hindering the binding site for substrate proteins [162,163]. Partner proteins may induce positive regulation by conformational changes leading to increased protein substrate binding [164], or by binding to both the HECT E3s and their substrate proteins in a ubiquitination-promoting complex structure [165]. Intramolecular binding between the HECT domain and other E3 domains may inhibit or increase the activity of HECT E3 ligases. The former outcome is caused by binding with the C2 domain [143], while the latter depends on binding with WW domains and protection from auto-ubiquitination-dependent HECT E3 degradation [166]. Partner proteins may inhibit the intramolecular WW-HECT domain interactions, and direct HECT Es towards proteasomal degradation [166].

RBR E3 ligases [108] are characterized by an ≈60 AA N-terminal RING domain that binds two Zn^{2+} ions, folds in a classical cross-braced ring finger structure and is named RING1 [167]. A second, ≈40 AA C-terminal RING2 domain binds either a single Zn^{2+} ion [168] or two Zn^{2+} ions [169], and has a hydrophobic core different from classical RING domains [167]. They are connected by another Cys-His ring finger structure, named in-between RING (IBR) domain [170]. The 13 human RBR E3 ligases show limited variability in the RBR region, especially in the Zn^{2+}-coordinating Cys-His residues of the RING1 and IBR domains [171]. Additional regions in specific RBR E3s include hydrophobic regions (ring finger protein 144A/RNF144A [172] and RNF19A/dorfin [173]), negative charge clusters (ariadne-1 homolog, ARIH1, HHARI [174] and ariadne-2 homolog/ARIH2/TRIAD1 [175]), RWD domains (RNF14, ARA54, TRIAD2 [176]), and UBQ-like domains (parkin [177], heme-oxidized IRP2 ubiquitin ligase 1/HOIL-1 [178]).

Although similar to RING domains of RING E3 ligases in terms of E2 binding [179], the RING1 domain of RBR E3s is not able, even in a construct containing also the IBR domain, to promote the transfer of UBQ from E2 enzymes to the Lys residue of a substrate protein [88]. The conserved nature of the IBR domain throughout RBR E3s would suggest its strong influence on substrate proteins [108]. Only an intradomain interaction between its C- and N-termini to bring near the RING1 and RING2 domains, though, is experimentally observed [180]. The Cys-His-rich AA sequence of the RING2 domain binds one (HHARI [168]) or two Zn^{2+} ions (parkin [169]). It does not engage all its conserved Cys residues in Zn^{2+} coordination. The parkin Cys431 residue [168] and the HHARI Cys357 residue [181] do not participate to intramolecular interactions, and are used as sites for the *trans*-thiolation reaction with UBQ-CO-S-E2 complexes [108].

RBR E3 ligases may catalyze mono- [182] and polyubiquitination of target substrates [183], and may be K63-linkage-selective [184]. LUBAC [69] is a 600 kDa E3 complex containing the RBR E3 ligases HOIL-1 [178] and HOIL-1 interacting protein (HOIP) [185], and the SH3 and multiple ankyrin repeat protein-associated RBCK1 homology domain-interacting protein (SHARPIN) [186]. It is the only known E3 ligase that assembles linear, Met1-connected polyUBQ chains. Its molecular mechanism of action is centered onto the RBR HOIP, which *per se* is a linear polyUBQ U3 ligase, but autoinhibits itself [187]. Binding with either HOIL-1 or SHARP-IN, or both, prevents auto-ubiquitination and establishes a fully functional LUBAC complex [187]. Negative regulation by auto-ubiquitination is observed also with parkin [188], dorfin [173], ARA54 [176], and ARIH2 [189]. In conclusion, the hybrid RING (presence of RING domains)-HECT (E3-UBQ intermediate thioester) mechanism is possibly conferring other unique, physiologically and/or pathologically relevant properties to RBR E3 ligases [108].

As ubiquitination is a reversible reaction, mono- and polyUBQ chains can be disassembled by an ≈100-membered class of isopeptide-specific *de-ubiquitinating enzymes (DUBs)* [190], belonging to one metalloprotease and five cysteine protease subfamilies [191,192]. Their roles in basal conditions are depicted in Figure 3.4.

UBQ is translated by the action of four genes, either as linear poly-UBQ (*Ubb* and *Ubc* genes [193], step 2a, Figure 3.4), or as a polyUBQ fusion with ribosomal proteins (UbL40, UbS27a [194], step 2b). In both cases, DUBs (step 1a) are essential to hydrolyze polyUBQ chains (step 1b) and to provide *free UBQ release* (path 2, Figure 3.4) [193] in the UBQ pool.

UBQ homeostasis is essential to mammalian cells. In addition to stress-induced higher expression of UBQ-producing genes [195], the UBQ pool is preserved by recycling UBQ before UPS-driven protein degradation [196].

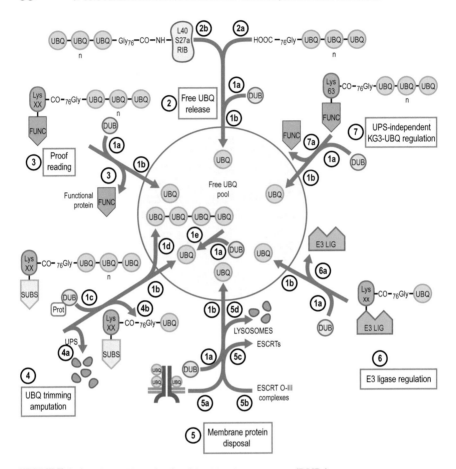

FIGURE 3.4   The UBQ cycle: de-ubiquitinating enzymes (DUBs).

At any point of the previously described UBQ cycle, the growth of a UPS-directed polyUBQ chain on a tagged substrate protein may be reverted by de-ubiquitination. This is useful to rescue a functional/mistagged substrate protein from degradation (*proofreading*, path 3, Figure 3.4 [196]). DUBs, then, dispose of the UBQ tag (step 1a) and release the functional protein (step 3) in a step-by-step, single UBQ-removing process, with the release of free UBQ (step 1b) in the UBQ pool.

Once UPS-targeted, polyUBQ-tagged proteins bind to the proteasome (a multisubunit proteolytic complex [197,198] described in detail later) through proteasomal UBQ receptors [199], the fate of the substrate protein is sealed (path 4, Figure 3.4). Proteasome-bound DUBs [200] cleave polyUBQ chains (step 1c) to yield free UBQ (step 1b) in the UBQ pool, and to enable proteasome-mediated protein degradation [201,202] that would be hindered by polyUBQ chains. UBQ chains can be removed step by step (path 4, *UBQ trimming*) or in a single step (path 4, *UBQ amputation*). Both processes are sketched in Figure 3.4. Proteasome-bound poly-UBQ proteins may be processed by two proteasome-associated trimming DUBs (step 1c) [203] that shorten polyUBQ chains in a step-by-step, single UBQ-removing process. Free UBQ (step 1b) and the shortened polyUBQ proteins from the UBQ receptors (step 4b) [201] are released. Conversely, unfolding and translocation of tagged substrate proteins is coupled with *en bloc* cleavage of the remaining UBQ residues by another proteasome-bound DUB (step 1c) [204]. The final UBQ cleavage produces untagged substrate proteins that are readily degraded by the proteasome (step 4a [205]). *En bloc* cleavage by a proteasome-bound DUB releases short, unattached polyUBQ chains (step 1d) that are processed by trimming DUBs (step 1e) to release once more free UBQ in the UBQ pool.

Membrane protein disposal happens (path 5, Figure 3.4) through UBQ/DUB-assisted receptor endocytosis [206]. Membrane proteins are mono- or oligo-ubiquitinated by specific E3 ligases [207]. They are then internalized into early endosomes [208] through binding to endosomal UBQ receptors in the early endosomal sorting complex required for transport machinery (ESCRT0). Then, they are progressed through the endocytic route by exchange of ESCRT components, up to the late ESCRTIII complex [209]. The multiple interactions between ubiquitinated membrane proteins and ES-CRTs are summarized in steps 5a and 5b. Late ESCRTIII complex-bearing endosomes then recruit specific DUBs (step 1a [210]). ESCRT III-bound DUBs release UBQ from membrane receptors for its recycling (step 1b) in the UBQ pool, and free the ESCRT machinery that re-enters the receptor endocytosis cycle (step 5c). "Naked" membrane receptors are finally degraded (step 5d) by sorting into multivesicular bodies (MVB) that eventually fuse with lysosomes [208].

DUBs have a role in UPS-independent mechanisms, by regulating the activity of ubiquitinated proteins or the activation of signaling pathways [196]. *E3 ligase regulation* (path 6), and specifically their auto-ubiquitination, often acts as a negative E3 regulator to prevent substrate protein degradation by the ligase [109]. Conversely, specific DUB–E3 ligase interactions/pairs (step 1a) de-ubiquitinate the E3 ligase (step 6a), release free UBQ (step 1b) in the UBQ pool, and restore E3 activity in the UBQ cycle [211].

DUBs may negatively regulate whole pathways through *UPS-independent K63-UBQ regulation* (path 7, Figure 3.4). The UBQ cycle may produce K63-UBQ-tagged proteins by activating NF-κB-dependent inflammation [212]. Several DUBs negatively regulate the NF-κB pathway (step 1a) by processing and de-ubiquitinating the K63-linked UBQ chains (step 7), releasing free UBQ in the UBQ pool (step 1b), and providing control on innate and adaptive immune responses (Figure 3.4) [213].

The vast majority of DUBs are cysteine proteases, and their mechanism is similar to the papain protease family [214]. A catalytic Cys residue is activated by a neighboring His for a nucleophilic attack on the carbonyl group of the isopeptide bond. The now charged His residue donates a proton to the ε-Lys leaving group, and a conserved Asp/Asn residue completes the catalytic Cys-His-Asp/Asn triad by orienting and polarizing the His group [202]. Notwithstanding the structural differences between the five DUB cysteine protease families, the tridimensional arrangement of their catalytic triad in the active conformation is almost superimposable [196,215].

Ubiquitin-specific proteases (*USP* [216]) are the largest family of DUBs, including more than 50 family members [192]. Their USP catalytic domain contains between ≈300 and ≈900 AAs, due to frequent insertions of large polypeptide sequences [192]. It is divided into three subdomains, resembling the palm, the thumb, and the fingers of a right hand [217]. The Cys residue is located in the thumb, and the His residue is in the palm. USPs are usually activated by polyUBQ binding, causing the fingers' domain to bind to the distal UBQ molecule. The induced conformational change allows the polyUBQ chain—more specifically the isopeptide bond to be cut by USPs—to be correctly positioned into the catalytic site [217]. Frequently recurring domains in USPs include ubiquitin-interacting motifs (UIMs) [218], zinc finger ubiquitin-specific protease (ZnF-UBP) [219], ubiquitin-associated [220] and ubiquitin-like (UBL) domains [221].

Ubiquitin C-terminal hydrolases (*UCH* [222]) are four Cys DUBs sharing an N-terminal, ≈230 AA catalytic domain. Two smaller family members, composed only by the catalytic domain, specifically act on monoUBQ substrates [223], including ester, thioester, and amide-UBQ conjugates [224]. Larger UCHs contain C-terminal extensions that promote the interaction with substrate proteins [225], and process also di- and polyUBQ chains [226]. The specificity for monoUBQ chains stems from a cross-over loop that covers the active site of smaller UCHs, and makes it difficult for polyUBQ chains to reach their catalytic site [226]. Conversely, the longer AA sequence of the loop in larger UCHs makes them more accessible for di- and polyUBQ chains [227].

The family of *ovarian tumor domain* (*OTU* [228]) DUBs contains ≈15 family members, divided into four subfamilies [196,228]. OTUs are characterized by a ≈180 AA catalytic domain [229] that does not seem to contain an obvious Asp/Asn residue, as is in the other Cys DUBs [230]. OTUs may contain only the OTU domain (otubain subfamily, two members; and otulin subfamily, one member), may include N- or C-terminus extensions with coiled coil regions, UIMs or UBLs (OTU subfamily, eight members), or may contain one or more zinc fingers (A20-like OTU subfamily, five members) [196]. OTUs show significant UBQ linkage preference throughout the family, with different specificities—see below [202]. Their specificity depends on the presence of properly positioned proximal UBQ in their catalytic site, and it is lost if the proximal UBQ molecule is modified [228].

Josephin domain proteins [231] are four Cys DUBs sharing an ≈180 AA catalytic domain [232]. Two of them (ataxin subfamily), in addition, contain one or more UIMs that promote UBQ binding, ubiquitination of ataxins and K63 linkage specificity [233]. Josephin domains contain an extended helical arm that regulates the access to the catalytic site, and hosts a second binding site that may contribute to their processing of long polyUBQ chains (>five residues) [234] by binding to two distant UBQ residues [235].

The family of *monocyte chemotactic protein-induced protein 1 (MCPIP1)-like proteins* [236] is a recent addition to Cys DUB families. At least seven MCPIP1-like DUBs should exist [190]. Their catalytic domain contains Cys and Asp residues, but the His residue is either absent or in another domain [236]. MCPIP1 contains an N-terminal ubiquitin-associated domain that assists in the interaction with UBQ but is not essential for DUB activity, a C-terminal Pro-rich region, a zinc finger, and an N-terminal conserved region among putative MCPIP1-like DUBs [236]. The last two domains are crucial for DUB activity.

A single metalloprotease family has DUB activity. The *JAB1/MPN/Mov34 (JAMM* [237]) metalloenzyme family includes seven functional DUBs, sharing a so-called JAMM domain. The domain contains a JAMM metal-binding motif (X_nHS/THX$_7$SXXD [238]). It acts by coordinating two Zn^{2+} ions and subsequently by activating an H_2O molecule to attack and hydrolyze the isopeptide of UBQ-containing substrates [239]. JAMM DUBs are often associated to multiprotein complexes, and preferentially act as *en bloc* DUBs, amputating the whole polyUBQ chain from a protein substrate [240].

DUBs display varying bond, substrate, and linkage selectivities. Most DUBs are true *isopeptidases* that preferentially process conformationally free ε-NH$_2$-COGly-UBQ isopeptide bonds with respect to conformationally and rotationally restricted peptide bonds [33]. Conversely, Met1 linkage selectivity requires a preference for peptide *vs.* isopeptide bond hydrolysis. JAMM, UCH, Josephin, and OTU DUBs (with an exception in the last subfamily) are true isopeptidases. The OTU otulin DUB shows complete Met1 linkage/peptide specificity [241]. Otulin binds to two UBQ molecules in a conformation that requires the proximal UBQ Gly16 to be positioned in such a way that only Met1-linked linear UBQ chains can assume [241]. USPs are more aspecific enzymes that cleave peptide bonds, although with lower efficiency, with a single exception [33]. The USP cylindromatosis (CYLD) DUB shows similar efficiency in peptide/Met1 linkage and isopeptide/K63 linkage hydrolysis [33].

UBQ-like proteins (ULPs) are protein modifiers used to tag proteins for degradation and/or cellular functions [242]. Their structure is similar to UBQ, but key residue changes in the C-terminal region

(Leu71-Arg72-Leu73-Arg74-Gly75-Gly76 for UBQ) confer extreme UBQ/ULQ selectivity for most DUBs [243]. Substrate-aspecific DUBs include USP21 (processing of UBQ- and NEDD8/Leu71-Ala72-Leu73-Arg74-Gly75-Gly76-tagged proteins [244]) and viral OTUs (processing of UBQ- and ISG15/UBQ-like C-terminus-tagged proteins [245]).

Lys *linkage selectivity* varies, and aspecific and specific DUBs are found in each of the six previously mentioned subfamilies. USP DUBs are aspecific isopeptidases, with some exceptions observed in a 12-member USP subfamily [246]. K48- and K63-linked UBQ chains are processed efficiently in general. A few examples (USP4 [247], CYLD [248]) are selective for K63-linked UBQ chains, while proteasome-associated USP14 is K48-selective [249]. K6- and K11-linked UBQ chains are similarly processed, while K29- and especially K27-linked UBQ chains are poorly hydrolyzed by USPs [246]. USP5 and USP13 are Lys linkage-aspecific, but selectively recognize and process unattached polyUBQ chains through their Gly75-Gly76 C-terminal dipeptide [250]. JAMM DUBS are generally K63-specific [196,251], while the josephin ataxin-3 cleaves preferentially K63-linked isopeptide bonds in mixed chains [234]. OTU DUBs are the most varied subfamily in terms of specificity [228]. In addition to the only Met1 linkage-selective DUB, they include four single Lys linkage-specific members (cezanne [252] and cezanne2 [253], K11-linked polyUBQ chains; OTUB1 [254], K48-linked polyUBQ chains; OTUD1 [255], K63-linked polyUBQ chains), four members with a double linkage specificity (OTUD3 [228], K6- and K11-linked polyUBQ chains; A20 [230] and valosin containing protein (p97)/p47 complex interacting protein 1-VCPIP [228], K11- and K48-linked polyUBQ chains; OTUD5 [256], K48- and K63-linked polyUBQ chains), and four OTUs which cleave three or more Lys linkages [228]. K63-selective DUBs take advantage of the extended conformation of K63-linked polyUBQ chains. For example, JAMM metalloprotease associated molecule with the SH3 domain of STAM-like protein (AMSH-LP) binds to the distal UBQ molecule, and to the Gln62 and Glu64 residues on the proximal UBQ [257]. Conversely, K48-linked polyUBQ chains are more resistant to the proteolytic action of many DUBs [258], to ensure that K48-UBQ-tagged substrate proteins are delivered to the UPS. K48-active DUBs are usually involved in post-proteasome binding polyUBQ trimming/cleavage/recycling [258].

Cleavage of the last UBQ molecule on a substrate protein, either after cleavage of distal UBQs—see below—or cleaving *monoubiquitinated* substrates, requires substrate ε-Lys-CONHGly76-UBQ specificity. Small UCH DUBs UCH-L1 and UCH-L3 are ideally suited for this task [196]. UCH-L1 and UCH-L3 also hydrolyze monoUBQ–nucleophile constructs accidentally formed in cellular environments [226]. Alternatively, aspecific DUBs such as USP9X and USP15 are known to de-ubiquitinate monoUBQ protein substrates [259].

PolyUBQ chains may be disassembled by processing one UBQ at a time from the distal UBQ residue (*exo specificity*), by isopeptide cleavage in the polyUBQ chain (*endo specificity*), or by cleavage of the substrate-proximal UBQ connection (*en bloc* cleavage) [196]. Proteasome-associated USP14 and UCH37 DUBs cleave K48-linked polyUBQ chains in an *exo*-single UBQ multistep process. The former binds the distal UBQ and uses the USP-typical fingers' subdomain to prevent the access of other UBQ residues [249]. The latter induces an unusual intramolecular salt bridge between Lys48 and Glu51 on the distal UBQ molecule and prevents the access of other UBQ residues to the catalytic site [260]. The USP CYLD is an *endo* isopeptidase that lacks the fingers' domain and preferentially binds two UBQ residues of a K63-linked UBQ chain [248]. AMSH-LP [257], A20 [230], and USP9X [261] are other *endo* isopeptidases. The constitutive proteasome DUB Rpn11 is an *en bloc* isopeptidase with higher affinity for K63-linked polyUBQ chains that—in addition to its main role in the cleavage of abundant K48-tagged UPS substrates, see below—acts as a proofreading mechanism to disassemble the UBQ tag from a few proteasome-bound, K63-tagged substrate proteins erroneously directed towards UPS degradation [262]. The aspecific USP USP5/ISOT recognizes the free C-terminus of an unattached polyUBQ chain—corresponding to the product of an *en bloc* cleavage by Rpn11—and disassembles the polyUBQ chain in a step-by-step process [263].

The *regulation of DUBs* depends on several factors. Phosphorylation may lead to DUB activation, as for OTUD5/DUBA [256] and for USP37 (by CK2 [264]), or to DUB stabilization by protein kinase B (Akt, PKB) phosphorylation, as for USP4 [265]. It may also lead to DUB inhibition, as for CYLD [266] and USP8 [267]. Oxidation of the catalytic Cys residue impairs the catalytic activity of many Cys DUBs, but does not prevent their binding to UBQ [268]. Oxidative inhibition of DUBs is reversible, and could be related to a response to acute oxidative stress. In fact, DNA damages and the related DNA replication impairment may be balanced by the inactivation of USP1, and the subsequent monoubiquitination/activation of the proliferating cell nuclear antigen (PCNA) that increases DNA replication [268]. Other post-translational modifications of DUBs, and their consequences, are documented (inactivation by auto-proteolysis, USP1 [269]; activation by ubiquitination, ataxin-3 [270]; inactivation by small ubiquitin-like modifier (SUMO)ylation, USP25 [218]). Non-catalytic UIM domains in DUBs may increase their efficiency [218,255,271]; binding with protein partners may cause DUB activation [272–276] or inhibition [277]; and may determine their linkage selectivity [234]. Finally, transmembrane domains (mitochondrial insertion, USP30 [278], and endoplasmic reticulum (ER) insertion, USP19 [279]) and subcellular localization (nucleolus, USP36 [280], and early endosomes, AMSH [272]) sometimes provide additional specificity to DUBs.

The 26S proteasome complex is the protein degradation terminal for UBQ-tagged proteins in eukaryotes [281]. It is made by a 20S barrel-shaped catalytic core particle (CP) composed of 28 subunits, structurally arranged in four, stacked, seven-membered rings [282]. Namely, two external rings are composed of α-type subunits, while two inner rings are made of β-type subunits [283]. The inner β1, β2, and β5 catalytic subunits act as N-terminal nucleophile (Ntn) hydrolases through their N-terminal Thr residue, and provide the CP, i.e., the whole UPS, with trypsin-, chemiotrypsin-, threonin-, and postglutamyl peptidase activities [284]. Two 19S regulatory particles (RPs), each composed of 19 subunits, interact with the CP. RPs are made by a 9 subunit lid, and a 10 subunit base structure [285]. The base structure contains a hexameric ring of six ATPases that interact through their C-terminus with binding pockets between the α-subunits of each external α-ring of the CP [286], and the scaffold proteins Rpn1 and Rpn2. The entire CP and the hexameric ATPase ring constitute an AAA+ (ATPases associated with various cellular activities) protease [287], the translocation–degradation core of the 26S proteasome. The RP lid contains various UBQ-interacting proteins, and shows direct interactions with the CP [288]. RP-CP-RP binding causes the opening of a channel in the CP that leads unfolded substrates to the catalytic UPS core [289]. Each CP-bound RP delimits a narrow area that ends with a substrate translocation channel (STC) made by the hexameric ATPase ring. The STC ends in correspondence of the CP catalytic channel [290]. RPs recognize UBQ-tagged proteins [41] and promote their DUB-mediated de-ubiquitination [291]. The AAA+ protease machinery unfolds polyUBQ-tagged substrates [292] and channels the "naked"-unfolded substrate proteins through the STC to the 20S catalytic chamber for degradation [293]. The energy provided by the hexameric ATPase ring in the RP base is used to unfold, de-ubiquitinate, and translocate the UPS substrates [294]. The CP may also act as a protease complex independently from the RPs, in an ATP- and ubiquitin-independent manner [295]—see below.

We now examine how a UBQ-tagged substrate protein (the result of the previously described UBQ/E1-E2-E3 cycle, and of DUB control/regulation/proofreading) is processed by the UPS, or—if the wrong UBQ-tagged substrate protein approaches the proteasome—how it is rescued. A UBQ-tagged substrate protein is at first recognized through its polyUBQ chain by UBQ receptors on the RPs, and through an intrinsically disordered region (IDR), with an unknown molecular mechanism. Once proteasome-bound, the UBQ-tagged substrate protein starts moving through the RP channel. Along the early/UBQ-interacting RP channel, the UBQ tag is de-ubiquitinated, either en bloc (UBQ amputation by Rpn11) or via a multistep process (UBQ trimming by USP14 and UCH37). UBQ trimming produces free UBQ to be recycled. UBQ amputation generates unattached polyUBQ chains that are then processed by UBQ trimming to yield once more free UBQ to be recycled. ATP-dependent protein substrate unfolding likely starts during the crossing of the late/STC-ATPase channel, but may initiate in the early RP channel. Unfolding of all protein regions, resulting from binding to, and ATPase activity of, the ATPase ring continue until translocation to the CP of the whole, untagged substrate proteins is completed. Finally, CP-internalized, unfolded proteins are processed by the β-type subunits of the inner heptameric rings.

Figure 3.5 provides a structural representation of the unfolding–deubiquitination–translocation process, by depicting the spatial relationship between a polyUBQ-tagged substrate, the RP-located UBQ receptors and DUB activities, and the AAA+ protease core.

The polyUBQ-tagged substrate at first binds through UBQ to either one of the lid RP UBQ receptors (S5a/hRpn10 [296] and adhesion regulating molecule 1/Ardm1/hRpn13 [297]). It binds also through an IDR of the substrate protein [298] that facilitates the process of the protein through the STC [299]. The UBQ chains and the disordered region must be separated by at least 30 AAs, i.e., by ≈60 Å [300], so that the latter may enter the STC and promote unfolding (see Figure 3.5) while the former is still bound to the UBQ receptors. The molecular basis of the interaction between the RP and the IDR of a substrate is not yet elucidated [201]. Conversely, five UPS-connected UBQ receptors are known (Figure 3.5). Rpn13 is located in the upper RP part, and binds UBQ through the atypical pleckstrin-like receptor for ubiquitin (Pru) domain [301]. Rpn10 is located close to the ATPase/unfoldase STC, and binds two adjacent UBQ molecules through two UIMs separated by a flexible linker [302]. Radiation-sensitive 23 (Rad23), DNA damage-inducible 1 (Ddi1),

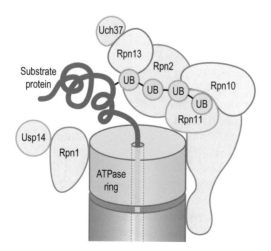

FIGURE 3.5   The proteasome: core and regulatory particles (CP and RPs).

and dominant suppressor of kar1 (Dsk2) are proteasome-associated UBQ receptors. They act as shuttling proteins that capture UPS substrates away from the proteasome, either as free UBQ-tagged substrates [303] or as freshly tagged, E3 ligase-bound proteins [304]. Then, they may bind to S5a and hand over their cargo [305], or may bind to the RP base scaffold protein Rpn1 [306].

Three proteasomal DUB enzymes free UPS substrate proteins from their tags (Figure 3.5). The ATP-dependent JAMM DUB Rpn11 [262] is strategically placed above the STC, and at a similar distance from the binding sites of all UBQ receptors—Rpn10, Rpn13 (70–80 Å [306]), and Rpn1-bound Ddi1, Rad23, and Dsk2 (80–120 Å [306]). One may assume that the initial binding of polyUBQ-tagged proteins with UBQ receptors, promoted by an IDR, causes the protein to start moving in an ATP-dependent manner towards the STC/CP channel. Once the distance to Rpn11 is covered (i.e., a tetraUBQ chain in an extended conformation covers ≈100 Å), Rpn11 cleaves the UBQ chain at the proximal UBQ residue, releasing the entire UBQ chain [204]. UBQ chain removal allows the protein to continue its ATP-dependent unfolding and translocation. Rpn11 de-ubiquitination does not cause the release of the substrate protein that has already entered the STC and approaches the CP catalytic channel [201]. Inactivation of the ATPases in the RP base prevents UBQ cleavage by Rpn11, possibly by stopping the movement of UBQ-tagged proteins [307]. Rpn11 activation needs the assembly of the whole RP that frees Rpn11 from active site-hindering inter-subunit interactions with Rpn9 and Rpn5. An inactive Rpn11 in the isolated RP lid prevents unspecific DUB activities, and provides additional selectivity to the UPS machinery [262].

The proteasome-associated USP14 (USP subfamily [308]) and UCH37 (UCH subfamily [309]) bind respectively to the scaffold protein Rpn1 [310] and to the UBQ receptor Rpn13 [311]. They are K48-specific *exo* isopeptidases, located at the entry of the whole proteasome. They may act as UBQ trimming enzymes when the substrate protein, and its polyUBQ tag, move towards the STC/CP channel [201]—especially when the polyUBQ chain consists of more than four UBQ molecules. They may also process the Rpn11-cleaved, unattached polyUBQ chains still bound by UBQ receptors, to free them for addition to the next UBQ-tagged substrate [312]. The common anchor point—Rpn1—for USP14 and UBQ receptors/shuttle proteins Rad23, Ddi1, and Dsk2 may suggest that USP14 is preferentially active on their UBQ-tagged cargos [307]. Surprisingly, UBQ chain trimming by USP14 and UCH37 may lead either to the inhibition [313] or to the stimulation [310] of proteasomal degradation. If UBQ trimming happens too fast, the affinity of the substrate protein for the RP lid decreases before it has reached the STC, and the substrate protein is released from the proteasome. If UBQ trimming is too slow, either unfolding of the moving, UBQ-tagged protein may slow down and the STC may get clogged, or the free polyUBQ chains still bound to UBQ receptors prevent the docking of a new UBQ-tagged substrate to the RP lid [201].

The ATPase ring is composed of six structurally different ATPase Rpt1–Rpt6 subunits, which arrange themselves in a ring with a defined Rpt1-Rpt2-Rpt6-Rpt3-Rpt4-Rpt5 order [314]. The C-terminus tails of Rpt2, Rpt3, and Rpt5 are solidly anchored respectively between the $\alpha3$–$\alpha4$, $\alpha1$–$\alpha2$, and $\alpha5$–$\alpha6$ subunits of the external heptameric CP ring [315]. Each C-terminus tail contains a hydrophobic Tyr-X (HbYX) sequence that determines, when docking is completed, the opening of the CP catalytic channel [286]. Most likely, the C-terminus tails of Rpt1, Rpt4, and Rpt6 also interact with the α-subunit CP ring in a more dynamic manner [316]. Surprisingly, the ATPase ring and the external α-type subunit CP ring are misaligned, with an $\approx10\,\text{Å}$ shift [317]. Unfolded protein must then "turn a corner," possibly reducing the tolerance for folded regions in a substrate protein engaged for UPS degradation. The strict requirement for unfolded proteins to cross the narrow STC/CP channels is coupled with a loose structural specificity that makes most of cytosolic proteins putative UPS substrates [287]. The substrate may enter the AAA+ protease machinery through its N- or C-terminus, or from internal sites, and may proceed in an N- or C-terminal direction [318]. D-amino acids, >10 internal consecutive Gly residues [319] and disulfide bonds [320], are among the tolerated structural features in UPS substrates. The narrow AAA+ channel is likely to be extremely flexible, undergoing punctual enlargements and contractions to adapt to unusual regions of UPS-targeted peptides [321].

Out of the six putative ATP binding sites, four are occupied in ATP-saturating conditions [322]. A single ATP hydrolysis event is enough to fuel conformational changes in the whole ATPase ring. The changes are then passed to the substrate protein through an aromatic/hydrophobic AA motif, whose mutation does not impair ATP hydrolysis but inhibits the overall AAA+ protease activity [323]. A dynamic ATP cycle switches between (1) four bound ATPs/hydrolysis, (2) three ATPs and one ADP/ADP dissociation, (3) three ATPs/ATP binding, and (4) four bound ATPs/cycle restart. A single ATP cycle is likely enough to progress the unfold–translocate–degrade process [324], although other outcomes are possible. The reluctance of a region of the substrate to unfold may cause a block of the process that can be resumed with an additional ATP cycle; or, in the presence of a major structural hurdle—i.e., a bound polyUBQ chain, or a compact substrate region—the substrate protein may dissociate from the proteasome [287]. The number of ATP cycles needed to degrade a given protein significantly varies depending on its folding status. The close AAA+ protease congener ClpXP needs ≈600 ATP cycles to process native titin-I27, while only ≈100 cycles are needed for an unfolded version of the same protein [325].

Substrate protein degradation is carried out in the CP by the three endopeptidase subunits of each inner β-type heptameric ring (β1—cleavage after acidic AAs; β2—cleavage after tryptic AAs; and β5—cleavage after hydrophobic AAs [281]). The functional catalytic subunits contain an N-terminal Tyr1 residue that is responsible for the nucleophilic attack on the peptide bond of the substrate protein [326]; an Asp17 residue and a Lys33 residue, which form a salt bridge and lower the pKa of the catalytic Tyr1 residue [283]; and two Ser (129 and 169) and an Asp166 residue, which contribute to keeping Tyr1 in the functional spatial orientation [327]. They are expressed as inactive propeptides that undergo activation through intramolecular autolysis [328]. The β3, β4, β6, and β7 subunits miss one or more of the key residues, and are catalytically inactive [329]. The different specificities for β1, β2, and β5 subunits mostly depend on the AA45 position—basic Arg45 in β1, favoring cleavage after an acidic residue; small Gly45 in β2, favoring trypsin-like cleavage after basic residues; and apolar Met45 in β3, favoring cleavage after hydrophobic residues [329]. Full processing of a polyUBQ-tagged substrate protein produces, in addition to several UBQ copies to be recycled in the UBQ cycle, a number of oligopeptides with an ≈7–8 average residue length. Such small peptides eventually exit the UPS and are recycled as peptide building blocks as such, or after the action of other peptidases [330].

The right equilibrium between recognition, de-ubiquitination (trimming and amputation), unfolding, translocation through the RP and the external CP, and proteolytic processing in the inner CP of an UPS substrate protein must be maintained in terms of timing and selectivity. One must consider that compact, slowly unfolding substrate proteins will require different overall residence times in the UPS machinery with respect to flexible, loosely folded substrate proteins. Thus, the UPS must possess an extreme adaptive capacity to cope with the wide range of its protein substrates.

3.2 UPS-MEDIATED DEGRADATION OF MISFOLDED PROTEINS IN NDDs

In the past, protein aggregation was seen as the main cause of several diseases, while the impairment of protein quality control (PQC) systems was considered as one of its consequences [331]. Now it is clear that in most cases—including tau [332]—protein aggregation is the result of impaired proteostasis. In other words, the failure lies in the impairment of UPS-driven disposal of pathologically misfolded, often abnormally post-translation-modified (i.e., abnormally phosphorylated) proteins [72]. Accordingly, their cellular concentration increases, eventually leading to significant cellular toxicity.

NDDs are often related to the impairment of protein degradation [333,334]. Proteasome dysfunctions and *Parkinson's disease* (PD) are connected, as depletion of 26S proteasome in transgenic (TG) mice leads to UBQ- and α-synuclein-containing intraneuronal inclusions with similarities to Lewy bodies in the brains of PD patients [335]. Accumulation of K48-connected polyUBQ chains, as a sign of impaired UPS activity, is observed in brains from TG mice models of *Huntington disease* (HD) and from HD patients [336]. TG mice with a pathologic mutation in the protein Cu-Zn superoxide dismutase (SOD1), recapitulating *amyothropic lateral sclerosis (ALS)*, show UPS impairment and reduced proteasome subunit expression in motor neurons [337]. The C-terminus elongated ubiquitin B + 1 (UBB^{+1}) version of UBQ is a potent, dose-dependent inhibitor of UPS. When overexpressed, it leads to neuronal cell death [338,339]. NDDs showing UPS-dependent impairment in proteostasis include Angelman syndrome [340], ataxia [341], and Wallerian degeneration [342].

Alzheimer's disease (AD) shows a plethora of abnormal events/proteins/regulations pertaining to protein and/or aggregate degradation. Several observations stem from the characterization of *post-mortem* AD brains, likely reflecting a therapeutical relevance for UPS-related molecular targets in AD and, more in general, in tauopathies. As to UPS, tau aggregates from AD human brains inhibit UPS *in vitro*, while non-aggregated tau from the same brains has no inhibitory effect on UPS—a "vicious circle" loop leading to an even higher UPS inhibition during AD-tauopathy progression may be inferred [343]. A similar, self-sustaining pathological effect is shown by the accumulation of misfolded, hyperphosphorylated (HP)- and polyUBQ tau at synapses in AD brains, when compared with control brains [344]. Such misdecorated protein copies have an inhibitory effect on the UPS system that increases with the progression of synaptic degeneration [344]. Ubiquitination of full-length (FL) tau in early stages of AD starts at the level of neurotoxic tau oligomers, and continues during their transition to neurofibrillary tangles (NFTs) [345]. It is inferred that polyUBQ chains are introduced on oligomers to facilitate their UPS degradation, but eventually lead to UPS inhibition due to their aggregation into insoluble NFTs [345]. The nature of soluble tau as a proteasomal substrate is supported by the inverse correlation between levels of tau and of 20S proteasome subunit in brain cytosolic fractions from the temporal cortex of AD patients [346]. The absence of tau in similar fractions from control patients underlines the association between AD and impaired UPS [346].

In vitro and *in vivo*/preclinical results related to UPS-directed compounds and targets are abundant in literature. They are described in the next paragraphs in some detail, mostly for their impact on tau-driven NDDs. It is important to anticipate that there is an interdependence among UPS- and autophagy-mediated processes (those last influencing the elimination of insoluble protein aggregates) in physiological and dys-regulated processes connected with NDDs. Their combined effects on tau depend on tau species, on the phosphorylation pattern of tau, and on the cellular environment. FL, soluble tau may be polyubiquitinated and degraded by the UPS [347], but may also be disposed of through autophagy ([348], see Chapter 4) or selective autophagy/aggrephagy ([349], see Chapter 5). As to UPS, phosphorylation of the KXGS motifs in the MT-binding repeats (MTBRs) of tau hinders ubiquitination of tau by the U-box E3 ligase C-terminus of Hsc70 interacting protein (CHIP, see Section 3.3.1) and proteasomal degradation [350]. Insoluble tau, as other insoluble protein aggregates, cannot physically access the site of UPS degradation due to its size [350]. Studies using engineered, tau-producing cell lines postulate a role for UBQ-independent UPS in soluble tau elimination [351,352].

An association is detected between a large tau fragment resulting from caspase cleavage at Asp421 [353], which is abundant in NFTs from AD brains, and polyubiquitination [354]. Ubiquitination of Asp421-truncated tau is an early pathological event in NFT formation, and promotes aggregation with FL tau to form NFTs [354]. Asp421-truncated tau is preferentially degraded through autophagy, when compared with FL tau—processed by UBQ-independent UPS [352]. Interestingly, both FL tau and its Asp421-truncated fragment are ubiquitinated in a cell-free system, but do not show ubiquitination in immortalized CN1.4 mouse cortical neurons [352].

In conclusion, theoretical and experimental evidence confirms that UPS-related and autophagy/aggrephagy-related mechanisms can be conceived in order to dispose of misfolded protein copies, oligomeric soluble species, and insoluble aggregates. In particular, this is true for misfolded tau, neurotoxic soluble tau oligomers, and insoluble tau NFTs.

3.3 UPS—TARGETS

Any UBQ-activating enzyme E1, UBQ-conjugating enzyme E2, UBQ ligase E3, and DUBs-deubiquitinase may be considered a suitable target to restore the UPS activity in an impaired cellular environment. Proteasome activity may also be targeted for a direct effect on UPS. NDDs in general, and tauopathies in particular, require the potentiation/restoring of cellular mechanisms leading to the elimination of misfolded/aggregated proteins. UPS inhibition may appear to lead in the opposite direction—decreasing UPS-mediated elimination of tau and other misfolded proteins. Certain enzymes in the UPS system, though, contribute to the elimination of proteins and hinder the rescuing–refolding of misfolded proteins. Their inhibition, thus, should be beneficial. Moreover, the intimate connection between UPS and autophagy may lead to the identification of targets causing synchronized UPS inhibition and autophagy induction (see also autophagy and aggrephagy, Chapters 4 and 5).

As to *E1-activating enzymes*, drug discovery efforts on them are advanced as they are popular targets in oncology. Two E1 inhibitor classes originated from high-throughput screening (HTS) campaigns are known. They are mechanism-based inhibitors, respectively acting through covalent inactivation of the E1 enzymes on a Cys residue [355], and by capturing the UBQ chain and blocking the adenylation domain of the E1 enzymes (substrate-assisted inhibition) [356]. The former inhibitors are selective *vs.* E2 and E3 enzymes, but significantly inhibit DUBs and raise questions about their overall effects on UPS [357]. The latter compounds inhibit the NEDD8 pathway, and specifically the NEDD8-activating enzyme (NAE) [358] out of the eight known E1 enzymes. Mechanistic studies with E1-aspecific, structurally related analogs show that substrate-assisted inhibition works with other E1 enzymes—including UBA1—and with UBQ as a tag protein [359]. Interestingly, substrate-assisted inhibitors induce autophagy in multiple cancer cell lines [360], supporting the hypothesis of E1 enzymes as targets for the elimination of dysfunctional proteins. The NEDD8 pathway itself may be a meaningful target for tauopathies, as NEDD8 is found together with UBQ in inclusion bodies and protein aggregates in AD brains [361].

As to *E2-conjugating enzymes*, most of them associate with UBA1 and transfer UBQ to E3 ligases. Their substrate-specific coupling with E3 ligases is appealing, as it should provide the needed efficacy and selectivity for correctly targeted inhibitors. Unfortunately, most of therapeutically meaningful E2–E3 couples are not yet characterized [81], and their potential as drug discovery targets is still to be determined. Among the few exceptions, small modulators [362–364] specifically inhibit the ubiquitin-conjugating enzyme E2N (UBE2N/Ubc13, a chain elongating E2 enzyme/oncology target involved in nuclear factor-κB/NF-κB activation [365]), showing promising, limited toxicity. Allosteric modulation of cdc34 (a chain elongating E2 enzyme/oncology target, which contributes to the elimination of cell-cycle regulatory proteins [366]) exploits an extremely selective interaction with a regulatory enzyme pocket, which may be common to other E2 enzymes [367].

UBE2K (E2-25K [368]) is a chain elongating E2 enzyme, which, together with the HECT E3 ligases thyroid hormone receptor interactor 12 (TRIP12) [369] or HUWE1 [370], ubiquitinates the neurogeneration-related mutant UBB$^+$ protein [338]. Ubiquitination of UBB$^+$ makes it a target for UPS degradation, but an excess of UBB$^+$ inhibits the UPS system with a positive feedback-self-sustaining mechanism [371]. Inhibition of the UBE2K–TRIP12 couple may produce a reduction of the observed, pathology-related UBB$^+$ accumulation in AD and other tauopathies [372]. Structural information about UBE2K complexed with UBQ is available through NMR [95], possibly facilitating the rational design of UBE2K inhibitors.

As to *E3 enzyme ligases*, several targets are being clinically pursued in oncology. Clinical trials targeting the E3 ligase mouse double minute 2 homolog (MDM2), to inhibit its action on/interaction with the pro-apoptotic tumor suppressor protein p53, involve at least two small molecules [373]. Seven clinical candidates target cellular Inhibitor of Apoptosis Proteins (cIAP-1 and cIAP-2), which possess an E3 ligase domain and regulate, among others, caspase-activated apoptosis and the canonical and non-canonical NF-κB pathway [374]. At least 10 more E3 ligases, out of the known ≈600 family members in the human genome, are validated oncology targets subjected to ongoing drug discovery efforts [375]. The role of E3 ligases in NDDs involving misfolded proteins—either as such or, more often, interacting with Hsp chaperones—is well recognized [376]. The RBR E3 ligase parkin, which has reduced solubility and co-localizes with HP-tau and amyloid Aβ in autophagic vacuoles (AVs) from AD brains, is involved in both UPS and autophagy [377]. Mutations in parkin-encoding gene PARK2 that translate a protein devoid of E3 activity are linked to autosomal recessive PD [378]. The HECT NEDD4 E3 ligase reverses SPT-phenotype protein 5 (Rsp5) is a K63-selective enzyme that contributes to endosomal trafficking [106]. The Rsp5 network has therapeutic potential against PD, and a small molecule identified through a large, phenotype-based screen [379] is able to reverse nitrosative stress, accumulation of endoplasmic reticulum (ER)-associated degradation substrates, and ER stress in neurons from PD patients [380]. Another E3 ligase candidate for therapeutic tauopathy targeting is discussed in the next section.

As to *proteasome inhibition*, it is a validated mechanism in oncology with a marketed drug and several clinical compounds [198]. Immunomodulation and parasite infestation are other applications for proteasome inhibitors [381]. Neurodegeneration has never been actively pursued, maybe due to the risk of considering the somewhat "dirty" adverse event (AE) profile of bortezomib [382] for chronic central nervous system (CNS) diseases. Selective approaches aiming to E1/E2/E3 modulators appear to be more promising, although a detailed profiling of second-generation, cleaner proteasome inhibitors in *in vitro* and *in vivo* models of tauopathies would be desirable.

As to *DUBs*, most of them are papain-like cysteine proteases—thus, they should be amenable to structure-based drug design [72]. In oncology, they may prevent—*via* de-ubiquitination and inhibition of UPS—the degradation of tumor suppressors [375]. Activators of tumor suppressor-processing DUBs would be beneficial, but they are not easy to find. DUB inhibitors, on the contrary, are known, and may be useful to promote UPS-mediated elimination of oncogenes [383]. They include pan-DUB inhibitors with moderate activity on several DUBs [384,385] and DUB-specific inhibitors against validated oncology targets—for example, against HDM2 elimination-promoting/p53-stabilizing and activating USP7/HAUSP [386–388]. Inhibitors of viral DUB enzymes are also known [389]. DUB inhibition may have therapeutic potential against NDDs [390]. At least 10 DUBs are connected with CNS physiology and pathology [391]. The DUB UCH-L1 is a validated target in PD and AD [392,393], as it is found in insoluble protein aggregates and inclusion bodies from AD and PD [394]. UCH-L1 mutations are associated to PD [395] and gracile axonal dystrophy (gad [396]). Reduced soluble UCH-L1 levels [397] and oxidatively modified UCH-L1 proteins [398] are observed in clinical AD samples and in TG mice models of AD [399]. Overexpression of functional UCH-L1 reverses behavioral impairments in the same AD models [399]. UCH-L1 regulates the levels of β-site amyloid precursor protein/β-APP cleaving enzyme 1 (BACE1), and its reduction in gad mice models causes an increase of neurotoxic Aβ peptides [400]. Selective UCH-L1 thieno-pyridinone inhibitors are known [401]. Altered expression levels of the endo isopeptidase USP9x are observed in a mouse model of PD [402]. The josephin DUB ataxin-3 is mutated by insertion of a polyglutamine (polyQ) repeat in spinocerebellar ataxia type 3 (SCA3 [403]). Expression of human wild-type (WT) ataxin-3 in a *Drosophila* polyQ model prevents neurodegeneration in a UBQ- and UPS-dependent manner [404]. A DUB candidate for therapeutic tauopathy targeting is discussed in Section 3.3.2.

3.3.1 CHIP

Carboxy-terminus of Hsp70-interacting protein (CHIP) [405] is an Hsp70/Hsp90 co-chaperone, due to three tandem tetratricopeptide (TPR) domains at its N-terminus [406]. It is also an E3 UBQ ligase, due to a U-box domain at its C-terminus [407]. CHIP and Hsp70/Hsp90 organizing protein (Hop) [408] are the only co-chaperones able to interact with both Hsp70 and Hsp90. Their role is strongly interdependent, as Hsp–Hop complexes lead towards protein rescuing-refolding, while Hsp–CHIP complexes "doom" client proteins towards degradation [409] (see also Chapter 2). Detailed mechanistic studies [410,411] prove that CHIP and Hop binding to Hsp70 and Hsp90 is mutually exclusive; that Hop can establish ternary Hsp70–Hop–Hsp90 hetero-complexes, while CHIP binds only to either Hsp70 or Hsp90; that the concentration of CHIP is strongly cell-dependent, but in general much lower ($\approx 0.1\ \mu M$) than the concentration of Hsp70 ($\approx 10\ \mu M$), Hsp90 ($\approx 5\ \mu M$), and Hop ($\approx 3\ \mu M$); that the dissociation constant of CHIP and Hop complexed with Hsp70 is in the low μM range; and that both co-chaperones have a lower, nM dissociation constant when complexed with Hsp90. Consequently, an ≈ 10-fold higher concentration of CHIP–Hsp70 than CHIP–Hsp90 is observed—pointing to a role for the former in the ubiquitination of client proteins. In physiological conditions, the Hsp90–Hop–Hsp70 ternary-refolding complex has an ≈10-fold higher concentration than the CHIP–Hsp70 binary-degradation complex—pointing to a privileged route towards protein recovery [410,411]. The low level of CHIP–Hsp70-promoted client protein degradation in basal conditions may be increased by CHIP overexpression, leading to a 4:1 excess of CHIP–Hsp70 over Hsp90–Hop–Hsp70 [412] and to an overall switching of the PQC machinery.

The relative ratio of CHIP- and Hop-containing complexes is the main, but not the only, determinant for the fate of a client protein. If the client protein can be easily refolded, refolding takes place rapidly and the functional protein is released from Hsp90–Hop–Hsp70. If the unfolding–refolding process is more complex, the dynamic equilibrium—client protein binding, ATP-dependent unfolding and refolding, nucleotide exchange, client protein release—may entail several binding–release cycles from Hsp90–Hop–Hsp70 before proper refolding is completed. Thus, the overall probability for a client protein to bind to the CHIP–Hsp70 degradation complex—even with its lower physiological abundance—increases, due to multiple binding–release cycles. A single CHIP–Hsp70 binding event is sufficient to direct misfolded proteins to ubiquitination and disposal by the UPS [413]. This complex framework facilitates the elimination of severely misfolded proteins, while ensuring proper refolding of minimally misfolded proteins.

CHIP is a key player in the management of misfolded proteins by the cell. Its role is even greater in several NDDs, due to client proteins of CHIP. PolyQ-containing, aggregation-prone *huntingtin* and *ataxin-3* are ubiquitinated in neuronal cells, while their WT, non-polyQ counterparts are not ubiquitinated [414]. CHIP immunoprecipitates with polyQ aggregates, enhances their ubiquitination rate in an Hsp70-dependent manner, and consequently increases their degradation rate by UPS. CHIP does not bind to/interact with WT ataxin-3 and huntingtin, and its overall neuroprotective effect is confirmed in CHIP- and Hsc70-overexpressing cells [414]. The chaperone-connected TPR domain, rather than the UPS-connected U-box domain of CHIP seems more relevant for the neuroprotective effect of CHIP on huntingtin [415]. The co-chaperone BAG-1 [416] (see also Chapter 2) has a role in huntingtin elimination by UPS, and shows neuroprotective effects, possibly acting in a concerted manner with CHIP [417]. A TG mouse model of SCA3 is exacerbated by genetic elimination of CHIP [418]. CHIP ablation does not modify soluble levels of ataxin-3, but increases the formation of neurotoxic, small microaggregates. Its overexpression reduces polyQ ataxin-3 levels, without affecting WT protein levels [418]. The effect of CHIP on polyQ ataxin-3 seems to be Hsp-independent, and is common to E4B (UFD2A [419]), another U-box E3 ligase. WT ataxin-3 has also a key role in the regulation of CHIP activity together with the monoubiquitinating E2 enzyme UBE2W [91]. The E2 enzyme monoubiquitinates CHIP, and monoUBQ CHIP binds to ataxin-3. UBE2W monoubiquitinates also ataxin-3, triggering its DUB activity. MonoUBQ ataxin-3 regulates the polyUBQ chain length of Hsp–CHIP client proteins, and de-ubiquitinates CHIP, to restart the UBE2W/CHIP/ataxin-3 cycle [91]. Neurotoxic, polyQ ataxin-3 has a higher affinity for CHIP, which leads to increased degradation of CHIP in SCA3 TG mice [91].

CHIP shows similar effects—co-immuno-precipitation, TPR domain- and Hsp70-dependent ubiquitination competence, overexpression-mediated neuroprotection—on *ataxin-1* in cells and in a *Drosophila* model of spinocerebellar ataxia type-1 (SCA1) [420]. Conversely, CHIP binds to, and ubiquitinates also, WT ataxin-1, as this protein—when expressed at high levels—is *per se* capable of neurotoxic aggregation [421]. The CHIP–ataxin-1 interaction leads to an increased amount of insoluble ataxin-1 aggregates in cells, reflecting the CHIP-mediated neuroprotective disposal of neurotoxic, soluble microaggregates *via* ubiquitination or macroaggregation [422]. Other E3 ligases, such as parkin and BAG-1, have no similar effects on ataxin-1 [422]. CHIP mutations are causative factors for autosomal recessive cerebellar ataxia (ARCA) [423].

CHIP is found in Lewy bodies together with *α-synuclein*, and it influences the morphology and the degradation of α-synuclein inclusions [424]. A TPR-/chaperone-dependent interaction influences UPS-mediated degradation, and a U-box/ubiquitination-dependent interaction directs α-synuclein inclusions towards autophagic degradation [424]. The most toxic, oligomeric α-synuclein species are the main CHIP targets in a TPR-/chaperone-dependent manner. An overall reduction of α-synuclein inclusions is observed [425], in contrast with the increase of ataxin-1 aggregates observed in SCA1 [422]. α-Synuclein inclusions-targeted effects are observed with TPR-deficient, Hsp70-independent CHIP constructs [425]. A ternary CHIP–Hsp70–BAG-5 complex influences the ubiquitination level of α-synuclein [426]. Namely, the nucleotide exchange factor (NEF) co-chaperone BAG-5 [427] reduces the levels of polyUBQ–α-synuclein species *via* negative modulation of CHIP. Thus, inhibiting the functions of BAG-5 may represent a therapeutic goal against PD [425]. Finally, CHIP associates to, and contributes to promote, the degradation of mutant *SOD1* through the UPS [428]. CHIP acts through a ubiquitination-independent mechanism, involving the Hsp70-dependent interaction with the proteasome and facilitation of mutant SOD1 translocation from the 19S RP to the 20S CP [429]. BAG-1 and the proteasome-associated AAA+ ATPase valosin-containing protein (VCP) participate with CHIP to mutant SOD1 regulation [429]. The activity of CHIP on mutant SOD1 includes its BAG-3-mediated elimination through autophagy [430].

The role of CHIP as an Hsp70-dependent, *tau*-ubiquitinating enzyme has been known for almost a decade [431,432]. CHIP immunoreactivity is observed in tau lesions from NDDs, with higher prevalence in 3R tauopathies such as Pick's disease [431]. Interactions between CHIP and tau involve the TPR and U-box domains of CHIP, and the MTBRs of tau. CHIP-mediated ubiquitination proceeds faster on 4R tau than on 3R isoforms [432]. CHIP levels are increased in AD brains, and aged CHIP knockout (KO) mice show increased insoluble tau aggregates, pointing to a pro-aggregation role for CHIP [433]. CHIP preferentially ubiquitinates HP-tau and drives it towards degradation in WT mice. The presence of polyUBQ tau in NFTs, though, indicates that CHIP may also stimulate the formation of insoluble tau aggregates from neurotoxic, HP-soluble tau oligomers when the UPS system capacity is overwhelmed [434].

CHIP-mediated tau degradation competes with refolding-targeted, Hsp-promoted tau processing. Inhibition of HSP90 using small molecules, or inhibition of the co-chaperone p23 and of peptidyl-prolyl cis-trans isomerase (Pin1, PPIase) with small interfering RNA (siRNA), switches tau towards UPS-mediated degradation [347]. HP-tau is a better ubiquitination substrate for the E3 activity of CHIP, but microtubule affinity-regulating kinase 2 (MARK2)-mediated tau phosphorylation on the MTB region (S252, S356) blocks the CHIP–tau interaction [347]. Conversely, CHIP binds to and processes HP-tau resulting from glycogen synthase kinase-3 (GSK-3) overactivation and protein phosphatase 2A (PP2A) inactivation [435].

The inter-Hsp70 class selectivity shown by CHIP causes either refolding or elimination of tau in physiological or pathological conditions [436]. Namely, constitutive Hsc70 slows tau clearance and promotes its CHIP-independent refolding, while stress-activated Hsp72 recruits CHIP and promotes UPS-dependent tau clearance. Both chaperones bind to the MTBR region of tau, but their structural differences (mostly in their C-terminal domains) determine the formation of tau refolding- (Hsc70) or tau clearance-directed (Hsp72) complexes. Increased Hsp72 levels induced by chronic stress in AD should be expected, in accordance with similar findings in other NDDs, but are not yet reported [436]. Thus, pharmacological enhancement of Hsp72 in AD and other tauopathies may lead to tau clearance and, eventually, to therapeutic benefits.

Reduced CHIP–tau binding, leading to a reduction of tau clearance and increased tau levels, is caused by endoplasmic reticulum (ER) stress [437]. Namely, ER stress increases the amount of unfolded protein substrates of CHIP, thus reducing the availability of CHIP in tau clearance-directed chaperone complexes [437]. CHIP efficiency in tau processing is also influenced by Aβ in double TG Aβ/tau mice [438]. Namely, Aβ decreases CHIP levels and promotes the development of tau pathology. Its effects may be prevented by CHIP overexpression, or by stimulation of the UPS machinery [438].

CHIP shows some redundancy with other E3 ligases. The Hsp70-binding RBR E3 ligase parkin shows a partial substrate overlap with CHIP [439]. A double TG, tau-overexpressing/parkin-null mice model unexpectedly shows tau-dependent neurodegeneration together with altered CHIP–Hsp70 levels and amyloidosis [440]. A "compensatory chaperone" effect, i.e., increased CHIP–Hsp70 levels, is observed in double TG, amyloid precursor protein (APP)-overexpressing/parkin-null mice [441]. The role of CHIP and parkin as E3 ligases involved in NDDs, including putative cooperative actions of the two enzymes, is reviewed [442].

CHIP-mediated tau clearance is regulated by the protein kinase Akt [443]. Akt regulation includes direct competition with CHIP for tau binding, and stimulation of previously mentioned MARK2 phosphorylation of tau at S252 and S356. The former effect reduces the levels of CHIP–tau levels, the latter increases the levels of CHIP ubiquitination-resistant HP-tau [443]. Akt is regulated by CHIP in a phosphorylation-dependent manner, as its doubly phosphorylated pT308 and pS473 form is preferentially ubiquitinated by CHIP and degraded through the UPS [444]. Overall, tau clearance should be promoted by Akt inhibition.

CHIP contributes to regulate the molecular switch between client protein refolding and elimination. CHIP also contributes to select the UBQ-dependent degradation pathway in a cell: proteasome-mediated, soluble misfolded protein digestion, or selective autophagy/aggrephagy-driven protein aggregate elimination (see Chapters 4 and 5). The formation of CHIP complexes containing chaperones (i.e., Hsp70, Hsp90) and co-chaperones (i.e., BAG family members) determines the client protein fate in a dynamic equilibrium, which is altered in disease-related conditions. Thus, carefully tailored CHIP modulation *in vivo* may have a great therapeutic potential against NDDs and tauopathies.

3.3.2 USP14

Three DUB enzymes act in proximity to the proteasome, either because they are structurally part of it—the JAMM metalloprotease Rpn11 [240]—or because they are activated as DUBs when associated with the proteasome—UCH37 [445] and USP14 [308]. The three DUBs are complementary in terms of their de-ubiquitinating mechanism, and of the fate of their substrates [203]. Rpn11 cleaves the whole polyUBQ chain from a given substrate—usually, four UBQ molecules or more—and, in so doing, permits the access of partially unfolded, de-ubiquitinated substrates to the proteasome CP [240] (see also Section 3.1). Acting as a UBQ trimming DUB at an early stage of the polyUBQ-substrate recognition process by the UPS, UCH37 causes the decrease of affinity of the ubiquitinated protein substrate for its proteasomal receptors, promoting either its dissociation from the proteasome and the prevention of its degradation (pro-survival action of UCH37 [203]), or the ATP-dependent movement of the substrate protein bearing a shortened UBQ tag towards the proteasome CP (pro-degradation activity, see below).

USP14 (USP subfamily) is an UBQ trimming DUB structurally different from UCH37 (UCH subfamily), which contributes to free UBQ recycling before substrate protein degradation by the UPS [201]. A pro-survival role at least partially overlapping with UCH37 is proven by the higher proteasomal degradation of ubiquitinated substrate proteins when USP14 and UCH37 are inactivated [446]. USP14 shows other proteasome-related functions. After Rpn10/Rpn13 binding, the polyUBQ substrate protein starts to move towards the STC/ATPase, while being de-ubiquitinated. USP14 trims polyUBQ chains, and may cause the dissociation of putative substrate proteins that have not established strong IDR–proteasome interactions. Conversely, if a polyUBQ substrate protein is strongly associated with the proteasome through an IDR, its binding to USP14 has a pro-degradation effect [447]. USP14 associates with the proteasome through the scaffold Rpn1 protein, and does not directly interact with the ATPase ring. The USP14-ubiquitinated substrate interaction, though, induces conformational changes leading to allosteric activation of the ATPase ring and opening of the CP channel [448]. USP14 stimulates proteasomal degradation in a DUB-independent manner [447]. UCH37 shows similar proteasome activity/ATPase activation-increasing effects. The site of proteasome association for UCH37—the scaffold protein Rpn2, located opposite to Rpn1—and the DUB-dependent nature of its proteasome activation suggest a different regulation mechanism for USP14 and UCH37 [448].

The role of USP14 in physiological and pathological events of the CNS is well known [390,391]. USP14 has a central role in synapse development and synaptic plasticity, as observed in a TG mice model of *ataxia (ax^J)* [449]. The *ax^J* phenotype is caused by a USP14 mutation expressing a shorter protein, missing the proteasome association domain. TG *ax^J* mice show aberrant neuronal and synaptic development (accumulation of phosphorylated neurofilaments, aberrant motor neuronal sprouting), with an ≈35% reduction in free UBQ in neurons [450], and eventually die between 6 and 10 weeks of age. Their neuromuscular junctions (NMJs) are poorly arborized and show reduced acetylcholine (AChE) release. USP14 deficiency in TG *ax^J* mice also dysregulates the cell surface distribution and the turnover of synaptic receptors, leading to reduced cerebellar output [451]. Either expression of FL USP14 or neuronal expression of free UBQ in *ax^J* TG mice rescue the defects of the *ax^J* phenotype, and prevent most biochemical abnormalities and early lethality [452,453].

A comparison of USP14 in mature T helper cells expressing the surface protein cluster of differentiation 4 (in short CD4⁺ T cells), from young and elderly humans shows an increased activity of USP14 coupled with a reduced proteasome activity in the latter samples [454]. A selective USP14 inhibitor is capable of increasing proteasomal activity through USP14 inhibition in cells from young adults, but cannot reach the same putative therapeutic result in elderly CD4⁺ T cells [454]. In detail, treatment with a selective USP14 inhibitor in young adult cells shows a significant disappearance of polyUBQ nuclear factor of kappa light polypeptide gene enhancer in B-cell inhibitor, alpha (IκBα), a proteasome substrate. The levels of the same polyubiquitinated protein are not decreased in elderly cells [454].

USP14 influences the clearance of polyubiquitinated tau and of 43K transactive response DNA-binding protein (TDP-43) in murine embryonic fibroblasts (MEFs) and in human embryonic kidney 293 (HEK293) cells [455]. USP14 overexpression stabilizes the two substrate proteins, but either a C114A mutation or a shorter USP14 version lacking its proteasome-association domain does not stabilize them. A USP14 inhibitor causes a significant reduction of tau and TDP-43 levels, while it is inactive in USP14$^{-/-}$ cells [455]. A later report [456] shows aberrant phosphorylation of tau in TG ax^J mice lacking USP14. The loss of USP14 does not influence the levels of tau or ataxin-3, and tau ablation does not reverse the diseased phenotype of TG ax^J mice [456].

Surprisingly, a proteasomal DUB-targeted double USP14–UCH37 inhibitor shows impaired rather than enhanced proteasomal activity. Most likely this is due to the accumulation of polyUBQ substrates (clogging the proteasome), and to the significant reduction of free UBQ levels caused by prevention of UBQ trimming [457]. The impact of USP14 inhibition on misfolded proteins in neurodegeneration, and in particular on tau in tauopathies, requires further investigation. The identification of proteasomal DUB inhibitors with varying levels of selectivity—clean USP14 inhibitors [455], USP14–UCH37 dual inhibitors [458], poly-DUB inhibitors [459]—should assist in identifying an ideal compromise between DUB potency/proteasomal activity restoration, and *in vitro* and *in vivo* tolerability.

3.4 DISEASE-MODIFYING COMPOUNDS

This chapter deals with mid steps leading to neuropathological alterations related to protein misfolding and aggregation in general, and to tau and/or tau-connected events in particular. A potential therapeutic mechanism is examined in detail, and two targets are chosen. Tens of other targets—some of which are validated and actively pursued by various labs—are neglected here, mostly due to space limitations. Thirty-nine compounds/scaffolds acting on selected targets are diffusely covered in a chemistry-oriented companion book [460] devoted to disease-modifying compounds, and are briefly summarized in Table 3.1. Each compound class is numbered as in the chemistry-oriented companion book, and its chemical core is structurally defined; its mechanism of action and molecular target are mentioned; the public or private laboratory that develops the compound is listed; and the development status—according to publicly available information—is provided.

TABLE 3.1   Compounds 3.1–3.37: Chemical Class, Target, Developing Organization, Development Status

Number	Chemical cpd./class	Target	Organization	Dev. status
3.1a,b	Thioflavin S	Hsp70–BAG-1	Cancer Research, UK	DD
3.2	Pifithrin-m, PES	Hsp70, plus others	University of Pennsylvania	LO
3.3	Gambogic acid, GA	Hsp70–, Hsp90–CHIP regulation	Jangsu University, China	TM
3.4	Methylene blue, MB	Hsp70–CHIP regulation	TauRX Therapeutics	Ph III
3.5	Apoptozole	Hsp70, ATPase inhibition	Yonsei University, South Korea	LO
3.6	Nutlin-3	HDM2-p53	Roche	Ph I
3.7	Serdemetan	HDM2-p53	Johnson & Johnson	Ph I
3.8	AT-406	IAPs	Ascenta	Ph I
3.9	GDC-0152	IAPs	Genentech	Ph I
3.10	LCL-161	IAPs	Novartis	Ph II
3.11	TL32711	IAPs	Tetralogic	Ph II
3.12	Diamines, compound A	Skp2	University of North Carolina	DD
3.13	Alkylidene thiazolidines, compound C1	Skp2	NY University	DD
3.14	Diacids, SKP-I2	Cdc4	Mount Sinai Hosp., Toronto, Canada	DD
3.15	Benzodiazepindiones, LS-101	Synoviolin	Tokyo Medical Univ.	LO
3.16	Triazines, LS-102	RING E3 ligases	Tokyo Medical Univ.	LO
3.17	Tetracycles, SMER3	Met30	UCLA	LO
3.18	Thalidomide	Cereblon	Tokyo Institute of Technology	Ph III
3.19	Curcumin	Pan-DUB inhibition	University of Utah	TM
3.20	Shikoccin	Pan-DUB inhibition	University of Utah	DD
3.21	Δ12-PGJ2	Pan-DUB inhibition	Karolinska Institute	DD

TABLE 3.1   Compounds 3.1–3.37: Chemical Class, Target, Developing Organization, Development Status *(cont.)*

Number	Chemical cpd./class	Target	Organization	Dev. status
3.22	Dienones, NSC 632839	Pan-DUB inhibition	Progenra	DD
3.23	Bis-isothiocyanate, PR-419	Pan-DUB inhibition	Oldenburg University, Germany	DD
3.24a,b	Tricyclic dinitriles, HBX-41,108 (3.24a)	USP DUBs	Hybrigenics	LO
3.25	Gold complexes	Pan-DUB inhibition	University of Hong Kong	DD
3.26	Tyrposthin-like WP-1130	USP5, USP9x, USP14, UCH-L1, UCH37	University of Michigan	LO
3.27	Alkyliden-pyrazolidindiones, PYR41	Cys DUBs	University of Michigan	DD
3.28	Betulinic acid	Pan-DUB inhibition	University of Miami	PE
3.29	Chalcones, RA-9	UPS2, UPS5, UPS8, UCH-L1, UCH-L3	University of Minnesota	DD
3.30	Phthalimide based, pimozide	UAF1, USP7	University of Delaware	DD
3.31	Spautin-1	USP10, USP13	Chinese Academy of Sciences	DD
3.32	Thiophene based, P5091	USP7, USP47	Harvard Med. School	LO
3.33	Thiophene based	USP7, USP47	Progenra	LO
3.34	Aminotetrahydroacri-dines, HBX-19,818	USP7	Hybrigenics	Ph I
3.35	Naphthylamides, GRL0617	PLpro	University of Illinois	LO
3.36	Electrophilic dienones, NSC687852/b-AP15	USP14	Karolinska Institute	LO
3.37	Pyrrole based, IU1	USP14	Harvard Med. School	LO

Not progressed, NP; early discovery, DD; lead optimization, LO; preclinical evaluation, PE; clinical Phase I-II-III, Ph I–Ph III; marketed, MKTD; traditional medicine, TM.

References

1. Hershko, A.; Ciechanover, A. The ubiquitin system. *Annu. Rev. Biochem.* **1998**, *67*, 425–479.
2. Weissman, A. M.; Shabek, N.; Ciechanover, A. The predator becomes the prey: regulating the ubiquitin system by ubiquitylation and degradation. *Nat. Rev. Mol. Cell Biol.* **2011**, *12*, 605–620.
3. Goldstein, G.; Scheid, M.; Hammerling, U.; Boyse, E. A.; Schlesinger, D. H.; Niall, H. D. Isolation of a polypeptide that has lymphocyte-differentiating properties and is probably represented universally in living cells. *Proc. Natl. Acad. Sci. U.S.A.* **1975**, *72*, 11–15.
4. Vijay-Kumar, S.; Bugg, C. E.; Cook, W. J. Structure of ubiquitin refined at 1.8 A resolution. *J. Mol. Biol.* **1987**, *194*, 531–544.
5. Hoege, C.; Pfander, B.; Moldovan, G. L.; Pyrowolakis, G.; Jentsch, S. RAD6-dependent DNA repair is linked to modification of PCNA by ubiquitin and SUMO. *Nature* **2002**, *419*, 135–141.
6. Carter, S.; Bischof, O.; Dejean, A.; Vousden, K. H. C-terminal modifications regulate MDM2 dissociation and nuclear export of p53. *Nat. Cell Biol.* **2007**, *9*, 428–435.
7. Haglund, K.; Sigismund, S.; Polo, S.; Szymkiewicz, I.; Di Fiore, P. P.; Dikic, I. Multiple monoubiquitination of RTKs is sufficient for their endocytosis and degradation. *Nat. Cell Biol.* **2003**, *5*, 461–466.
8. Sun, L.; Chen, Z. J. The novel functions of ubiquitination in signaling. *Curr. Opin. Cell Biol.* **2004**, *16*, 119–126.
9. Greer, S. F.; Zika, E.; Conti, B.; Zhu, X. -S.; Ting, J. P. -Y. Enhancement of CIITA transcriptional function by ubiquitin. *Nat. Immunol.* **2003**, *4*, 1074–1082.
10. Freudenthal, B. D.; Gakhar, L.; Ramaswamy, S.; Washington, M. T. Structure of monoubiquitinated PCNA and implications for translesion synthesis and DNA polymerase exchange. *Nat. Struct. Mol. Biol.* **2010**, *17*, 479–484.
11. Hoeller, D.; Crosetto, N.; Blagoev, B.; Raiborg, C.; Tikkanen, R.; Wagner, S., et al. Regulation of ubiquitin-binding proteins by monoubiquitination. *Nat. Cell Biol.* **2006**, *8*, 163–169.
12. Stringer, D. K.; Piper, R. C. A single ubiquitin is sufficient for cargo protein entry into MVBs in the absence of ESCRT ubiquitination. *J. Cell Biol.* **2011**, *192*, 229–242.
13. Ulrich, H. D.; Walden, D. Ubiquitin signalling in DNA replication and repair. *Nat. Rev. Mol. Cell Biol.* **2010**, *11*, 479–489.
14. Weinmaster, G.; Fischer, J. A. Notch ligand ubiquitylation: what is it good for? *Dev. Cell* **2011**, *21*, 134–144.
15. Kravtsova-Ivantsiv, Y.; Cohen, S.; Ciechanover, A. Modification by single ubiquitin moieties rather than polyubiquitination is sufficient for proteasomal processing of the p105 NF-kappaB precursor. *Mol. Cell* **2009**, *33*, 496–504.
16. Li, M.; Brooks, C. L.; Wu-Baer, F.; Chen, D.; Baer, R.; Gu, W. Mono- versus polyubiquitination: differential control of p53 fate by Mdm2. *Science* **2003**, *302*, 1972–1975.
17. Komander, D.; Rape, R. The ubiquitin code. *Annu. Rev. Biochem.* **2012**, *81*, 203–229.
18. Sloper-Mould, K. E.; Jemc, J. C.; Pickart, C. M.; Hicke, L. Distinct functional surface regions on ubiquitin. *J. Biol. Chem.* **2001**, *276*, 30483–30489.
19. Kamadurai, H. B.; Souphron, J.; Scott, D. C.; Duda, D. M.; Miller, D. J.; Stringer, D., et al. Insights into ubiquitin transfer cascades from a structure of a UbcH5B~ubiquitin-HECTNEDD4L complex. *Mol. Cell* **2009**, *36*, 1095–1102.
20. Dikic, I.; Wakatsuki, S.; Walters, K. J. Ubiquitin-binding domains—from structures to functions. *Nat. Rev. Mol. Cell Biol.* **2009**, *10*, 659–671.
21. Jin, L.; Williamson, A.; Banerjee, S.; Philipp, I.; Rape, M. Mechanism of ubiquitin-chain formation by the human anaphase-promoting complex. *Cell* **2008**, *133*, 653–665.
22. Penengo, L.; Mapelli, M.; Murachelli, A. G.; Confalonieri, S.; Magri, L.; Musacchio, A., et al. Crystal structure of the ubiquitin binding domains of Rabex-5 reveals two modes of interaction with ubiquitin. *Cell* **2006**, *124*, 1183–1195.

23. Winget, J. M.; Mayor, T. The diversity of ubiquitin recognition: hot spots and varied specificity. *Mol. Cell* **2010**, *38*, 627–635.

24. Cook, W. J.; Jeffrey, L. C.; Carson, M.; Chen, Z.; Pickart, C. M. Structure of a diubiquitin conjugate and a model for interaction with ubiquitin conjugating enzyme (E2). *J. Biol. Chem.* **1992**, *267*, 16467–16471.

25. Ryabov, Y.; Fushman, D. Interdomain mobility in di-ubiquitin revealed by NMR. *Proteins* **2006**, *63*, 787–796.

26. Eddins, M. J.; Varadan, R.; Fushman, D.; Pickart, C. M.; Wolberger, C. Crystal structure and solution NMR studies of Lys48-linked tetraubiquitin at neutral pH. *J. Mol. Biol.* **2007**, *367*, 204–211.

27. Virdee, S.; Ye, Y.; Nguyen, D. P.; Komander, D.; Chin, J. W. Engineered diubiquitin synthesis reveals Lys29-isopeptide specificity of an OTU deubiquitinase. *Nat. Chem. Biol.* **2010**, *6*, 750–757.

28. Bremm, A.; Freund, S. M.; Komander, D. Lys11-linked ubiquitin chains adopt compact conformations and are preferentially hydrolyzed by the deubiquitinase Cezanne. *Nat. Struct. Mol. Biol.* **2010**, *17*, 939–947.

29. Matsumoto, M. L.; Wickliffe, K. E.; Dong, K. C.; Yu, C.; Bosanac, I.; Bustos, D., et al. K11-linked polyubiquitination in cell cycle control revealed by a K11 linkage-specific antibody. *Mol. Cell* **2010**, *39*, 477–484.

30. Bremm, A.; Komander, D. Emerging roles for Lys11-linked polyubiquitin in cellular regulation. *Trends Biochem. Sci.* **2011**, *36*, 355–363.

31. Varadan, R.; Assfalg, M.; Haririnia, A.; Raasi, S.; Pickart, C.; Fushman, D. Solution conformation of Lys63-linked di-ubiquitin chain provides clues to functional diversity of polyubiquitin signaling. *J. Biol. Chem.* **2004**, *279*, 7055–7063.

32. Sims, J. J.; Cohen, R. E. Linkage-specific avidity defines the lysine 63-linked polyubiquitin-binding preference of rap80. *Mol. Cell* **2009**, *33*, 775–783.

33. Komander, D.; Reyes-Turcu, F.; Licchesi, J. D.; Odenwaelder, P.; Wilkinson, K. D.; Barford, D. Molecular discrimination of structurally equivalent Lys 63-linked and linear polyubiquitin chains. *EMBO Rep.* **2009**, *10*, 466–473.

34. Datta, A. B.; Hura, G. L.; Wolberger, C. The structure and conformation of Lys63–linked tetraubiquitin. *J. Mol. Biol.* **2009**, *392*, 1117–1124.

35. Rohaim, A.; Kawasaki, M.; Kato, R.; Dikic, I.; Wakatsuki, S. Structure of a compact conformation of linear diubiquitin. *Acta Crystallogr. D Biol. Crystallogr.* **2012**, *68*, 102–108.

36. Fushman, D.; Walker, O. Exploring the linkage dependence of polyubiquitin conformations using molecular modeling. *J. Mol. Biol.* **2010**, *395*, 803–814.

37. Husnjak, K.; Dikic, I. Ubiquitin-binding proteins: decoders of ubiquitin-mediated cellular functions. *Annu. Rev. Biochem.* **2012**, *81*, 291–322.

38. Xu, M.; Skaug, B.; Zeng, W.; Chen, Z. J. A ubiquitin replacement strategy in human cells reveals distinct mechanisms of IKK activation by TNFalpha and IL-1beta. *Mol. Cell* **2009**, *36*, 302–314.

39. Ben-Saadon, R.; Zaaroor, D.; Ziv, T.; Ciechanover, A. The polycomb protein Ring1B generates self atypical mixed ubiquitin chains required for its in vitro histone H2A ligase activity. *Mol. Cell* **2006**, *24*, 701–711.

40. Kaiser, S. E.; Riley, B. E.; Shaler, T. A.; Trevino, R. S.; Becker, C. H.; Schulman, H.; Kopito, R. R. Protein standard absolute quantification (PSAQ) method for the measurement of cellular ubiquitin pools. *Nat. Methods* **2011**, *8*, 691–696.

41. Finley, D. Recognition and processing of ubiquitin-protein conjugates by the proteasome. *Annu. Rev. Biochem.* **2009**, *78*, 477–513.

42. Kim, W.; Bennett, E. J.; Huttlin, E. L.; Guo, A.; Li, J.; Possemato, A., et al. Systematic and quantitative assessment of the ubiquitin-modified proteome. *Mol. Cell* **2011**, *44*, 325–340.

43. Finley, D. Inhibition of proteolysis and cell cycle progression in a multiubiquitination-deficient yeast mutant. *Mol. Cell. Biol.* **1994**, *14*, 5501–5509.

44. Baboshina, O. V.; Haas, A. L. Novel multiubiquitin chain linkages catalyzed by the conjugating enzymes E2EPF and RAD6 are recognized by 26 S proteasome subunit 5. *J. Biol. Chem.* **1996**, *271*, 2823–2831.
45. Glickman, M. H.; Ciechanover, A. The ubiquitin–proteasome proteolytic pathway: destruction for the sake of construction. *Physiol. Rev.* **2002**, *82*, 373–428.
46. Wickliffe, K. E.; Williamson, A.; Meyer, H.-J.; Kelly, A.; Rape, M. K11-linked ubiquitin chains as novel regulators of cell division. *Trends Cell Biol.* **2011**, *21*, 656–663.
47. Wagner, S. A.; Beli, P.; Weinert, B. T.; Nielsen, M. L.; Cox, J.; Mann, M.; Choudhary, C. A proteome-wide, quantitative survey of in vivo ubiquitylation sites reveals widespread regulatory roles. *Mol. Cell Proteomics* **2011**, *10*, M111.013284.
48. Xu, P.; Duong, D. M.; Seyfried, N. T.; Cheng, D.; Xie, Y.; Robert, J., et al. Quantitative proteomics reveals the function of unconventional ubiquitin chains in proteasomal degradation. *Cell* **2009**, *137*, 133–145.
49. Goto, E.; Yamanaka, Y.; Ishikawa, A.; Aoki-Kawasumi, M.; Mito-Yoshida, M.; Ohmura-Hoshino, M., et al. Contribution of lysine 11-linked ubiquitination to MIR2-mediated major histocompatibility complex class I internalization. *J. Biol. Chem.* **2010**, *285*, 35311–35319.
50. Dynek, J. N.; Goncharov, T.; Dueber, E. C.; Fedorova, A. V.; Izrael-Tomasevic, A.; Phu, L., et al. c-IAP1 and UbcH5 promote K11-linked polyubiquitination of RIP1 in TNF signalling. *EMBO J.* **2010**, *29*, 4198–4209.
51. Xu, P.; Peng, J. Dissecting the ubiquitin pathway by mass spectrometry. *Biochim. Biophys. Acta* **1764**, *2006*, 1940–1947.
52. Kulahtu, Y.; Komander, D. Atypical ubiquitylation—the unexplored world of polyubiquitin beyond Lys48 and Lys63 linkages. *Nat. Rev. Mol. Cell Biol.* **2012**, *13*, 508–523.
53. Wu-Baer, F.; Lagrazon, K.; Yuan, W.; Baer, R. The BRCA1/BARD1 heterodimer assembles polyubiquitin chains through an unconventional linkage involving lysine residue K6 of ubiquitin. *J. Biol. Chem.* **2003**, *278*, 34743–34746.
54. Sato, K. Nucleophosmin/B23 is a candidate substrate for the BRCA1–BARD1 ubiquitin ligase. *J. Biol. Chem.* **2004**, *279*, 30919–30922.
55. Wu, W.; Nishikawa, H.; Hayami, R.; Sato, K.; Honda, A.; Aratani, S., et al. BRCA1 ubiquitinates RPB8 in response to DNA damage. *Cancer Res.* **2007**, *67*, 951–958.
56. Glauser, L.; Sonnay, S.; Stafa, K.; Moore, D. J. Parkin promotes the ubiquitination and degradation of the mitochondrial fusion factor mitofusin 1. *J. Neurochem.* **2011**, *118*, 636–645.
57. Peng, D.-J.; Zeng, M.; Muromoto, R.; Matsuda, T.; Shimoda, K.; Subramaniam, M., et al. Noncanonical K27-linked polyubiquitination of TIEG1 regulates Foxp3 expression and tumor growth. *J. Immunol.* **2011**, *186*, 5638–5647.
58. Koegl, M.; Hoppe, T.; Schlenker, S.; Ulrich, H. D.; Mayer, T. U.; Jentsch, S. A novel ubiquitination factor, E4, is involved in multiubiquitin chain assembly. *Cell* **1999**, *96*, 635–644.
59. Chastagner, P.; Israël, A.; Brou, C. Itch/AIP4 mediates deltex degradation through the formation of K29-linked polyubiquitin chains. *EMBO Rep.* **2006**, *7*, 1147–1153.
60. Peng, J.; Schwartz, D.; Elias, J. E.; Thoreen, C. C.; Cheng, D.; Marsischky, G.; Roelofs, J.; Finley, D.; Gygi, S. P. A proteomics approach to understanding protein ubiquitination. *Nat. Biotechnol.* **2003**, *21*, 921–926.
61. Huang, H.; Jeon, M. S.; Liao, L.; Yang, C.; Elly, C.; Yates, J. R.; Liu, Y. C. K33-linked polyubiquitination of T-cell receptor-zeta regulates proteolysis-independent T cell signalling. *Immunity* **2010**, *33*, 60–70.
62. Kodadek, T. No splicing, no dicing: non-proteolytic roles of the ubiquitin-proteasome system in transcription. *J. Biol. Chem.* **2010**, *285*, 2221–2226.
63. Bellare, P.; Small, E. C.; Huang, X.; Wohlschlegel, J. A.; Staley, J. P.; Sontheimer, E. J. A role for ubiquitin in the spliceosome assembly pathway. *Nat. Struct. Mol. Biol.* **2008**, *15*, 444–451.
64. Spence, J.; Gali, R. R.; Dittmar, G.; Sherman, F.; Karin, M.; Finley, D. Cell cycle-regulated modification of the ribosome by a variant multiubiquitin chain. *Cell* **2000**, *102*, 67–76.

65. Wang, C.; Deng, L.; Hong, M.; Akkaraju, G. R.; Inoue, J.; Chen, Z. J. TAK1 is a ubiquitin-dependent kinase of MKK and IKK. *Nature* **2001**, *412*, 346–351.
66. Al-Hakim, A.; Escribano-Diaz, C.; Landry, M. C.; O'Donnell, L.; Panier, S.; Szilard, R. K.; Durocher, D. The ubiquitous role of ubiquitin in the DNA damage response. *DNA Repair* **2010**, *9*, 1229–1240.
67. Xia, Z. P.; Sun, L.; Chen, X.; Pineda, G.; Jiang, X.; Adhikari, A., et al. Direct activation of protein kinases by unanchored polyubiquitin chains. *Nature* **2009**, *461*, 114–119.
68. Tan, J. M.; Wong, E. S.; Dawson, V. L.; Dawson, T. M.; Lim, K. L. Lysine 63-linked poly-ubiquitin potentially partners with p62 to promote the clearance of protein inclusions by autophagy. *Autophagy* **2008**, *4*, 251–253.
69. Kirisako, T.; Kamei, K.; Murata, S.; Kato, M.; Fukumoto, H.; Kanie, M., et al. A ubiquitin ligase complex assembles linear polyubiquitin chains. *EMBO J.* **2006**, *25*, 4877–4887.
70. Gerlach, B.; Cordier, S. M.; Schmukle, A. C.; Emmerich, C. H.; Rieser, E.; Haas, T. L., et al. Linear ubiquitination prevents inflammation and regulates immune signalling. *Nature* **2011**, *471*, 591–596.
71. Ikeda, F.; Deribe, Y. L.; Skanland, S. S.; Stieglitz, B.; Grabbe, C.; Franz-Wachtel, M., et al. SHARPIN forms a linear ubiquitin ligase complex regulating NF-kappaB activity and apoptosis. *Nature* **2011**, *471*, 637–641.
72. Bedford, L.; Lowe, J.; Dick, L. R.; Mayer, R. J.; Brownell, J. E. Ubiquitin-like protein con-jugation and the ubiquitin-proteasome system as drug targets. *Nat. Rev. Drug Discov.* **2011**, *10*, 29–46.
73. Kulkarni, M.; Smith, H. E. E1 ubiquitin-activating enzyme UBA-1 plays multiple roles throughout *C. elegans* development. *PLoS Genet.* **2008**, *4*, e1000131.
74. Gavin, J. M.; Chen, J. J.; Liao, H.; Rollins, N.; Yang, X.; Xu, Q., et al. Mechanistic studies on activation of ubiquitin and di-ubiquitin-like protein, FAT10, by ubiquitin-like modi-fier activating enzyme 6, Uba6. *J. Biol. Chem.* **2012**, *287*, 15512–15522.
75. Bohnsack, R. N.; Haas, A. L. Conservation in the mechanism of NEDD8 activation by the human AppBp1–Uba3 heterodimer. *J. Biol. Chem.* **2003**, *278*, 26823–26830.
76. Haas, A. L.; Rose, I. A. The mechanism of ubiquitin-activating enzyme. A kinetic and equilibrium analysis. *J. Biol. Chem.* **1982**, *257*, 10329–10337.
77. Jin, J.; Li, X.; Gygi, S. P.; Harper, J. W. Dual E1 activation systems for ubiquitin differen-tially regulate E2 enzyme charging. *Nature* **2007**, *447*, 1135–1138.
78. Lee, P. C. W.; Dodart, J.-C.; Aron, L.; Finley, L. W.; Bronson, R. T.; Haigis, M. C., et al. Altered social behavior and neuronal development in mice lacking the Uba6-Use1 ubiq-uitin transfer system. *Mol. Cell* **2013**, *50*, 172–184.
79. Schulman, B. A.; Harper, J. W. Ubiquitin-like protein activation by E1 enzymes: the apex for downstream signalling pathways. *Nat. Rev. Mol. Cell Biol.* **2009**, *10*, 319–331.
80. Burroughs, A. M.; Jaffee, M.; Iyer, L. M.; Aravind, L. Anatomy of the E2 ligase fold: implications for enzymology and evolution of ubiquitin/Ub-like protein conjugation. *J. Struct. Biol.* **2008**, *162*, 205–218.
81. van Wijk, S. J.; Timmers, H. T. The family of ubiquitin-conjugating enzymes (E2s): decid-ing between life and death of proteins. *FASEB J.* **2010**, *24*, 981–993.
82. Summers, M. K.; Pan, B.; Mukhyala, K.; Jackson, P. K. The unique N terminus of the UbcH10 E2 enzyme controls the threshold for APC activation and enhances checkpoint regulation of the APC. *Mol. Cell* **2008**, *31*, 544–556.
83. Sadowski, M.; Mawson, A.; Baker, R.; Sarcevic, B. Cdc34 C-terminal tail phosphorylation regulates Skp1/cullin/F-box (SCF)-mediated ubiquitination and cell cycle progression. *Biochem. J.* **2007**, *405*, 569–581.
84. Hao, Y.; Sekine, K.; Kawabata, A.; Nakamura, H.; Ishioka, T.; Ohata, H., et al. Apollon ubiquitinates SMAC and caspase-9, and has an essential cytoprotection function. *Nat. Cell Biol.* **2004**, *6*, 849–860.
85. Haldeman, M. T.; Xia, G.; Kasperek, E. M.; Pickart, C. M. Structure and function of ubiq-uitin conjugating enzyme E2-25K: the tail is a core-dependent activity element. *Biochem-istry* **1997**, *36*, 10526–10537.

86. Ye, Y.; Rape, M. Building ubiquitin chains: E2 enzymes at work. *Nat. Rev. Mol. Cell Biol.* **2009**, *10*, 755–764.
87. Wenzel, D. M.; Stoll, K. E.; Klevit, R. E. E2s: structurally economical and functionally replete. *Biochem. J.* **2010**, *433*, 31–42.
88. Wenzel, D. M.; Lissounov, A.; Brzovic, P. S.; Klevit, R. E. UBCH7 reactivity profile reveals parkin and HHARI to be RING/HECT hybrids. *Nature* **2011**, *474*, 105–108.
89. Hochstrasser, M. Lingering mysteries of ubiquitin chain assembly. *Cell* **2006**, *124*, 27–34.
90. Alpi, A. F.; Pace, P. E.; Babu, M. M.; Patel, K. J. Mechanistic insight into site-restricted mono-ubiquitination of FANCD2 by Ube2t, FANCL, and FANCI. *Mol. Cell* **2008**, *32*, 767–777.
91. Scaglione, K. M.; Zavodszky, E.; Todi, S. V.; Patury, S.; Xu, P.; Rodrıguez-Lebron, E., et al. Ube2w and ataxin-3 coordinately regulate the ubiquitin ligase CHIP. *Mol. Cell* **2011**, *43*, 599–612.
92. Hao, Z.; Zhang, H.; Cowell, J. Ubiquitin-conjugating enzyme UBE2C: molecular biology, role in tumorigenesis, and potential as a biomarker. *Tumor Biology* **2012**, *33*, 723–730.
93. Windheim, M.; Peggie, M.; Cohen, P. Two different classes of E2 ubiquitin-conjugating enzymes are required for the monoubiquitination of proteins and elongation by poly-ubiquitin chains with a specific topology. *Biochem. J.* **2008**, *409*, 723–729.
94. Sakata, E.; Satoh, T.; Yamamoto, S.; Yamaguchi, Y.; Yagi-Utsumi, M.; Kurimoto, E., et al. Crystal structure of UbcH5b~ubiquitin intermediate: insight into the formation of the self-assembled E2~Ub conjugates. *Structure* **2010**, *18*, 138–147.
95. Wilson, R. C.; Edmondson, S. P.; Flatt, J. W.; Helms, K.; Twigg, P. D. The E2-25K ubiquitin-associated (UBA) domain aids in polyubiquitin chain synthesis and linkage specificity. *Biochem. Biophys. Res. Commun.* **2011**, *405*, 662–666.
96. Ziemba, A.; Hill, S.; Sandoval, D.; Webb, K.; Bennett, E. J.; Kleiger, G. Multimodal mechanism of action for the Cdc34 acidic loop. *J. Biol. Chem.* **2013**, *288*, 34882–34896.
97. Semplici, F.; Meggio, F.; Pinna, L. A.; Oliviero, S. CK2-dependent phosphorylation of the E2 ubiquitin conjugating enzyme UBC3B induces its interaction with beta-TrCP and enhances beta-catenin degradation. *Oncogene* **2002**, *21*, 3978–3987.
98. Haghikia, A.; Missol-Kolka, E.; Tsikas, D.; Venturini, L.; Brundiers, S.; Castoldi, M., et al. Signal transducer and activator of transcription 3-mediated regulation of miR-199a-5p links cardiomyocyte and endothelial cell function in the heart: a key role for ubiquitin-conjugating enzymes. *Eur. Heart J.* **2011**, *32*, 1287–1297.
99. Bocik, W. E.; Sircar, A.; Gray, J. J.; Tolman, J. R. Mechanism of polyubiquitin chain recognition by the human ubiquitin conjugating enzyme Ube2g2. *J. Biol. Chem.* **2011**, *286*, 3981–3991.
100. Wickliffe, K. E.; Lorenz, S.; Wemmer, D. E.; Kuriyan, J.; Rape, M. The mechanism of linkage-specific ubiquitin chain elongation by a single-subunit E2. *Cell* **2011**, *144*, 769–781.
101. Lim, G. G. Y.; Chew, K. C. M.; Ng, X.-H.; Henry-Basil, A.; Sim, R. W. X.; Tan, J. M. M., et al. Proteasome inhibition promotes parkin-Ubc13 interaction and lysine 63-linked ubiquitination. *PLoS One* **2013**, *8*, e73235.
102. Pelzer, L.; Pastushok, L.; Moraes, T.; Mark Glover, J. N.; Ellison, M. J.; Ziola, B.; Xiao, W. Biological significance of structural differences between two highly conserved Ubc variants. *Biochem. Biophys. Res. Commun.* **2009**, *378*, 563–568.
103. Lewis, M. J.; Saltibus, L. F.; Hau, D. D.; Xiao, W.; Spyracopoulos, L. Structural basis for non-covalent interaction between ubiquitin and the ubiquitin conjugating enzyme variant human MMS2. *J. Biomol. NMR* **2006**, *34*, 89–100.
104. Eddins, M. J.; Carlile, C. M.; Gomez, K. M.; Pickart, C. M.; Wolberger, C. Mms2–Ubc13 covalently bound to ubiquitin reveals the structural basis of linkage-specific polyubiquitin chain formation. *Nature Struct. Mol. Biol.* **2006**, *13*, 915–920.
105. Li, W.; Bengtson, M. H.; Ulbrich, A.; Matsuda, A.; Reddy, V. A.; Orth, A., et al. Genome-wide and functional annotation of human E3 ubiquitin ligases identifies MULAN, a mitochondrial E3 that regulates the organelle's dynamics and signaling. *PLoS One* **2008**, *3*, e1487.

106. Rotin, D.; Kumar, S. Physiological functions of the HECT family of ubiquitin ligases. *Nat. Rev. Mol. Cell. Biol.* **2009**, *10*, 398–409.

107. Budhidarmo, R.; Nakatani, Y.; Day, C. L. RINGs hold the key to ubiquitin transfer. *Trends Biochem. Sci.* **2012**, *37*, 58–65.

108. Wenzel, D. M.; Klevit, R. E. Following Ariadne's thread: a new perspective on RBR ubiquitin ligases. *BMC Biol.* **2012**, *10*, 24.

109. Deshaies, R. J.; Joazeiro, C. A. P. Ring domain E3 ubiquitin ligases. *Annu. Rev. Biochem.* **2009**, *78*, 399–434.

110. Borden, K. L.; Boddy, M. N.; Lally, J.; O'Reilly, N. J.; Martin, S.; Howe, K., et al. The solution structure of the RING finger domain from the acute promyelocytic leukaemia proto-oncoprotein PML. *EMBO J.* **1995**, *14*, 1532–1541.

111. Sarikas, A.; Hartmann, T.; Pan, Z.-Q. The cullin protein family. *Genome Biol.* **2011**, *12*, 220.

112. Aravind, L.; Koonin, E. V. The U box is a modified RING finger—a common domain in ubiquitination. *Curr. Biol.* **2000**, *10*, R132–R134.

113. Metzger, M. B.; Hristova, V. A.; Weissman, A. M. HECT and RING finger families of E3 ubiquitin ligases at a glance. *J. Cell Sci.* **2012**, *125*, 531–537.

114. Park, Y.; Burkitt, V.; Villa, A.; Tong, L.; Wu, H. Structural basis for self-association and receptor recognition of human TRAF2. *Nature* **1999**, *398*, 533–538.

115. Kozlov, G.; Peschard, P.; Zimmerman, B.; Lin, T.; Moldoveanu, T.; Mansur-Azzam, N., et al. Structural basis for UBA-mediated dimerization of c-Cbl ubiquitin ligase. *J. Biol. Chem.* **2007**, *282*, 27547–27555.

116. Poyurovsky, M. V.; Priest, C.; Kentsis, A.; Borden, K. L.; Pan, Z. Q.; Pavletich, N.; Prives, C. The Mdm2 RING domain C-terminus is required for supramolecular assembly and ubiquitin ligase activity. *EMBO J.* **2007**, *26*, 90–101.

117. Tang, X.; Orlicky, S.; Lin, Z.; Willems, A.; Neculai, D.; Ceccarelli, D., et al. Suprafacial orientation of the SCFCdc4 dimer accommodates multiple geometries for substrate ubiquitination. *Cell* **2007**, *129*, 1165–1176.

118. Zheng, N.; Wang, P.; Jeffrey, P. D.; Pavletich, N. P. Structure of a c-Cbl-UbcH7 complex: RING domain function in ubiquitin-protein ligases. *Cell* **2000**, *102*, 533–539.

119. Zheng, N.; Schulman, B. A.; Song, L.; Miller, J. J.; Jeffrey, P. D.; Wang, P., et al. Structure of the Cul1-Rbx1-Skp1-F boxSkp2 SCF ubiquitin ligase complex. *Nature* **2002**, *416*, 703–709.

120. Seol, J. H.; Feldman, R. M.; Zachariae, W.; Shevchenko, A.; Correll, C. C.; Lyapina, S., et al. Cdc53/cullin and the essential Hrt1 RING-H2 subunit of SCF define a ubiquitin ligase module that activates the E2 enzyme Cdc34. *Genes Dev.* **1999**, *13*, 1614–1626.

121. Duda, D. M.; Borg, L. A.; Scott, D. C.; Hunt, H. W.; Hammel, M.; Schulman, B. A. Structural insights into NEDD8 activation of cullin-RING ligases: conformational control of conjugation. *Cell* **2008**, *134*, 995–1006.

122. Petroski, M. D.; Deshaies, R. J. Mechanism of lysine 48-linked ubiquitin-chain synthesis by the cullin-RING ubiquitin-ligase complex SCF/Cdc34. *Cell* **2005**, *123*, 1107–1120.

123. Petroski, M. D.; Deshaies, R. J. Context of multiubiquitin chain attachment influences the rate of Sic1 degradation. *Mol. Cell* **2003**, *11*, 1435–1444.

124. Feldman, R. M.; Correll, C. C.; Kaplan, K. B.; Deshaies, R. J. A complex of Cdc4p, Skp1p, and Cdc53p/cullin catalyzes ubiquitination of the phosphorylated CDK inhibitor Sic1p. *Cell* **1997**, *91*, 221–230.

125. Hofmann, R. M.; Pickart, C. M. Noncanonical MMS2-encoded ubiquitin-conjugating enzyme functions in assembly of novel polyubiquitin chains for DNA repair. *Cell* **1999**, *96*, 645–653.

126. Li, W.; Tu, D.; Brunger, A. T.; Ye, Y. A ubiquitin ligase transfers preformed polyubiquitin chains from a conjugating enzyme to a substrate. *Nature* **2000**, *446*, 333–337.

127. Orlicky, S.; Tang, X.; Willems, A.; Tyers, M.; Sicheri, F. Structural basis for phosphodependent substrate selection and orientation by the SCFCdc4 ubiquitin ligase. *Cell* **2003**, *112*, 243–256.

128. Skowyra, D.; Craig, K. L.; Tyers, M.; Elledge, S. J.; Harper, J. W. F-box proteins are receptors that recruit phosphorylated substrates to the SCF ubiquitin-ligase complex. *Cell* **1997**, *91*, 209–219.
129. Yoshida, Y.; Chiba, T.; Tokunaga, F.; Kawasaki, H.; Iwai, K.; Suzuki, T., et al. E3 ubiquitin ligase that recognizes sugar chains. *Nature* **2002**, *418*, 438–442.
130. Ivan, M.; Kondo, K.; Yang, H.; Kim, W.; Valiando, J.; Ohh, M., et al. HIFalpha targeted for VHL-mediated destruction by proline hydroxylation: implications for O2 sensing. *Science* **2001**, *292*, 464–468.
131. Rudner, A. D.; Murray, A. W. Phosphorylation by Cdc28 activates the Cdc20-dependent activity of the anaphase-promoting complex. *J. Cell Biol.* **2000**, *149*, 1377–1390.
132. Zachariae, W.; Schwab, M.; Nasmyth, K.; Seufert, W. Control of cyclin ubiquitination by CDK regulated binding of Hct1 to the anaphase promoting complex. *Science* **1998**, *282*, 1721–1724.
133. Zheng, J.; Yang, X.; Harrell, J. M.; Ryzhikov, S.; Shim, E. H. CAND1 binds to unneddylated CUL1 and regulates the formation of SCF ubiquitin E3 ligase complex. *Mol. Cell* **2002**, *10*, 1519–1526.
134. Choi, E.; Dial, J. M.; Jeong, D. E.; Hall, M. C. Unique Dbox and KEN box sequences limit ubiquitination of Acm1 and promote pseudosubstrate inhibition of the anaphase-promoting complex. *J. Biol. Chem.* **2008**, *283*, 23701–23710.
135. Tan, X.; Calderon-Villalobos, L. I.; Sharon, M.; Zheng, C.; Robinson, C. V.; Estelle, M.; Zheng, N. Mechanism of auxin perception by the TIR1 ubiquitin ligase. *Nature* **2007**, *446*, 640–645.
136. Rachakonda, G.; Xiong, Y.; Sekhar, K. R.; Stamer, S. L.; Liebler, D. C.; Freeman, M. L. Covalent modification at Cys151 dissociates the electrophile sensor Keap1 from the ubiquitin ligase CUL3. *Chem. Res. Toxicol.* **2008**, *21*, 705–710.
137. Galan, J. M.; Peter, M. Ubiquitin-dependent degradation of multiple F-box proteins by an autocatalytic mechanism. *Proc. Natl. Acad. Sci. U.S.A.* **1999**, *96*, 9124–9129.
138. Li, Y.; Gazdoiu, S.; Pan, Z. Q.; Fuchs, S. Y. Stability of homologue of Slimb F-box protein is regulated by availability of its substrate. *J. Biol. Chem.* **2004**, *279*, 11074–11080.
139. Lamothe, B.; Besse, A.; Campos, A. D.; Webster, W. K.; Wu, H.; Darnay, B. G. Site-specific Lys-63-linked tumor necrosis factor receptor-associated factor 6 auto-ubiquitination is a critical determinant of I kappa B kinase activation. *J. Biol. Chem.* **2007**, *282*, 4102–4112.
140. Kawakami, T.; Chiba, T.; Suzuki, T.; Iwai, K.; Yamanaka, K.; Minato, N., et al. NEDD8 recruits E2-ubiquitin to SCF E3 ligase. *EMBO J.* **2001**, *20*, 4003–4012.
141. Kumar, S.; Tomooka, Y.; Noda, M. Identification of a set of genes with developmentally down-regulated expression in the mouse brain. *Biochem. Biophys. Res. Commun.* **1992**, *185*, 1155–1161.
142. Morrione, A.; Plant, P.; Valentinis, B.; Staub, O.; Kumar, S.; Rotin, D.; Baserga, R. mGrb10 interacts with Nedd4. *J. Biol. Chem.* **1999**, *274*, 24094–24099.
143. Wiesner, S.; Ogunjimi, A. A.; Wang, H. R.; Rotin, D.; Sicheri, F.; Wrana, J. L.; Forman-Kay, J. D. Autoinhibition of the HECT-type ubiquitin ligase Smurf2 through its C2 domain. *Cell* **2007**, *130*, 651–662.
144. Dunn, R.; Klos, D. A.; Adler, A. S.; Hicke, L. The C2 domain of the Rsp5 ubiquitin ligase binds membrane phosphoinositides and directs ubiquitination of endosomal cargo. *J. Cell Biol.* **2004**, *165*, 135–144.
145. Ding, Y.; Zhang, Y.; Xu, C.; Tao, Q.-H.; Chen, Y.-G. HECT domain-containing E3 ubiquitin ligase NEDD4L negatively regulates Wnt signaling by targeting dishevelled for proteasomal egradation. *J. Biol. Chem.* **2013**, *288*, 8289–8298.
146. Gautam, V.; Trinidad, J. C.; Rimerman, R. A.; Costa, B. M.; Burlingame, A. L.; Monaghan, D. T. Nedd4 is a specific E3 ubiquitin ligase for the NMDA receptor subunit GluN2D. *Neuropharmacology* **2013**, *74*, 96–107.
147. Garcia-Gonzalo, F. R.; Rosa, J. L. The HERC proteins: functional and evolutionary insights. *Cell. Mol. Life Sci.* **2005**, *62*, 1826–1838.

148. Renault, L.; Kuhlmann, J.; Henkel, A.; Wittinghofer, A. Structural basis for guanine nu-cleotide exchange on Ran by the regulator of chromosome condensation (RCC1). *Cell* **2001**, *105*, 245–255.

149. Al-Hakim, A. K.; Bashkurov, M.; Gingras, A.-C.; Durocher, D.; Pelletier, L. Interaction proteomics identify NEURL4 and the HECT E3 ligase HERC2 as novel modulators of centrosome architecture. *Mol. Cell. Proteom.* **2012**, *11*, M111.014233.

150. D'Arca, D.; Zhao, X.; Xu, W.; Ramirez-Martinez, N. C.; Iavarone, A.; Lasorella, A. Huwe1 ubiquitin ligase is essential to synchronize neuronal and glial differentiation in the de-veloping cerebellum. *Proc. Natl. Acad. Sci. U.S.A.* **2010**, *107*, 5875–5880.

151. Torrino, S.; Visvikis, O.; Doye, A.; Boyer, L.; Stefani, C.; Munro, P., et al. The E3 ubiquitin-ligase HACE1 catalyzes the ubiquitylation of active Rac1. *Dev. Cell* **2011**, *21*, 959–965.

152. Brooks, W. S.; Helton, E. S.; Banerjee, S.; Venable, M.; Johnson, L.; Schoeb, T. R., et al. G2E3 is a dual function ubiquitin ligase required for early embryonic development. *J. Biol. Chem.* **2008**, *283*, 22304–22315.

153. Verdecia, M. A.; Joazeiro, C. A. P.; Wells, N. J.; Ferrer, J.-L.; Bowman, M. E.; Hunter, T.; Noel, J. P. Conformational flexibility underlies ubiquitin ligation mediated by the WWP1 HECT domain E3 ligase. *Mol. Cell* **2003**, *11*, 249–259.

154. Ogunjimi, A. A.; Briant, D. J.; Pece-Barbara, N.; Le Roy, C.; Di Guglielmo, G. M.; Kavsak, P., et al. Regulation of Smurf2 ubiquitin ligase activity by anchoring the E2 to the HECT domain. *Mol. Cell* **2005**, *19*, 297–308.

155. Huang, L.; Kinnucan, E.; Wang, G.; Beaudenon, S.; Howley, P. M.; Huibregtse, J. M.; Pavletich, N. P. Structure of an E6AP–UbcH7 complex: insights into ubiquitination by the E2-E3 enzyme cascade. *Science* **1999**, *286*, 1321–1326.

156. Salvat, C.; Wang, G.; Dastur, A.; Lyon, N.; Huibregtse, J. M. The -4 phenylalanine is required for substrate ubiquitination catalyzed by HECT ubiquitin ligases. *J. Biol. Chem.* **2004**, *279*, 18935–18943.

157. Kim, H. C.; Huibregtse, J. M. Polyubiquitination by HECT E3s and the determinants of chain type specificity. *Mol. Cell. Biol.* **2009**, *29*, 3307–3318.

158. Wang, M.; Pickart, C. M. Different HECT domain ubiquitin ligases employ distinct mechanisms of polyubiquitin chain synthesis. *EMBO J.* **2005**, *24*, 4324–4333.

159. Wang, M.; Cheng, D.; Peng, J.; Pickart, C. M. Molecular determinants of polyubiquitin linkage selection by an HECT ubiquitin ligase. *EMBO J.* **2006**, *25*, 1710–1719.

160. Galan, J. M.; Haguenauer-Tsapis, R. Ubiquitin Lys63 is involved in ubiquitination of a yeast plasma membrane protein. *EMBO J.* **1997**, *16*, 5847–5854.

161. Maspero, E.; Valentini, E.; Mari, S.; Cecatiello, V.; Soffientini, P.; Pasqualato, S.; Polo, S. Structure of a ubiquitin-loaded HECT ligase reveals the molecular basis for catalytic priming. *Nature Struct. Mol. Biol.* **2013**, *20*, 696–701.

162. Snyder, P. M.; Olson, D. R.; Thomas, B. C. Serum and glucocorticoid-regulated kinase modulates Nedd4-2-mediated inhibition of the epithelial Na⁺ channel. *J. Biol. Chem.* **2002**, *277*, 5–8.

163. Oberst, A.; Malatesta, M.; Aqeilan, R. I.; Rossi, M.; Salomoni, P.; Murillas, R., et al. The Nedd4-binding partner 1 (N4BP1) protein is an inhibitor of the E3 ligase Itch. *Proc. Natl Acad. Sci. U.S.A.* **2007**, *104*, 11280–11285.

164. Lu, K.; Yin, X.; Weng, T.; Xi, S.; Li, L.; Xing, G., et al. Targeting WW domains linker of HECT-type ubiquitin ligase Smurf1 for activation by CKIP-1. *Nature Cell Biol.* **2008**, *10*, 994–1002.

165. McGill, M. A.; McGlade, C. J. Mammalian numb proteins promote Notch1 receptor ubiquitination and degradation of the Notch1 intracellular domain. *J. Biol. Chem.* **2003**, *278*, 23196–23203.

166. Bruce, M. C.; Kanelis, V.; Fouladkou, F.; Debonneville, A.; Staub, O.; Rotin, D. Regula-tion of Nedd4-2 selfubiquitination and stability by a PY motif located within its HECT-domain. *Biochem. J.* **2008**, *415*, 155–163.

167. Eisenhaber, B.; Chumak, N.; Eisenhaber, F.; Hauser, M. T. The ring between ring fingers (RBR) protein family. *Genome Biol.* **2007**, *8*, 209.

168. Capili, A. D.; Edghill, E. L.; Wu, K.; Borden, K. L. Structure of the C-terminal RING finger from a RING-IBR-RING/TRIAD motif reveals a novel zinc-binding domain distinct from a RING. *J. Mol. Biol.* **2004**, *340*, 1117–1129.
169. Hristova, V. A.; Beasley, S. A.; Rylett, R. J.; Shaw, G. S. Identification of a novel Zn2+-binding domain in the autosomal recessive juvenile Parkinson-related E3 ligase parkin. *J. Biol. Chem.* **2009**, *284*, 14978–14986.
170. Marin, I.; Ferrus, A. Comparative genomics of the RBR family, including the Parkinson's disease-related gene parkin and the genes of the Ariadne subfamily. *Mol. Biol. Evol.* **2002**, *19*, 2039–2050.
171. Marin, I.; Lucas, J. I.; Gradilla, A. C.; Ferrus, A. Parkin and relatives: the RBR family of ubiquitin ligases. *Physiol. Genomics* **2004**, *17*, 253–263.
172. Marzook, H.; Li, D.-Q.; Nair, V. S.; Mudvari, P.; Reddy, S. D. N.; Pakala, S. B., et al. Metastasis-associated protein 1 drives tumor cell migration and invasion through transcriptional repression of RING finger protein 144A. *J. Biol. Chem.* **2012**, *287*, 5615–5626.
173. Niwa, J.; Ishigaki, S.; Doyu, M.; Suzuki, T.; Tanaka, K.; Sobue, G. A novel centrosomal ring-finger protein, dorfin, mediates ubiquitin ligase activity. *Biochem. Biophys. Res. Commun.* **2001**, *281*, 706–713.
174. Ardley, H. C.; Tan, N. G.; Rose, S. A.; Markham, A. F.; Robinson, P. A. Features of the parkin/ariadne-like ubiquitin ligase, HHARI, that regulate its interaction with the ubiquitin-conjugating enzyme, Ubch7. *J. Biol. Chem.* **2001**, *276*, 19640–19647.
175. van der Reijden, B. A.; Erpelinck-Verschueren, C. A.; Lowenberg, B.; Jansen, J. H. TRIADs: a new class of proteins with a novel cysteine-rich signature. *Protein Sci.* **1999**, *8*, 1557–1561.
176. Ito, K.; Adachi, S.; Iwakami, R.; Yasuda, H.; Muto, Y.; Seki, N.; Okano, Y. N-Terminally extended human ubiquitin-conjugating enzymes (E2s) mediate the ubiquitination of RING-finger proteins, ARA54 and RNF8. *Eur. J. Biochem.* **2001**, *268*, 2725–2732.
177. Byrd, R. A.; Weissman, A. M. Compact Parkin only: insights into the structure of an autoinhibited ubiquitin ligase. *EMBO J.* **2013**, *32*, 2087–2089.
178. Sato, Y.; Fujita, H.; Yoshikawa, A.; Yamashita, M.; Yamagata, A.; Kaiser, S. E., et al. Specific recognition of linear ubiquitin chains by the Npl4 zinc finger (NZF) domain of the HOIL-1L subunit of the linear ubiquitin chain assembly complex. *Proc. Natl. Acad. Sci. U.S.A.* **2011**, *108*, 20520–20525.
179. Brzovic, P. S.; Keeffe, J. R.; Nishikawa, H.; Miyamoto, K.; Fox, D., 3rd.; Fukuda, M., et al. Binding and recognition in the assembly of an active BRCA1/BARD1 ubiquitin-ligase complex. *Proc. Natl. Acad. Sci. U.S.A.* **2003**, *100*, 5646–5651.
180. Beasley, S. A.; Hristova, V. A.; Shaw, G. S. Structure of the Parkin in-between-ring domain provides insights for E3-ligase dysfunction in autosomal recessive Parkinson's disease. *Proc. Natl. Acad. Sci. U.S.A.* **2007**, *104*, 3095–3100.
181. Rankin, C. A.; Roy, A.; Zhang, Y.; Richter, M. Parkin, a top level manager in the cell's sanitation department. *Open Biochem. J.* **2011**, *5*, 9–26.
182. Hampe, C.; Ardila-Osorio, H.; Fournier, M.; Brice, A.; Corti, O. Biochemical analysis of Parkinson's disease-causing variants of Parkin, an E3 ubiquitin-protein ligase with monoubiquitylation capacity. *Hum. Mol. Genet.* **2006**, *15*, 2059–2075.
183. Matsuda, N.; Kitami, T.; Suzuki, T.; Mizuno, Y.; Hattori, N.; Tanaka, K. Diverse effects of pathogenic mutations of Parkin that catalyze multiple monoubiquitylation in vitro. *J. Biol. Chem.* **2006**, *281*, 3204–3209.
184. Olzmann, J. A.; Li, L.; Chudaev, M. V.; Chen, J.; Perez, F. A.; Palmiter, R. D.; Chin, L. S. Parkin-mediated K63-linked polyubiquitination targets misfolded DJ-1 to aggresomes via binding to HDAC6. *J. Cell Biol.* **2007**, *178*, 1025–1038.
185. Smit, J. J.; Monteferrario, D.; Noordermeer, S. M.; van Dijk, W. J.; van der Reijden, B. A.; Sixma, T. K. The E3 ligase HOIP specifies linear ubiquitin chain assembly through its RING-IBR-RING domain and the unique LDD extension. *EMBO J.* **2012**, *31*, 3833–3844.

186. Tokunaga, F.; Nakagawa, T.; Nakahara, M.; Saeki, Y.; Taniguchi, M.; Sakata, S., et al. SHARPIN is a component of the NF-kappaB-activating linear ubiquitin chain assembly complex. *Nature* **2011**, *471*, 633–636.

187. Stieglitz, B.; Morris-Davies, A. C.; Koliopoulos, M. G.; Christodoulou, E.; Rittinger, K. LUBAC synthesizes linear ubiquitin chains via a thioester intermediate. *EMBO Rep.* **2012**, *13*, 840–846.

188. Chaugule, V. K.; Burchell, L.; Barber, K. R.; Sidhu, A.; Leslie, S. J.; Shaw, G. S.; Walden, H. Autoregulation of Parkin activity through its ubiquitin-like domain. *EMBO J.* **2011**, *30*, 2853–2867.

189. Marteijn, J. A.; van Emst, L.; Erpelinck-Verschueren, C. A.; Nikoloski, G.; Menke, A.; de Witte, T., et al. The E3 ubiquitin-protein ligase Triad1 inhibits clonogenic growth of primary myeloid progenitor cells. *Blood* **2005**, *106*, 4114–4123.

190. Fraile, J. M.; Quesada, V.; Rodríguez, D.; Freije, J. M. P.; Lopez-Otín, C. Deubiquitinases in cancer: new functions and therapeutic options. *Oncogene* **2012**, *31*, 2373–2388.

191. Nijman, S. M. B.; Luna-Vargas, M. P. A.; Velds, A.; Brummelkamp, T. R.; Dirac, A. M. G.; Sixma, T. K.; Bernards, R. A genomic and functional inventory of deubiquitinating enzymes. *Cell* **2005**, *123*, 773–786.

192. Ye, Y.; Scheel, H.; Hofmann, K.; Komander, D. Dissection of USP catalytic domains reveals five common insertion points. *Mol. Biosyst.* **2009**, *5*, 1797–1808.

193. Kimura, Y.; Tanaka, K. Regulatory mechanisms involved in the control of ubiquitin homeostasis. *J. Biochem.* **2010**, *147*, 793–798.

194. Finley, D.; Bartel, B.; Varshavsky, A. The tails of ubiquitin precursors are ribosomal proteins whose fusion to ubiquitin facilitates ribosome biogenesis. *Nature* **1989**, *338*, 394–401.

195. Fornace, A. J., Jr.; Alamo, I., Jr.; Hollander, M. C.; Lamoreaux, E. Ubiquitin mRNA is a major stress-induced transcript in mammalian cells. *Nucleic Acids Res.* **1989**, *17*, 1215–1230.

196. Komander, D.; Clague, M. J.; Urbe, S. Breaking the chains: structure and function of the deubiquitinases. *Nat. Rev. Mol. Cell Biol.* **2009**, *10*, 550–563.

197. Bhaumik, S. R.; Malik, S. Diverse regulatory mechanisms of eukaryotic transcriptional activation by the proteasome complex. *Crit. Rev. Biochem. Mol. Biol.* **2008**, *43*, 419–433.

198. Frankland-Searby, S.; Bhaumik, S. R. The 26S proteasome system: an attractive target for cancer therapy. *Bioch. Biophys. Acta* **2012**, *1825*, 64–76.

199. Young, P.; Deveraux, Q.; Beal, R. E.; Pickart, C. M.; Rechsteiner, M. Characterization of two polyubiquitin binding sites in the 26 S protease subunit 5a. *J. Biol. Chem.* **1998**, *273*, 5461–5467.

200. Koulich, E.; Li, X.; DeMartino, G. N. Relative structural and functional roles of multiple deubiquitylating proteins associated with mammalian 26S proteasome. *Mol. Biol. Cell* **2008**, *19*, 1072–1082.

201. Liu, C.-W.; Jacobson, A. D. Functions of the 19S complex in proteasomal degradation. *Trends Biochem. Sci.* **2013**, *38*, 103–110.

202. Eletr, Z. M.; Wilkinson, K. D. Regulation of proteolysis by human deubiquitinating enzymes. *Biochim. Biophys. Acta* **1843**, *2014*, 114–128.

203. Lee, M. J.; Lee, B.-H.; Hanna, J.; King, R. W.; Finley, D. Trimming of ubiquitin chains by proteasome associated deubiquitinating enzymes. *Mol. Cell. Proteomics* **2011**, *10* R110.003871.

204. Yao, T.; Cohen, R. E. A cryptic protease couples deubiquitination and degradation by the proteasome. *Nature* **2002**, *419*, 403–407.

205. Lecker, S. H.; Goldberg, A. L.; Mitch, W. E. Protein degradation by the ubiquitin–proteasome pathway in normal and disease states. *J. Am. Soc. Nephrol.* **2006**, *17*, 1807–1819.

206. Mukhopadhyay, D.; Riezman, H. Proteasome-independent functions of ubiquitin in endocytosis and signaling. *Science* **2007**, *315*, 201–205.

207. Haglund, K.; Dikic, I. Ubiquitylation and cell signaling. *EMBO J.* **2005**, *24*, 3353–3359.
208. Grabbe, C.; Husnjak, K.; Dikic, I. The spatial and temporal organization of ubiquitin networks. *Nat. Rev. Mol. Cell Biol.* **2011**, *12*, 295–307.
209. Raiborg, C.; Stenmark, H. The ESCRT machinery in endosomal sorting of ubiquitylated membrane proteins. *Nature* **2009**, *458*, 445–452.
210. Clague, M. J.; Urbe, S. Endocytosis: the DUB version. *Trends Cell Biol.* **2006**, *16*, 551–559.
211. Lu, Y.; Adegoke, O. A. J.; Nepveu, A.; Nakayama, K. I.; Bedard, N.; Cheng, D., et al. USP19 deubiquitinating enzyme supports cell proliferation by stabilizing KPC1, a ubiquitin ligase for p27Kip1. *Mol. Cell Biol.* **2009**, *29*, 547–558.
212. Chiu, Y. H.; Zhao, M.; Chen, Z. J. Ubiquitin in NF-κB signaling. *Chem. Rev.* **2009**, *4*, 1549–1560.
213. Sun, S. C. Deubiquitylation and regulation of the immune response. *Nat. Rev. Immunol.* **2008**, *8*, 501–511.
214. Storer, A. C.; Menard, R. Catalytic mechanism in papain family of cysteine peptidases. *Methods Enzymol.* **1994**, *244*, 486–500.
215. Amerik, A. Y.; Hochstrasser, M. Mechanism and function of deubiquitinating enzymes. *Biochim. Biophys. Acta* **1695**, *2004*, 189–207.
216. Komander, D. Mechanism, specificity and structure of the deubiquitinases. *Subcell. Biochem.* **2010**, *54*, 69–87.
217. Hu, M.; Li, P.; Li, M.; Li, W.; Yao, T.; Wu, J. W., et al. Crystal structure of a UBP-family deubiquitinating enzyme in isolation and in complex with ubiquitin aldehyde. *Cell* **2002**, *111*, 1041–1054.
218. Meulmeester, E.; Kunze; Marion, H.; He, H.; Urlaub, H.; Melchior, F. Mechanism and consequences for paralog-specific sumoylation of Ubiquitin-Specific Protease 25. *Mol. Cell* **2008**, *30*, 610–619.
219. Avvakumov, G. V.; Walker, J. R.; Xue, S.; Allali-Hassani, A.; Asinas, A.; Nair, U. B., et al. Two ZnF-UBP domains in isopeptidase T (USP5). *Biochemistry* **2012**, *51*, 1188–1198.
220. Reyes-Turcu, F. E.; Shanks, J. R.; Komander, D.; Wilkinson, K. D. Recognition of polyubiquitin isoforms by the multiple ubiquitin binding modules of isopeptidase T. *J. Biol. Chem.* **2008**, *283*, 19581–19592.
221. Edelmann, M. J.; Kessler, B. M. Ubiquitin and ubiquitin-like specific proteases targeted by infectious pathogens: emerging patterns and molecular principles. *Biochim. Biophys. Acta* **1782**, *2008*, 809–816.
222. Fang, Y.; Fu, D.; Shen, X.-Z. The potential role of ubiquitin c-terminal hydrolases in oncogenesis. *Biochim. Biophys. Acta* **1806**, *2010*, 1–6.
223. Johnston, S. C.; Riddle, S. M.; Cohen, R. E.; Hill, C. P. Structural basis for the specificity of ubiquitin C-terminal hydrolases. *EMBO J.* **1999**, *18*, 3877–3887.
224. Reyes-Turcu, F. E.; Ventii, K. H.; Wilkinson, K. D. Regulation and cellular roles of ubiquitin-specific deubiquitinating enzymes. *Annu. Rev. Biochem.* **2009**, *78*, 363–397.
225. Yu, H.; Mashtalir, N.; Daou, S.; Hammond-Martel, I.; Ross, J.; Sui, G., et al. The ubiquitin carboxyl hydrolase BAP1 forms a ternary complex with YY1 and HCF-1 and is a critical regulator of gene expression. *Mol. Cell. Biol.* **2010**, *30*, 5071–5085.
226. Larsen, C. N.; Krantz, B. A.; Wilkinson, K. D. Substrate specificity of deubiquitinating enzymes: ubiquitin C-terminal hydrolases. *Biochemistry* **1998**, *37*, 3358–3368.
227. Machida, Y. J.; Machida, Y.; Vashisht, A. A.; Wohlschlegel, J. A.; Dutta, A. The deubiquitinating enzyme BAP1 regulates cell growth via interaction with HCF-1. *J. Biol. Chem.* **2009**, *284*, 34179–34188.
228. Mevissen, T. E. T.; Hospenthal, M. K.; Geurink, P. P.; Elliott, P. R.; Akutsu, M.; Arnaudo, N., et al. OTU deubiquitinases reveal mechanisms of linkage specificity and enable ubiquitin chain restriction analysis. *Cell* **2013**, *154*, 169–184.
229. Makarova, K. S.; Aravind, L.; Koonin, E. V. A novel superfamily of predicted cysteine proteases from eukaryotes, viruses and *Chlamydia pneumoniae*. *Trends Biochem. Sci.* **2000**, *25*, 50–52.

230. Lin, S. C.; Chung, J. Y.; Lamothe, B.; Rajashankar, K.; Lu, M.; Lo, Y. C., et al. Molecular basis for the unique deubiquitinating activity of the NF-kappaB inhibitor A20. *J. Mol. Biol.* **2008**, *376*, 526–540.

231. Tzvetkov, N.; Breuer, P. Josephin domain-containing proteins from a variety of species are active de-ubiquitination enzymes. *Biol. Chem.* **2007**, *388*, 973–978.

232. Nicastro, G.; Todi, S. V.; Karaca, E.; Bonvin, A. M. J. J.; Paulson, H. L.; Pastore, A. Understanding the role of the Josephin domain in the polyUb binding and cleavage properties of ataxin-3. *PLoS One* **2010**, *5*, e12430.

233. Berke, S. J.; Chai, Y.; Marrs, G. L.; Wen, H.; Paulson, H. L. Defining the role of ubiquitin-interacting motifs in the polyglutamine disease protein, ataxin-3. *J. Biol. Chem.* **2005**, *280*, 32026–32034.

234. Winborn, B. J.; Travis, S. M.; Todi, S. V.; Scaglione, K. M.; Xu, P.; Williams, A. J., et al. The deubiquitinating enzyme ataxin-3, a polyglutamine disease protein, edits Lys63 linkages in mixed linkage ubiquitin chains. *J. Biol. Chem.* **2008**, *283*, 26436–26443.

235. Nicastro, G.; Masino, L.; Esposito, V.; Menon, R. P.; De Simone, A.; Fraternali, F.; Pastore, A. Josephin domain of ataxin-3 contains two distinct ubiquitin-binding sites. *Biopolymers* **2009**, *91*, 1203–1214.

236. Liang, J.; Saad, Y.; Lei, T.; Wang, J.; Qi, D.; Yang, Q., et al. MCP induced protein 1 deubiquitinates TRAF proteins and negatively regulates JNK and NF-kappaB signaling. *J. Exp. Med.* **2010**, *207*, 2959–2973.

237. Ambroggio, X. I.; Rees, D. C.; Deshaies, R. J. JAMM: A metalloprotease-like zinc site in the proteasome and signalosome. *PLoS Biol.* **2004**, *2*, 0113–0119.

238. Cope, G. A.; Suh, G. S.; Aravind, L.; Schwarz, S. E.; Zipursky, S. L.; Koonin, E. V.; Deshaies, R. J. Role of predicted metalloprotease motif of Jab1/Csn5 in cleavage of Nedd8 from Cul1. *Science* **2002**, *298*, 608–611.

239. Tran, H. J.; Allen, M. D.; Lowe, J.; Bycroft, M. Structure of the Jab1/MPN domain and its implications for proteasome function. *Biochemistry* **2003**, *42*, 11460–11465.

240. Verma, R.; Aravind, L.; Oania, R.; McDonald, W. H.; Yates, J. R., III.; Koonin, E. V.; Deshaies, R. J. Role of Rpn11 metalloprotease in deubiquitination and degradation by the 26S proteasome. *Science* **2002**, *298*, 611–615.

241. Keusekotten, K.; Elliott, P. R.; Glockner, L.; Fiil, B. K.; Damgaard, R. B.; Kulathu, Y., et al. OTULIN antagonizes LUBAC signaling by specifically hydrolyzing Met1-linked polyubiquitin. *Cell* **2013**, *153*, 1312–1326.

242. Fu, Q.-S.; Song, A.-X.; Hu, H.-Y. Structural aspects of ubiquitin binding specificities. *Curr. Prot. Pept. Sci.* **2012**, *13*, 482–489.

243. Drag, M.; Mikolajczyk, J.; Bekes, M.; Reyes-Turcu, F. E.; Ellman, J. A.; Wilkinson, K. D.; Salvesen, G. S. Positional-scanning fluorigenic substrate libraries reveal unexpected specificity determinants of DUBs (deubiquitinating enzymes). *Biochem. J.* **2008**, *415*, 367–375.

244. Gong, L.; Kamitani, T.; Millas, S.; Yeh, E. T. Identification of a novel isopeptidase with dual specificity for ubiquitin- and NEDD8-conjugated proteins. *J. Biol. Chem.* **2000**, *275*, 14212–14216.

245. Catic, A.; Fiebiger, E.; Korbel, G. A.; Blom, D.; Galardy, P. J.; Ploegh, H. L. Screen for ISG15-crossreactive deubiquitinases. *PLoS One* **2007**, *2*, e679.

246. Faesen, A. C.; Luna-Vargas, M. P. A.; Geurink, P. P.; Clerici, M.; Merkx, R.; van Dijk, W. J., et al. The differential modulation of USP activity by internal regulatory domains, interactors and eight ubiquitin chain types. *Chem. Biol.* **2011**, *18*, 1550–1561.

247. Luna-Vargas, M. P.; Faesen, A. C.; van Dijk, W. J.; Rape, M.; Fish, A.; Sixma, T. K. Ubiquitin-specific protease 4 is inhibited by its ubiquitin-like domain. *EMBO Rep.* **2011**, *12*, 365–372.

248. Komander, D.; Lord, C. J.; Scheel, H.; Swift, S.; Hofmann, K.; Ashworth, A.; Barford, D. The structure of the CYLD USP domain explains its specificity for Lys63-linked polyubiquitin and reveals a B box module. *Mol. Cell* **2008**, *29*, 451–464.

249. Hu, M.; Li, P.; Song, L.; Jeffrey, P. D.; Chernova, T. A.; Wilkinson, K. D., et al. Structure and mechanisms of the proteasome-associated deubiquitinating enzyme USP14. *EMBO J.* **2005**, *24*, 3747–3756.

250. Reyes-Turcu, F. E.; Horton, J. R.; Mullally, J. E.; Heroux, A.; Cheng, X.; Wilkinson, K. D. The ubiquitin binding domain ZnF UBP recognizes the C-terminal diglycine motif of unanchored ubiquitin. *Cell* **2006**, *124*, 1197–1208.

251. Cooper, E. M.; Cutcliffe, C.; Kristiansen, T. Z.; Pandey, A.; Pickart, C. M.; Cohen, R. E. K63-specific deubiquitination by two JAMM/MPN+ complexes: BRISC-associated Brcc36 and proteasomal Poh1. *EMBO J.* **2009**, *28*, 621–631.

252. Bremm, A.; Freund, S. M. V.; Komander, D. Lys11-linked ubiquitin chains adopt compact conformations and are preferentially hydrolyzed by the deubiquitinase Cezanne. *Nature Struct. Mol. Biol.* **2010**, *17*, 939–947.

253. Xu, Z.; Pei, L.; Wang, L.; Zhang, F.; Hu, X.; Gui, Y. Snail1-dependent transcriptional repression of Cezanne2 in hepatocellular carcinoma. *Oncogene* **2013**, doi: 10.1038/onc.2013.243.

254. Wang, T.; Yin, L.; Cooper, E. M.; Lai, M.-Y.; Dickey, S.; Pickart, C. M., et al. Evidence for bidentate substrate binding as the basis for the K48 linkage specificity of otubain 1. *J. Mol. Biol.* **2009**, *386*, 1011–1023.

255. Kayagaki, N.; Phung, Q.; Chan, S.; Chaudhari, R.; Quan, C.; O'Rourke, K. M., et al. DUBA: a deubiquitinase that regulates type I interferon production. *Science* **2007**, *318*, 1628–1632.

256. Huang, O. W.; Ma, X.; Yin, J. P.; Flinders, J.; Maurer, T.; Kayagaki, N., et al. Phosphorylation-dependent activity of the deubiquitinase DUBA. *Nature Struct. Mol. Biol.* **2012**, *19*, 171–175.

257. Sato, Y.; Yoshikawa, A.; Yamagata, A.; Mimura, H.; Yamashita, M.; Ookata, K., et al. Structural basis for specific cleavage of Lys 63-linked polyubiquitin chains. *Nature* **2008**, *455*, 358–362.

258. Schaefer, J. B.; Morgan, D. O. Protein-linked ubiquitin chain structure restricts activity of deubiquitinating enzymes. *J. Biol. Chem.* **2011**, *286*, 45186–45196.

259. Dupont, S.; Inui, M.; Newfeld, S. J. Regulation of TGF-β signal transduction by mono- and deubiquitylation of Smads. *FEBS Lett.* **2012**, *586*, 1913–1920.

260. Morrow, M. E.; Kim, M.-I.; Ronau, J. A.; Sheedlo, M. J.; White, R. R.; Chaney, J., et al. Stabilization of an unusual salt bridge in ubiquitin by the extra C-terminal domain of the proteasome-associated deubiquitinase UCH37 as a mechanism of its exo specificity. *Biochemistry* **2013**, *52*, 3564–3578.

261. Al-Hakim, A. K.; Zagorska, A.; Chapman, L.; Deak, M.; Peggie, M.; Alessi, D. R. Control of AMPK-related kinases by USP9X and atypical Lys29/Lys33-linked polyubiquitin chains. *Biochem. J.* **2008**, *411*, 249–260.

262. Verma, R.; Aravind, L.; Oania, R.; McDonald, W. H.; Yates, J. R., 3rd.; Koonin, E. V.; Deshaies, R. J. Role of Rpn11 metalloprotease in deubiquitination and degradation by the 26S proteasome. *Science* **2002**, *298*, 611–615.

263. Wilkinson, K. D.; Tashayev, V. L.; O'Connor, L. B.; Larsen, C. N.; Kasperek, E.; Pickart, C. M. Metabolism of the polyubiquitin degradation signal: structure, mechanism, and role of isopeptidase T. *Biochemistry* **1995**, *34*, 14535–14546.

264. Huang, X. D.; Summers, M. K.; Pham, V.; Lill, J. R.; Liu, J.; Lee, G., et al. Deubiquitinase USP37 is activated by CDK2 to antagonize APCCDH1 and promote S phase entry. *Mol. Cell* **2011**, *42*, 511–523.

265. Zhang, L.; Zhou, F. F.; Drabsch, Y.; Gao, R.; Snaar-Jagalska, B. E.; Mickanin, C., et al. USP4 is regulated by AKT phosphorylation and directly deubiquitylates TGF-β type I receptor. *Nature Cell Biol.* **2012**, *14*, 717–726.

266. Reiley, W.; Zhang, M.; Wu, X.; Granger, E.; Sun, S. C. Regulation of the deubiquitinating enzyme CYLD by IκB kinase gamma-dependent phosphorylation. *Mol. Cell. Biol.* **2005**, *25*, 3886–3895.

267. Mizuno, E.; Kitamura, N.; Komada, M. 14-3-3-dependent inhibition of the deubiquitinating activity of UBPY and its cancellation in the M phase. *Exp. Cell Res.* **2007**, *313*, 3624–3634.

268. Lee, J.-G.; Baek, K.; Soetandyo, N.; Ye, Y. Reversible inactivation of deubiquitinases by reactive oxygen species in vitro and in cells. *Nature Commun.* **2013**, *4*, 2532.

269. Huang, T. T.; Nijman, S. M. B.; Galardy, P. J.; Cohn, M. A.; Haas, W.; Gygi, S. P., et al. Regulation of monoubiquitinated PCNA by DUB autocleavage. *Nature Cell Biol.* **2006**, *8*, 341–347.

270. Todi, S. V.; Winborn, B. J.; Scaglione, K. M.; Blount, J. R.; Travis, S. M.; Paulson, H. L. Ubiquitination directly enhances activity of the deubiquitinating enzyme ataxin-3. *EMBO J.* **2009**, *28*, 372–382.

271. Mao, Y.; Senic-Matuglia, F.; Di Fiore, P. P.; Polo, S.; Hodsdon, M. E.; De Camilli, P. Deubiquitinating function of ataxin-3: insights from the solution structure of the Josephin domain. *Proc. Natl Acad. Sci. U.S.A.* **2005**, *102*, 12700–12705.

272. McCullough, J.; Row, P. E.; Lorenzo, O.; Doherty, M.; Beynon, R.; Clague, M. J.; Urbé, S. Activation of the endosome-associated ubiquitin isopeptidase AMSH by STAM, a component of the multivesicular body-sorting machinery. *Curr. Biol.* **2006**, *16*, 160–165.

273. Iha, H.; Peloponese, J. M.; Verstrepen, L.; Zapart, G.; Ikeda, F.; Smith, C. D., et al. Inflammatory cardiac valvulitis in TAX1BP1-deficient mice through selective NF-κB activation. *EMBO J.* **2008**, *27*, 629–641.

274. Wagner, S.; Carpentier, I.; Rogov, V.; Kreike, M.; Ikeda, F.; Loehr, F., et al. Ubiquitin binding mediates the NF-κB inhibitory potential of ABIN proteins. *Oncogene* **2008**, *27*, 3739–3745.

275. Cohn, M. A.; Kowal, P.; Yang, K.; Haas, W.; Huang, T. T.; Gygi, S. P.; D'Andrea, A. D. A UAF1-containing multisubunit protein complex regulates the Fanconi anemia pathway. *Mol. Cell* **2007**, *28*, 786–797.

276. Cohn, M. A.; Kee, Y.; Haas, W.; Gygi, S. P.; D'Andrea, A. D. UAF1 is a subunit of multiple deubiquitinating enzyme complexes. *J. Biol. Chem.* **2008**, *8*, 5343–5351.

277. Yao, T.; Song, L.; Jin, J.; Cai, Y.; Takahashi, H.; Swanson, S. K., et al. Distinct modes of regulation of the Uch37 deubiquitinating enzyme in the proteasome and in the Ino80 chromatin-remodeling complex. *Mol. Cell* **2008**, *31*, 909–917.

278. Nakamura, N.; Hirose, S. Regulation of mitochondrial morphology by USP30, a deubiquitinating enzyme present in the mitochondrial outer membrane. *Mol. Biol. Cell* **2008**, *19*, 1903–1911.

279. Lee, J.-G.; Kim, W.; Gygi, S.; Ye, Y. Characterization of the deubiquitinating activity of USP19 and its role in endoplasmic reticulum-associated degradation. *J. Biol. Chem.* **2014**, doi: 10.1074/jbc.M113.538934.

280. Endo, A.; Matsumoto, M.; Inada, T.; Yamamoto, A.; Nakayama, K. I.; Kitamura, N.; Komada, M. Nucleolar structure and function are regulated by the deubiquitylating enzyme USP36. *J. Cell Sci.* **2009**, *122*, 678–686.

281. Gallastegui, N.; Groll, M. The 26S proteasome: assembly and function of a destructive machine. *Trends Biochem. Sci.* **2010**, *35*, 634–642.

282. Groll, M.; Dtizel, L.; Lowe, J.; Stock, D.; Bochtler, M.; Wolf, D. H.; Huber, R. Structure of the 20S proteasome from yeast at 2.4 A resolution. *Nature* **1997**, *386*, 463–471.

283. Borissenko, L.; Groll, M. 20S proteasome and its inhibitors: crystallographic knowledge for drug development. *Chem. Rev.* **2007**, *107*, 687–717.

284. Groll, M.; Heinemeyer, W.; Jager, S.; Ulrich, T.; Bochtler, M.; Wolf, D. H.; Huber, R. The catalytic sites of 20S proteasomes and their role in subunit maturation: a mutational and crystallographic study. *Proc. Natl. Acad. Sci. U.S.A.* **1999**, *96*, 10975–10983.

285. Glickman, M. H.; Rubin, D. M.; Coux, O.; Wefes, I.; Pfeifer, G.; Cjeka, Z., et al. A subcomplex of the proteasome regulatory particle required for ubiquitin-conjugate degradation and related to the COP9-signalosome and eIF3. *Cell* **1998**, *94*, 615–623.

286. Smith, D. M.; Chang, S.-C.; Park, S.; Finley, D.; Cheng, Y.; Goldberg, A. L. Docking of the proteasomal ATPases' carboxyl termini in the 20S proteasome's alpha ring opens the gate for substrate entry. *Mol. Cell* **2007**, *27*, 731–744.

287. Sauer, R. T.; Baker, T. A. AAA+ proteases: ATP-fueled machines of protein destruction. *Annu. Rev. Biochem.* **2011**, *80*, 587–612.

288. Tanaka, K.; Mizushima, T.; Saeki, Y. The proteasome: molecular machinery and pathophysiological roles. *Biol. Chem.* **2012**, *393*, 217–234.

289. Rabl, J.; Smith, D. M.; Yu, Y.; Chang, S.-C.; Goldberg, A. L.; Cheng, Y. Mechanism of gate opening in the 20S proteasome by the proteasomal ATPases. *Mol. Cell* **2008**, *30*, 360–368.

290. Rosenzweig, R.; Osmulski, P. A.; Gaczynska, M.; Glickman, M. H. The central unit within the 19S regulatory particle of the proteasome. *Nat. Struct. Mol. Biol.* **2008**, *15*, 573–580.

291. D'Arcy, P.; Linder, S. Proteasome deubiquitinases as novel targets for cancer therapy. *Int. J. Biochem. Cell Biol.* **2012**, *44*, 1729–1738.

292. Sharon, M.; Taverner, T.; Ambroggio, X. I.; Deshaies, R. J.; Robinson, C. V. Structural organization of the 19S proteasome lid: insights from MS of intact complexes. *PLoS Biology* **2006**, *4*, 1314–1323.

293. Lipson, C.; Alalouf, G.; Bajorek, M.; Rabinovich, E.; Atir-Lande, A.; Glickman, M.; Bar-Nun, S. A proteasomal ATPase contributes to dislocation of endoplasmic reticulum-associated degradation (ERAD) substrates. *J. Biol. Chem.* **2008**, *283*, 7166–7175.

294. Liu, C. W.; Li, X.; Thompson, D.; Wooding, K.; Chang, T. L.; Tang, Z., et al. ATP binding and ATP hydrolysis play distinct roles in the function of 26S proteasome. *Mol. Cell* **2006**, *24*, 39–50.

295. Jariel-Encontre, I.; Bossis, G.; Piechaczyk, M. Ubiquitin-independent degradation of proteins by the proteasome. *Biochim. Biophys. Acta* **1786**, *2008*, 153–177.

296. Deveraux, Q.; Ustrell, V.; Pickart, C. M.; Rechsteiner, M. A 26 S protease subunit that binds ubiquitin conjugates. *J. Biol. Chem.* **1994**, *269*, 7059–7061.

297. Husnjak, K.; Elsasser, S.; Zhang, N.; Chen, X.; Randles, L.; Shi, Y., et al. Proteasome subunit Rpn13 is a novel ubiquitin receptor. *Nature* **2008**, *453*, 481–488.

298. Zhao, M.; Zhang, N.-Y.; Zurawel, A.; Hansen, K. C.; Liu, C.-W. Degradation of some polyubiquitinated proteins requires an intrinsic proteasomal binding element in the substrates. *J. Biol. Chem.* **2010**, *285*, 4771–4780.

299. Peth, A.; Uchiki, T.; Goldberg, A. L. ATP-dependent steps in the binding of ubiquitin conjugates to the 26S proteasome that commit to degradation. *Mol. Cell* **2010**, *40*, 671–681.

300. Inobe, T.; Fishbain, S.; Prakash, S.; Matouschek, A. Defining the geometry of the two-component proteasome degron. *Nat. Chem. Biol.* **2011**, *7*, 161–167.

301. Schreiner, P.; Chen, X.; Husnjak, K.; Randles, L.; Zhang, N.; Elsasser, S., et al. Ubiquitin docking at the proteasome through a novel pleckstrin-homology domain interaction. *Nature* **2008**, *453*, 548–552.

302. Wang, Q.; Young, P.; Walters, K. J. Structure of S5a bound to monoubiquitin provides a model for polyubiquitin recognition. *J. Mol. Biol.* **2005**, *348*, 727–739.

303. Wang, Q.; Goh, A. M.; Howley, P. M.; Walters, K. J. Ubiquitin recognition by the DNA repair protein hHR23a. *Biochemistry* **2003**, *42*, 13529–13535.

304. Kim, I.; Mi, K.; Rao, H. Multiple interactions of Rad23 suggest a mechanism for ubiquitylated substrate delivery important in proteolysis. *Mol. Biol. Cell* **2004**, *15*, 3357–3365.

305. Hiyama, H.; Yokoi, M.; Masutani, C.; Sugasawa, K.; Maekawa, T.; Tanaka, K., et al. Interaction of hHR23 with S5a. The ubiquitin-like domain of hHR23 mediates interaction with S5a subunit of 26 S proteasome. *J. Biol. Chem.* **1999**, *274*, 28019–28025.

306. Lander, G. C.; Estrin, E.; Matyskiela, M. E.; Bashore, C.; Nogales, E.; Martin, A. Complete subunit architecture of the proteasome regulatory particle. *Nature* **2012**, *482*, 186–191.

307. Matyskiela, M. E.; Martin, A. Design principles of a universal protein degradation machine. *J. Mol. Biol.* **2013**, *425*, 199–213.

308. Hanna, J.; Hathaway, N. A.; Tone, Y.; Elsasser, S.; Kirkpatrick, D. S.; Leggett, D. S., et al. Deubiquitinating enzyme Ubp6 functions noncatalytically to delay proteasomal degradation. *Cell* **2006**, *127*, 99–111.

309. Lam, Y. A.; Xu, W.; DeMartino, G. N.; Cohen, R. E. Editing of ubiquitin conjugates by an isopeptidase in the 26S proteasome. *Nature* **1997**, *385*, 737–740.

310. Leggett, D. S.; Hanna, J.; Borodovsky, A.; Crosas, B.; Schmidt, M.; Baker, R. T., et al. Multiple associated proteins regulate proteasome structure and function. *Mol. Cell* **2002**, *10*, 495–507.

311. Qiu, X. B.; Ouyang, S. Y.; Li, C. J.; Miao, S.; Wang, L.; Goldberg, A. L. hRpn13/ADRM1/ GP110 is a novel proteasome subunit that binds the deubiquitinating enzyme, UCH37. *EMBO J.* **2006**, *25*, 5742–5753.

312. Thrower, J. S.; Hoffman, L.; Rechsteiner, M.; Pickart, C. M. Recognition of the polyubiquitin proteolytic signal. *EMBO J.* **2000**, *19*, 94–102.

313. Guterman, A.; Glickman, M. H. Complementary roles for Rpn11 and Ubp6 in deubiquitination and proteolysis by the proteasome. *J. Biol. Chem.* **2004**, *279*, 1729–1738.

314. Tomko, R. J., Jr.; Funakoshi, M.; Schneider, K.; Wang, J.; Hochstrasser, M. Heterohexameric ring arrangement of the eukaryotic proteasomal ATPases: implications for proteasome structure and assembly. *Mol. Cell* **2010**, *38*, 393–403.

315. Gillette, T. G.; Kumar, B.; Thompson, D.; Slaughter, C. A.; DeMartino, G. N. Differential roles of the COOH termini of AAA subunits of PA700 (19 S regulator) in asymmetric assembly and activation of the 26 S proteasome. *J. Biol. Chem.* **2008**, *283*, 31813–31822.

316. Tian, G.; Park, S.; Lee, M. J.; Huck, B.; McAllister, F.; Hill, C. P., et al. An asymmetric interface between the regulatory and core particles of the proteasome. *Nat. Struct. Mol. Biol.* **2011**, *18*, 1259–1267.

317. Nickell, S.; Beck, F.; Scheres, S. H. W.; Korinek, A.; Förster, F.; Lasker, K., et al. Insights into the molecular architecture of the 26S proteasome. *Proc. Natl Acad. Sci. U.S.A.* **2009**, *106*, 11943–11947.

318. Lee, C.; Schwartz, M. P.; Prakash, S.; Iwakura, M.; Matouschek, A. ATP-dependent proteases degrade their substrates by processively unraveling them from the degradation signal. *Mol. Cell* **2001**, *7*, 627–637.

319. Barkow, S. R.; Levchenko, I.; Baker, T. A.; Sauer, R. T. Polypeptide translocation by the AAA+ ClpXP protease machine. *Chem. Biol.* **2009**, *16*, 605–612.

320. Lee, C.; Prakash, S.; Matouschek, A. Concurrent translocation of multiple polypeptide chains through the proteasomal degradation channel. *J. Biol. Chem.* **2002**, *277*, 34760–34765.

321. Glynn, S. E.; Martin, A.; Nager, A. R.; Baker, T. A.; Sauer, R. T. Crystal structures of asymmetric ClpX hexamers reveal nucleotide-dependent motions in a AAA+ protein-unfolding machine. *Cell* **2009**, *139*, 744–756.

322. Hersch, G. L.; Burton, R. E.; Bolon, D. N.; Baker, T. A.; Sauer, R. T. Asymmetric interactions of ATP with the AAA+ ClpX6 unfoldase: allosteric control of a protein machine. *Cell* **2005**, *121*, 1017–1027.

323. Song, H. K.; Hartmann, C.; Ramachandran, R.; Bochtler, M.; Behrendt, R.; Moroder, L.; Huber, R. Mutational studies on HslU and its docking mode with HslV. *Proc. Natl. Acad. Sci. U.S.A.* **2000**, *97*, 14103–14108.

324. Martin, A.; Baker, T. A.; Sauer, R. T. Rebuilt AAA+ motors reveal operating principles for ATP-fuelled machines. *Nature* **2005**, *437*, 1115–1120.

325. Kenniston, J. A.; Baker, T. A.; Fernandez, J. M.; Sauer, R. T. Linkage between ATP consumption and mechanical unfolding during the protein processing reactions of an AAA+ degradation machine. *Cell* **2003**, *114*, 511–520.

326. Brannigan, J. A.; Dodson, G.; Duggleby, H. J.; Moody, P. C.; Smith, J. L.; Tomchick, D. R.; Murzin, A. G. A protein catalytic framework with an N-terminal nucleophile is capable of self-activation. *Nature* **1995**, *378*, 416–419.

327. Seemuller, E.; Lupas, A.; Stock, D.; Lowe, J.; Huber, R.; Baumeister, W. Proteasome from *Thermoplasma acidophilum*: a threonine protease. *Science* **1995**, *268*, 579–582.

328. Arendt, C. S.; Hochstrasser, M. Identification of the yeast 20S proteasome catalytic centers and subunit interactions required for active-site formation. *Proc. Natl. Acad. Sci. U.S.A.* **1997**, *94*, 7156–7161.

329. Marques, A. J.; Palanimurugan, R.; Matias, A. C.; Ramos, P. C.; Dohmen, R. J. Catalytic mechanism and assembly of the proteasome. *Chem. Rev.* **2009**, *109*, 1509–1536.
330. Komlosh, A.; Momburg, F.; Weinschenk, T.; Emmerich, N.; Schild, H.; Nadav, E., et al. A role for a novel luminal endoplasmic reticulum aminopeptidase in final trimming of 26S proteasome-generated major histocompatability complex class I antigenic peptides. *J. Biol. Chem.* **2001**, *276*, 30050–30056.
331. Tyedmers, J.; Moegk, A.; Bukau, B. Cellular strategies for controlling protein aggregation. *Nat. Rev. Mol. Cell Biol.* **2010**, *11*, 777–788.
332. Wang, Y.; Mandelkow, E. Degradation of tau protein by autophagy and proteasomal pathways. *Bioch. Soc. Trans.* **2012**, *40*, 644–652.
333. Hegde, A. N.; Upadhya, S. C. Role of ubiquitin-proteasome mediated proteolysis in nervous system disease. *Biochim. Biophys. Acta* **1809**, *2011*, 128–140.
334. Baptista, M. S.; Duarte, C. B.; Maciel, P. Role of the ubiquitin-proteasome system in nervous system function and disease: using *C. elegans* as a dissecting tool. *Cell. Mol. Life Sci.* **2012**, *69*, 2691–2715.
335. Bedford, L.; Hay, D.; Devoy, A.; Paine, S.; Powe, D. G.; Seth, R., et al. Depletion of 26S proteasomes in mouse brain neurons causes neurodegeneration and Lewy-like inclusions resembling human pale bodies. *J. Neurosci.* **2008**, *28*, 8189–8198.
336. Bennett, E. J.; Shaler, T. A.; Woodman, B.; Ryu, K. Y.; Zaitseva, T. S.; Becker, C. H., et al. Global changes to the ubiquitin system in Huntington's disease. *Nature* **2007**, *448*, 704–708.
337. Cheroni, C.; Marino, M.; Tortarolo, M.; Veglianese, P.; De, B. S.; Fontana, E., et al. Functional alterations of the ubiquitin–proteasome system in motor neurons of a mouse model of familial amyotrophic lateral sclerosis. *Hum. Mol. Genet.* **2009**, *18*, 82–96.
338. Tan, Z.; Sun, X.; Hou, F. S.; Oh, H. W.; Hilgenberg, L. G.; Hol, E. M., et al. Mutant ubiquitin found in Alzheimer's disease causes neuritic beading of mitochondria in association with neuronal degeneration. *Cell Death Differ.* **2007**, *14*, 1721–1732.
339. Bertram, L.; Hiltunen, M.; Parkinson, M.; Ingelsson, M.; Lange, C.; Ramasamy, K., et al. Family-based association between Alzheimer's disease and variants in UBQLN1. *N. Engl. J. Med.* **2005**, *352*, 884–894.
340. Matsuura, T.; Sutcliffe, J. S.; Fang, P.; Galjaard, R. J.; Jiang, Y. H.; Benton, C. S., et al. De novo truncating mutations in E6-AP ubiquitin–protein ligase gene (UBE3A) in Angelman syndrome. *Nat. Genet.* **1997**, *15*, 74–77.
341. Crimmins, S.; Jin, Y.; Wheeler, C.; Huffman, A. K.; Chapman, C.; Dobrunz, L. E.; Levey, A., et al. Transgenic rescue of ataxia mice with neuronal-specific expression of ubiquitin-specific protease 14. *J. Neurosci.* **2006**, *26*, 11423–11431.
342. Zhai, Q.; Wang, J.; Kim, A.; Liu, Q.; Watts, R.; Hoopfer, E., et al. Involvement of the ubiquitin–proteasome system in the early stages of Wallerian degeneration. *Neuron* **2003**, *39*, 217–225.
343. Keck, S.; Nitsch, R.; Grune, T.; Ullrich, O. Proteasome inhibition by paired helical filament-tau in brains of patients with Alzheimer's disease. *J. Neurochem.* **2003**, *85*, 115–122.
344. Tai, H.-C.; Serrano-Pozo, A.; Hashimoto, T.; Frosch, M. P.; Spires-Jones, T. L.; Hyman, B. T. The synaptic accumulation of hyperphosphorylated tau oligomers in Alzheimer disease is associated with dysfunction of the ubiquitin-proteasome system. *Am. J. Pathol.* **2012**, *181*, 1426–1435.
345. Lasagna-Reeves, C. A.; Castillo-Carranza, D. L.; Sengupta, U.; Sarmiento, J.; Troncoso, J.; Jackson, G. R.; Kayed, R. Identification of oligomers at early stages of tau aggregation in Alzheimer's disease. *FASEB J.* **2012**, *26*, 1946–1959.
346. Yen, S. S. Proteasome degradation of brain cytosolic tau in Alzheimer's disease. *Int. J. Clin. Exp. Pathol.* **2011**, *4*, 385–402.
347. Dickey, C. A.; Kamal, A.; Lundgren, K.; Klosak, N.; Bailey, R. M.; Dunmore, J., et al. The high-affinity HSP90-CHIP complex recognizes and selectively degrades phosphorylated tau client proteins. *J. Clin. Invest.* **2007**, *117*, 648–658.

348. Wang, Y.; Martinez-Vicente, M.; Kruger, U.; Kaushik, S.; Wong, E.; Mandelkow, E. M., et al. Tau fragmentation, aggregation and clearance: the dual role of lysosomal processing. *Hum. Mol. Genet.* **2009**, *18*, 4153–4170.

349. Ramesh Babu, J.; Lamar Seibenhener, M.; Peng, J.; Strom, A. L.; Kemppainen, R.; Cox, N., et al. Genetic inactivation of p62 leads to accumulation of hyperphosphorylated tau and neurodegeneration. *J. Neurochem.* **2008**, *106*, 107–120.

350. Rubinsztein, D. C. The roles of intracellular protein-degradation pathways in neurodegeneration. *Nature* **2006**, *443*, 780–786.

351. Grune, T.; Botzen, D.; Engels, M.; Voss, P.; Kaiser, B.; Jung, T., et al. Tau protein degradation is catalyzed by the ATP/ubiquitin-independent 20S proteasome under normal cell conditions. *Arch. Biochem. Biophys.* **2010**, *500*, 181–188.

352. Dolan, P. J.; Johnson, G. V. A caspase cleaved form of tau is preferentially degraded through the autophagy pathway. *J. Biol. Chem.* **2010**, *285*, 21978–21987.

353. Gamblin, T. C.; Chen, F.; Zambrano, A.; Abraha, A.; Lagalwar, S.; Guillozet, A. L., et al. Caspase cleavage of tau: linking amyloid and neurofibrillary tangles in Alzheimer's disease. *Proc. Natl. Acad. Sci. U.S.A.* **2003**, *100*, 10032–10037.

354. García-Sierra, F.; Jarero-Basulto, J. J.; Kristofikova, Z.; Majer, E.; Binder, L. I.; Ripova, D. Ubiquitin is associated with early truncation of tau protein at aspartic acid[421] during the maturation of neurofibrillary tangles in Alzheimer's disease. *Brain Pathol.* **2012**, *22*, 240–250.

355. Yang, Y.; Kitagaki, J.; Dai, R.-M.; Tsai, Y. C.; Lorick, K. L.; Ludwig, R. L., et al. Inhibitors of ubiquitin-activating enzyme (E1), a new class of potential cancer therapeutics. *Cancer Res.* **2007**, *67*, 9472–9481.

356. Soucy, T. A.; Smith, P. G.; Milhollen, M. A.; Berger, A. J.; Gavin, J. M.; Adhikari, S., et al. An inhibitor of NEDD8-activating enzyme as a new approach to treat cancer. *Nature* **2009**, *458*, 732–736.

357. Kapuria, V.; Peterson, L. F.; Showalter, H. D. H.; Kirchhoff, P. D.; Talpaz, M.; Donato, N. J. Protein cross-linking as a novel mechanism of action of a ubiquitin-activating enzyme inhibitor with anti-tumor activity. *Biochem. Pharmacol.* **2011**, *82*, 341–349.

358. Pan, Z. Q.; Kentsis, A.; Dias, D. C.; Yamoah, K.; Wu, K. NEDD8 on Cullin: building an expressway to protein destruction. *Oncogene* **2004**, *23*, 1985–1997.

359. Brownell, J. E.; Sintchak, M. D.; Gavin, J. M.; Liao, H.; Bruzzese, F. J.; Bump, N. J., et al. Substrate-assisted inhibition of ubiquitin-like protein-activating enzymes: the NEDD8 E1 inhibitor MLN4924 forms a NEDD8-AMP mimetic in situ. *Mol. Cell* **2010**, *37*, 102–111.

360. Zhao, Y.; Xiong, X.; Jia, L.; Sun, Y. Targeting Cullin-RING ligases by MLN4924 induces autophagy via modulating the HIF1-REDD1-TSC1-mTORC1-DEPTOR axis. *Cell Death Dis.* **2012**, *3*, e386.

361. Dil, K. A.; Kito, K.; Abe, Y.; Shin, R. W.; Kamitani, T.; Ueda, N. NEDD8 protein is involved in ubiquitinated inclusion bodies. *J. Pathol.* **2003**, *199*, 259–266.

362. Scheper, J.; Guerra-Rebollo, M.; Sanclimens, G.; Moure, A.; Masip, I.; Gonzalez-Ruiz, D., et al. Protein-protein interaction antagonists as novel inhibitors of non-canonical poly-ubiquitylation. *PloS One* **2010**, *5*, e11403.

363. Tsukamoto, S.; Takeuchi, T.; Rotinsulu, H.; Mangindaan, R. E. P.; van Soest, R. W. M.; Ukai, K., et al. Leucettamol A: a new inhibitor of Ubc13-Uev1A interaction isolated from a marine sponge, Leucetta aff. microrhaphis. *Bioorg. Med. Chem. Lett.* **2008**, *18*, 6319–6320.

364. Pulvino, M.; Liang, Y.; Oleksyn, D.; DeRan, M.; Van Pelt, E.; Shapiro, J., et al. Inhibition of proliferation and survival of diffuse large B-cell lymphoma cells by a small-molecule inhibitor of the ubiquitin-conjugating enzyme Ubc13-Uev1A. *Blood* **2012**, *120*, 1668–1677.

365. Yamamoto, M.; Okamoto, T.; Takeda, K.; Sato, S.; Sanjo, H.; Uematsu, S., et al. Key function for the Ubc13 E2 ubiquitin-conjugating enzyme in immune receptor signaling. *Nat. Immunol.* **2006**, *7*, 962–970.

366. Kleiger, G.; Saha, A.; Lewis, S.; Kuhlman, B.; Deshaies, R. J. Rapid E2-E3 assembly and disassembly enable processive ubiquitylation of cullin-RING ubiquitin ligase substrates. *Cell* **2009**, *139*, 957–968.

367. Ceccarelli, D. F.; Tang, X.; Pelletier, B.; Orlicky, S.; Xie, W.; Plantevin, V., et al. An allosteric inhibitor of the human cdc34 ubiquitin-conjugating enzyme. *Cell* **2011**, *145*, 1075–1087.
368. Chen, Z.; Pickart, C. M. A 25-kilodalton ubiquitin carrier protein (E2) catalyzes multiubiquitin chain synthesis via lysine 48 of ubiquitin. *J. Biol. Chem.* **1990**, *265*, 21835–21842.
369. Park, Y.; Yoon, S. K.; Yoon, J. B. The HECT domain of TRIP12 ubiquitinates substrates of the ubiquitin fusion degradation pathway. *J. Biol. Chem.* **2009**, *284*, 1540–1549.
370. Poulsen, E. G.; Steinhauer, C.; Lees, M.; Lauridsen, A. M.; Ellgaard, L.; Hartmann-Petersen, R. HUWE1 and TRIP12 collaborate in degradation of ubiquitin-fusion proteins and misframed ubiquitin. *PloS One* **2012**, *7*, e50548.
371. Ko, S.; Kang, G. B.; Song, S. M.; Lee, J. G.; Shin, D. Y.; Yun, J. H., et al. Structural basis of E2-25K/UBB+1 interaction leading to proteasome inhibition and neurotoxicity. *J. Biol. Chem.* **2010**, *285*, 36070–36080.
372. van Leeuwen, F. W.; de Kleijn, D. P.; van den Hurk, H. H.; Neubauer, A.; Sonnemans, M. A.; Sluijs, J. A., et al. Frameshift mutants of beta amyloid precursor protein and ubiquitin-B in Alzheimer's and Down patients. *Science* **1998**, *279*, 242–247.
373. Suzuki, K.; Matsubara, H. Recent advances in p53 research and cancer treatment. *J. Biomed. Biotechnol.* **2011**, *2011*, 978312.
374. Varfolomeev, E.; Vucic, D. Inhibitors of apoptosis proteins: fascinating biology leads to attractive tumor therapeutic targets. *Future Oncol.* **2011**, *7*, 633–648.
375. Shi, D.; Grossman, S. R. Ubiquitin becomes ubiquitous in cancer: emerging roles of ubiquitin ligases and deubiquitinases in tumorigenesis and as therapeutic targets. *Cancer Biol. Ther.* **2010**, *10*, 737–747.
376. Chhangani, D.; Ranjan Jana, N.; Mishra, A. Misfolded proteins recognition strategies of E3 ubiquitin ligases and neurodegenerative diseases. *Mol. Neurobiol.* **2013**, *47*, 302–312.
377. Lonskaya, I.; Shekoyan, A. R.; Hebron, M. L.; Desforges, N.; Algarzae, N. K.; Moussa, C. E.-H. Diminished parkin solubility and co-localization with intraneuronal amyloid-β are associated with autophagic defects in Alzheimer's disease. *J. Alzheimer's Dis.* **2013**, *33*, 231–247.
378. Dawson, T. M.; Dawson, V. L. The role of parkin in familial and sporadic Parkinson's disease. *Mov. Disord.* **2010**, *25*, S32–S39.
379. Tardiff, D. F.; Jui, N. T.; Khurana, V.; Tambe, M. A.; Thompson, M. L.; Chung, C. Y., et al. Yeast reveals a "druggable" Rsp5/Nedd4 network that ameliorates a-synuclein toxicity in neurons. *Science* **2013**, *342*, 979–983.
380. Chung, C. Y.; Khurana, V.; Auluck, P. K.; Tardiff, D. F.; Mazzulli, J. R.; Soldner, F., et al. Identification and rescue of a-synuclein toxicity in Parkinson patient-derived neurons. *Science* **2013**, *342*, 983–987.
381. Graewert, M. A.; Groll, M. Exploiting nature's rich source of proteasome inhibitors as starting points in drug development. *Chem. Commun.* **2012**, *48*, 1364–1378.
382. Cavaletti, G.; Jakubowiak, A. J. Peripheral neuropathy during bortezomib treatment of multiple myeloma: a review of recent studies. *Leuk. Lymphoma* **2010**, *51*, 1178–1187.
383. Mattern, M. R.; Wu, J.; Nicholson, B. Ubiquitin-based anticancer therapy: carpet bombing with proteasome inhibitors vs surgical strikes with E1, E2, E3, or DUB inhibitors. *Biochim. Biophys. Acta* **1823**, *2012*, 2014–2021.
384. Mullally, J. E.; Moos, P. J.; Edes, K.; Fitzpatrick, F. A. Cyclopentenone prostaglandins of the J series inhibit the ubiquitin isopeptidase activity of the proteasome pathway. *J. Biol. Chem.* **2001**, *276*, 30366–30373.
385. Altun, M.; Kramer, H. B.; Willems, L. I.; McDermott, J. L.; Leach, C. A.; Goldenberg, S. J., et al. Activity-based chemical proteomics accelerates inhibitor development for deubiquitylating enzymes. *Chem. Biol.* **2011**, *18*, 1401–1412.
386. Cummins, J. M.; Rago, C.; Kohli, M.; Kinzler, K. W.; Lengauer, C.; Vogelstein, B. Tumour suppression: disruption of HAUSP gene stabilizes p53. *Nature* **2004**, *428*, 486.
387. Li, M.; Brooks, C. L.; Kon, N.; Gu, W. A dynamic role of HAUSP in the p53–Mdm2 pathway. *Mol. Cell* **2004**, *13*, 879–886.

388. Reverdy, C.; Conrath, S.; Lopez, R.; Planquette, C.; Atmanene, C.; Collura, V., et al. Discovery of specific inhibitors of human USP7/HAUSP deubiquitinating enzyme. *Chem. Biol.* **2012**, *19*, 467–477.

389. Ratia, K.; Pegan, S.; Takayama, J.; Sleeman, K.; Coughlin, M.; Baliji, S., et al. A noncovalent class of papain-like protease/deubiquitinase inhibitors blocks SARS virus replication. *Proc. Natl. Acad. Sci. U.S.A.* **2008**, *105*, 16119–16124.

390. Todi, S. V.; Paulson, H. L. Balancing act: deubiquitinating enzymes in the nervous system. *Tr. Neurosci.* **2011**, *34*, 370–382.

391. Kowalski, J. R.; Juo, P. The role of deubiquitinating enzymes in synaptic function and nervous system diseases. *Neural Plastic.* **2012**, *2012*, 892749.

392. Setsuie, R.; Wada, K. The functions of UCH-L1 and its relation to neurodegenerative diseases. *Neurochem. Int.* **2007**, *51*, 105–111.

393. Liu, Z.; Meray, R. K.; Grammatopoulos, T. N.; Fredenburg, R. A.; Cookson, M. A.; Liu, Y., et al. Membrane-associated farnesylated UCH-L1 promotes α-synuclein neurotoxicity and is a therapeutic target for Parkinson's disease. *Proc. Natl. Acad. Sci. U.S.A.* **2009**, *106*, 4635–4640.

394. Lowe, J.; McDermott, H.; Landon, M.; Mayer, R. J.; Wilkinson, K. D. Ubiquitin carboxylterminal hydrolase (PGP 9.5) is selectively present in ubiquitinated inclusion bodies characteristic of human neurodegenerative diseases. *J. Pathol.* **1990**, *161*, 153–160.

395. Setsuie, R.; Wang, Y. L.; Mochizuki, H.; Osaka, H.; Hayakawa, H.; Ichihara, N., et al. Dopaminergic neuronal loss in transgenic mice expressing the Parkinson's disease-associated UCH-L1 I93M mutant. *Neurochem. Int.* **2007**, *50*, 119–129.

396. Mukoyama, M.; Yamazaki, K.; Kikuchi, T.; Tomita, T. Neuropathology of gracile axonal dystrophy (GAD) mouse. An animal model of central distal axonopathy in primary sensory neurons. *Acta Neuropathol.* **1989**, *79*, 294–299.

397. Choi, J.; Levey, A. I.; Weintraub, S. T.; Rees, H. D.; Gearing, M.; Chin, L. S.; Li, L. Oxidative modifications and down-regulation of ubiquitin carboxylterminal hydrolase L1 associated with idiopathic Parkinson's and Alzheimer's diseases. *J. Biol. Chem.* **2004**, *279*, 13256–13264.

398. Castegna, A.; Aksenov, M.; Aksenova, M.; Thongboonkerd, V.; Klein, J. B.; Pierce, W. M., et al. Proteomic identification of oxidatively modified proteins in Alzheimer's disease brain. Part I: creatine kinase BB, glutamine synthase, and ubiquitin carboxy-terminal hydrolase L-1. *Free Radic. Biol. Med.* **2002**, *33*, 562–571.

399. Gong, B.; Cao, Z.; Zheng, P.; Vitolo, O. V.; Liu, S.; Staniszewski, A., et al. Ubiquitin hydrolase Uch-L1 rescues beta-amyloid-induced decreases in synaptic function and contextual memory. *Cell* **2006**, *126*, 775–788.

400. Zhang, M.; Deng, Y.; Luo, Y.; Zhang, S.; Zou, H.; Cai, F., et al. Control of BACE1 degradation and APP processing by ubiquitin carboxylterminal hydrolase L1. *J. Neurochem.* **2012**, *120*, 1129–1138.

401. Mermerian, A. H.; Case, A.; Stein, R. L.; Cuny, G. D. Structure-activity relationship, kinetic mechanism, and selectivity for a new class of ubiquitin C-terminal hydrolase-L1 (UCH-L1) inhibitors. *Bioorg. Med. Chem. Lett.* **2007**, *17*, 3729–3732.

402. Zhang, X.; Zhou, J.; Chin, M. H.; Schepmoes, A. A.; Petyuk, V. A.; Weitz, K. K., et al. Region-specific protein abundance changes in the brain of MPTP-induced Parkinson's disease mouse model. *J. Proteome Res.* **2010**, *9*, 1496–1509.

403. Williams, A. J.; Paulson, H. L. Polyglutamine neurodegeneration: protein misfolding revisited. *Trends Neurosci.* **2008**, *31*, 521–528.

404. Warrick, J. M.; Morabito, L. M.; Bilen, J.; Gordesky-Gold, B.; Faust, L. Z.; Paulson, H. L.; Bonini, N. M. Ataxin-3 suppresses polyglutamine neurodegeneration in Drosophila by a ubiquitin-associated mechanism. *Mol. Cell* **2005**, *18*, 37–48.

405. Ballinger, C. A.; Connell, P.; Wu, Y.; Hu, Z.; Thompson, L. J.; Yin, L. Y.; Patterson, C. Identification of CHIP, a novel tetratricopeptide repeat-containing protein that interacts with heat shock proteins and negatively regulates chaperone functions. *Mol. Cell. Biol.* **1999**, *19*, 4535–4545.

406. Connell, P.; Ballinger, C. A.; Jiang, J.; Wu, Y.; Thompson, L. J.; Hohfeld, J.; Patterson, C. The co-chaperone CHIP regulates protein triage decisions mediated by heat-shock proteins. *Nat. Cell Biol.* **2001**, *3*, 93–96.
407. Murata, S.; Minami, Y.; Minami, M.; Chiba, T.; Tanaka, K. CHIP is a chaperone-dependent E3 ligase that ubiquitylates unfolded protein. *EMBO Rep.* **2001**, *2*, 1133–1138.
408. Scheufler, C.; Brinker, A.; Bourenkov, G.; Pegoraro, S.; Moroder, L.; Bartunik, H., et al. Structure of TPR domain-peptide complexes: critical elements in the assembly of the Hsp70-Hsp90 multichaperone machine. *Cell* **2000**, *101*, 199–210.
409. Hoehfeld, J.; Cyr, D. M.; Patterson, C. From the cradle to the grave: molecular chaperones that may choose between folding and degradation. *EMBO Rep.* **2001**, *2*, 885–890.
410. Kundrat, L.; Regat, L. Balance between folding and degradation for Hsp90-dependent client proteins: a key role for CHIP. *Biochemistry* **2010**, *49*, 7428–7438.
411. Stankiewicz, M.; Nikolay, R.; Rybin, V.; Mayer, M. P. CHIP participates in protein triage decisions by preferentially ubiquitinating Hsp70-bound substrates. *FEBS J.* **2010**, *277*, 3353–3367.
412. Meacham, G. C.; Patterson, C.; Zhang, W.; Younger, J. M.; Cyr, D. M. The Hsc70 co-chaperone CHIP targets immature CFTR for proteasomal degradation. *Nat. Cell Biol.* **2001**, *3*, 100–105.
413. Min, J. N.; Whaley, R. A.; Sharpless, N. E.; Lockyer, P.; Portbury, A. L.; Patterson, C. CHIP deficiency decreases longevity, with accelerated aging phenotypes accompanied by altered protein quality control. *Mol. Cell. Biol.* **2001**, *28*, 4018–4025.
414. Jana, N. R.; Dikshit, P.; Goswami, A.; Kotliarova, S.; Murata, S.; Tanaka, K.; Nukina, N. Co-chaperone CHIP associates with expanded polyglutamine protein and promotes their degradation by proteasomes. *J. Biol. Chem.* **2005**, *280*, 11635–11640.
415. Miller, V. M.; Nelson, R. F.; Gouvion, C. M.; Williams, A.; Rodriguez-Lebron, E.; Harper, S. Q., et al. CHIP suppresses polyglutamine aggregation and toxicity *in vitro* and *in vivo*. *J. Neurosci.* **2005**, *25*, 9152–9161.
416. Takayama, S.; Sato, T.; Krajewski, S.; Kochel, K.; Irie, S.; Millan, J. A.; Reed, J. C. Cloning and functional analysis of BAG-1: a novel Bcl-2-binding protein with anti-cell death activity. *Cell* **1995**, *80*, 279–284.
417. Sroka, K.; Voigt, A.; Deeg, S.; Reed, J. C.; Schulz, J. B.; Baehr, M.; Kermer, P. BAG1 modulates huntingtin toxicity, aggregation, degradation, and subcellular distribution. *J. Neurochem.* **2009**, *111*, 801–807.
418. Williams, A. J.; Knutson, T. M.; Colomer Gould, V. F.; Paulson, H. L. In vivo suppression of polyglutamine neurotoxicity by C-terminus of Hsp70-interacting protein (CHIP) supports an aggregation model of pathogenesis. *Neurobiol. Dis.* **2009**, *33*, 342–353.
419. Matsumoto, M.; Yada, M.; Hatakeyama, S.; Ishimoto, H.; Tanimura, T.; Tsuji, S., et al. Molecular clearance of ataxin-3 is regulated by a mammalian E4. *EMBO J.* **2004**, *23*, 659–669.
420. Al-Ramahi, I.; Lam, Y. C.; Chen, H.-K.; de Gouyon, B.; Zhang, M.; Perez, A. M., et al. CHIP protects from the neurotoxicity of expanded and wild-type ataxin-1 and promotes their ubiquitination and degradation. *J. Biol. Chem.* **2006**, *281*, 26714–26724.
421. Fernandez-Funez, P.; Nino-Rosales, M. L.; de Gouyon, B.; She, W. C.; Luchak, J. M.; Martinez, P., et al. Identification of genes that modify ataxin-1-induced neurodegeneration. *Nature* **2000**, *408*, 101–106.
422. Choi, J. Y.; Ryu, J. H.; Kim, H. S.; Park, S. G.; Bae, K.-H.; Kang, S., et al. Co-chaperone CHIP promotes aggregation of ataxin-1. *Mol. Cell. Neurosci.* **2007**, *34*, 69–79.
423. Shi, Y.; Wang, J.; Li, J. -D.; Ren, H.; Guan, W.; He, M., et al. Identification of CHIP as a novel causative gene for autosomal recessive cerebellar ataxia. *PloS One* **2013**, *8*, e81884.
424. Shin, Y.; Klucken, J.; Patterson, C.; Hyman, B. T.; McLean, P. J. The cochaperone carboxyl terminus of Hsp70-interacting protein (CHIP) mediates alpha-synuclein degradation decisions between proteasomal and lysosomal pathways. *J. Biol. Chem.* **2005**, *280*, 23727–23734.
425. Tetzlaff, J. E.; Putcha, P.; Outeiro, T. F.; Ivanov, A.; Berezovska, O.; Hyman, B. T.; McLean, P. J. CHIP targets toxic alpha-synuclein oligomers for degradation. *J. Biol. Chem.* **2008**, *283*, 17962–17968.

426. Kalia, L. V.; Kalia, S. K.; Chau, H.; Lozano, A. M.; Hyman, B. T.; McLean, P. J. Ubiquitinylation of α-synuclein by carboxyl terminus Hsp70-interacting protein (CHIP) is regulated by Bcl-2-associated athanogene 5 (BAG5). *PLoS ONE* **2011**, *6*, e14695.
427. Arakawa, A.; Handa, N.; Ohsawa, N.; Shida, M.; Kigawa, T.; Hayashi, F., et al. The C-terminal BAG domain of BAG5 induces conformational changes of the Hsp70 nucleotide-binding domain for ADP-ATP exchange. *Structure* **2010**, *18*, 309–319.
428. Choi, J. S.; Cho, S.; Park, S. G.; Park, B. C.; Lee, D. H. Cochaperone CHIP associates with mutant Cu/Zn-superoxide dismutase proteins linked to familial amyotrophic lateral sclerosis and promotes their degradation by proteasomes. *Biochem. Biophys. Res. Commun.* **2004**, *321*, 574–583.
429. Choi, J.-S.; Lee, D. H. CHIP promotes the degradation of mutant SOD1 by reducing its interaction with VCP and S6/S6' subunits of 26S proteasome. *Animal Cells Syst.* **2010**, *14*, 1–10.
430. Gamerdinger, M.; Carra, S.; Behl, C. Emerging roles of molecular chaperones and cochaperones in selective autophagy: focus on BAG proteins. *J. Mol. Med.* **2011**, *89*, 1175–1182.
431. Petrucelli, L.; Dickson, D.; Kehoe, K.; Taylor, J.; Snyder, H.; Grover, A., et al. CHIP and Hsp70 regulate tau ubiquitination, degradation and aggregation. *Human Mol. Genet.* **2004**, *13*, 703–714.
432. Hatakeyama, S.; Matsumoto, M.; Kamura, T.; Murayama, M.; Chui, D.-H.; Planel, E., et al. U-box protein carboxyl terminus of Hsc70-interacting protein (CHIP) mediates polyUbiquitylation preferentially on four-repeat Tau and is involved in neurodegeneration of tauopathy. *J. Neurochem.* **2004**, *91*, 299–307.
433. Sahara, N.; Murayama, M.; Mizoroki, T.; Urushitani, M.; Imai, Y.; Takahashi, R., et al. In vivo evidence of CHIP up-regulation attenuating tau aggregation. *J. Neurochem.* **2005**, *94*, 1254–1263.
434. Dickey, C. A.; Yue, M.; Lin, W.-L.; Dickson, D. W.; Dunmore, J. H.; Lee, W. C., et al. Deletion of the ubiquitin ligase CHIP leads to the accumulation, but not the aggregation, of both endogenous phosphor and caspase-3-cleaved tau species. *J. Neurosci.* **2006**, *26*, 6985–6996.
435. Zhang, Y.-J.; Xu, Y.-F.; Liu, X.-H.; Li, D.; Yin, J.; Liu, Y.-H., et al. Carboxyl terminus of heat-shock cognate 70-interacting protein degrades tau regardless its phosphorylation status without affecting the spatial memory of the rats. *J. Neural Transm.* **2008**, *115*, 483–491.
436. Jinwal, U. K.; Akoury, E.; Abisambra, J. F.; O'Leary, J. C., III.; Thompson, A. D.; Blair, L. J., et al. Imbalance of Hsp70 family variants fosters tau accumulation. *FASEB J.* **2013**, *27*, 1–10.
437. Sakagami, Y.; Kudo, T.; Tanimukai, H.; Kanayama, D.; Omi, T.; Horiguchi, K., et al. Involvement of endoplasmic reticulum stress in tauopathy. *Biochem. Biophys. Res. Commun.* **2013**, *430*, 500–504.
438. Oddo, S.; Caccamo, A.; Tseng, B.; Cheng, D.; Vasilevko, V.; Cribbs, D. H.; LaFerla, F. M. Blocking Aβ42 accumulation delays the onset and progression of tau pathology via the C terminus of heat shock protein70-interacting protein: a mechanistic link between Aβ and tau pathology. *J. Neurosci.* **2008**, *28*, 12163–12175.
439. Morishima, Y.; Wang, A. M.; Yu, Z.; Pratt, W. B.; Osawa, Y.; Lieberman, A. P. CHIP deletion reveals functional redundancy of E3 ligases in promoting degradation of both signalling proteins and expanded glutamine proteins. *Human Mol. Gen.* **2008**, *17*, 3942–3952.
440. Rodrıguez-Navarro, J. A.; Gomez, A.; Rodal, I.; Perucho, J.; Martinez, A.; Furio, V., et al. Parkin deletion causes cerebral and systemic amyloidosis in human mutated tau overexpressing mice. *Human Mol. Genet.* **2008**, *17*, 3128–3143.
441. Perucho, J.; Casarejos, M. J.; Rubio, I.; Rodriguez-Navarro, J. A.; Gómez, A.; Ampuero, I., et al. The effects of parkin suppression on the behaviour, amyloid processing, and cell survival in APP mutant transgenic mice. *Exp. Neurol.* **2010**, *221*, 54–67.
442. Kumar, P.; Pradhan, K.; Karunya, R.; Ambasta, R. K.; Querfurth, R. W. Cross-functional E3 ligases parkin and C-terminus Hsp70-interacting protein in neurodegenerative diseases. *J. Neurochem.* **2012**, *120*, 350–370.

443. Dickey, C. A.; Koren, J.; Zhang, Y.-J.; Xu, Y.-F.; Jinwal, U. K.; Birnbaum, M. J., et al. Akt and CHIP coregulate tau degradation through coordinated interactions. *Proc. Natl. Acad. Sci. U.S.A.* **2008**, *105*, 3622–3627.
444. Su, C.-H.; Wang, C.-Y.; Lan, K.-H.; Li, C.-P.; Chao, Y.; Lin, H.-C., et al. Akt phosphorylation at Thr308 and Ser473 is required for CHIP-mediated ubiquitination of the kinase. *Cell. Signall.* **2011**, *23*, 1824–1830.
445. Yao, T.; Song, L.; Xu, W.; DeMartino, G. N.; Florens, L.; Swanson, S. K., et al. Proteasome recruitment and activation of the Uch37 deubiquitinating enzyme by Adrm1. *Nat. Cell Biol.* **2006**, *8*, 994–1002.
446. Kirkpatrick, D. S.; Hathaway, N. A.; Hanna, J.; Elsasser, S.; Rush, J.; Finley, D., et al. Quantitative analysis of in vitro ubiquitinated cyclin B1 reveals complex chain topology. *Nat. Cell Biol.* **2006**, *8*, 700–710.
447. Peth, A.; Besche, H. C.; Goldberg, A. L. Ubiquitinated proteins activate the proteasome by binding to Usp14/Ubp6, which causes 20S gate opening. *Mol. Cell* **2009**, *36*, 794–804.
448. Peth, A.; Kukushkin, N.; Bossé, M.; Goldberg, A. L. Ubiquitinated proteins activate the proteasomal ATPases by binding to Usp14 or Uch37 homologs. *J. Biol. Chem.* **2013**, *288*, 7781–7790.
449. Wilson, S. M.; Bhattacharyya, B.; Rachel, R. A.; Coppola, V.; Tessarollo, L.; Householder, D. B., et al. Synaptic defects in ataxia mice result from a mutation in Usp14, encoding a ubiquitin-specific protease. *Nat. Gen.* **2002**, *32*, 420–425.
450. Anderson, C.; Crimmins, S.; Wilson, J. A.; Korbel, G. A.; Ploegh, H. L.; Wilson, S. M. Loss of Usp14 results in reduced levels of ubiquitin in ataxia mice. *J. Neurochem.* **2005**, *95*, 724–731.
451. Lappe-Siefke, C.; Loebrich, S.; Hevers, W.; Waidmann, O. B.; Schweizer, M.; Fehr, S., et al. The ataxia (axJ) mutation causes abnormal GABA(A) receptor turnover in mice. *PLoS Genet.* **2009**, *5*, e1000631.
452. Crimmins, S.; Jin, Y.; Wheeler, C.; Huffman, A. K.; Chapman, C.; Dobrunz, L. E., et al. Transgenic rescue of ataxia mice with neuronal-specific expression of ubiquitin-specific protease 14. *J. Neurosci.* **2006**, *26*, 11423–11431.
453. Chen, P. C.; Bhattacharyya, B. J.; Hanna, J.; Minkel, H.; Wilson, J. A.; Finley, D., et al. Ubiquitin homeostasis is critical for synaptic development and function. *J. Neurosci.* **2011**, *31*, 17505–17513.
454. Ponnappan, S.; Palmieri, M.; Sullivan, D. H.; Ponnappan, U. Compensatory increase in USP14 activity accompanies impaired proteasomal proteolysis during aging. *Mechan. Ageing Devel.* **2013**, *134*, 53–59.
455. Lee, B.-H.; Lee, M. J.; Park, S.; Oh, D.-C.; Elsasser, S.; Chen, P. C., et al. Enhancement of proteasome activity by a small-molecule inhibitor of USP14. *Nature* **2010**, *467*, 179–184.
456. Jin, Y. N.; Chen, P.-C.; Watson, J. A.; Walters, B. J.; Phillips, S. E.; Green, K., et al. Usp14 deficiency increases tau phosphorylation without altering tau degradation or causing tau-dependent deficits. *PLoS One* **2012**, *7*, e47884.
457. D'Arcy, P.; Brnjic, S.; Hägg Olofsson, M.; Fryknäs, M.; Lindsten, K.; De Cesare, M., et al. Inhibition of proteasome deubiquitinating activity as a new cancer therapy. *Nat. Med.* **2011**, *17*, 1636–1640.
458. D'Arcy, P.; Linder, S. Proteasome deubiquitinases as novel targets for cancer therapy. *Int. J. Biochem. Cell Biol.* **2012**, *44*, 1729–1738.
459. Kapuria, V.; Peterson, L. F.; Fang, D.; Bornmann, W. G.; Talpaz, M.; Donato, N. J. Deubiquitinase inhibition by small-molecule WP1130 triggers aggresome formation and tumor cell apoptosis. *Cancer Res.* **2010**, *70*, 9265–9276.
460. Seneci, P. Chemical modulators of protein misfolding and neurodegenerative disease. Elsevier, accepted for publication, **2015**.

Unselective Disposal of Cellular Aggregates

Engulf, Devour and Digest to Recycle

4.1 AUTOPHAGY-MEDIATED DEGRADATION OF PROTEIN AGGREGATES

Two main elimination pathways leading to protein elimination in cells must be considered. The *ubiquitin-proteasome system* (*UPS*) [1,2] deals with most regulated proteolytic events on soluble proteins. It usually targets relatively small and short-lived proteins that can access the 20S catalytic subunit of the proteasome [3]. UPS is described in detail in Chapter 3.

Autophagy is a self-degradation process of cellular components—from proteins to whole organelles—known for many years [4]. Its crucial role, though, has been fully appreciated for less than a decade [5,6]. Two non-selective autophagic processes are known, where lysosomal degradation of cellular components is triggered by various mechanisms (Figure 4.1).

The most common is *macroautophagy* (*MA* from now on, Figure 4.1, left [7]). MA is a non-selective cellular process, induced by stress stimuli that include, among others, nutrient stress and hypoxia [8]. In MA, a limited portion of cytoplasm is surrounded by an isolation membrane/phagophore (step 1). Building of the isolation membrane starts at the so-called phagophore assembly site (PAS). The isolation membrane gradually grows and curves around a cytoplasmic region (step 2). The growing membrane evolves into a double-layer barrier, confining the soon-to-be-degraded cytoplasmic region into an organelle called autophagosome (AP, step 3 [9]). The outer AP membrane then merges with either endosomes (ES), creating an amphisome [10], and/or with lysosomes (LS, step 4), creating an early autolysosome that initiates the digestion of cellular components by lysosomal enzymes [11]. Large autolysosomes eventually evolve into more densely packed LSs (step 5), which complete cytoplasmic protein and/or organelle degradation [11] (step 6). *Microautophagy* (Figure 4.1,

Molecular Targets in Protein Misfolding and Neurodegenerative Disease. DOI: 10.1016/B978-0-12-800186-8.00004-3

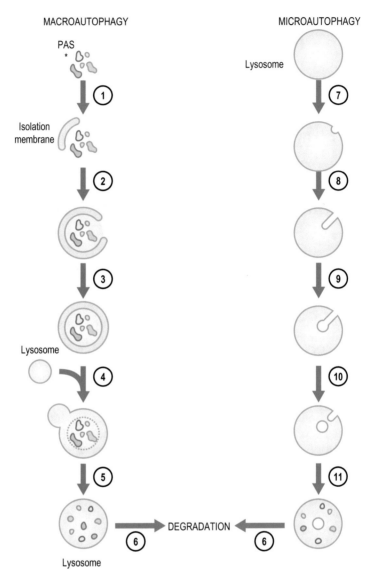

FIGURE 4.1   Macroautophagy and microautophagy: main features.

right [12]) is a poorly characterized cellular process initiated by direct engulfment of cytoplasmic portions—smaller than in MA—by the lysosomal membrane, leading to their rapid degradation [13]. Namely, the initial invagination assumes a controlled tubular form (autophagic tube, step 7 [14]) that grows steadily towards the LS center in an ATP-dependent manner (step 8 [15]), increasing the lipid content and reducing the relative content of intra-membranous proteins. Vesicle formation and expansion (step 9) [12] then take place at the top of the autophagic tube, leading ultimately to its detachment from the tube itself (step 10) in an enzyme-assisted process [16]. Once free from the tube, microautophagy vesicles move freely into the LS (step 11) and are degraded as their macrocounterparts in MA (step 6, Figure 4.1) [17].

Due to its impact on neurodegenerative diseases (NDDs) in general and tauopathies in particular, MA is the autophagic process mainly discussed here. MA is considered a non-selective, starvation-responding process aimed at replenishing the reservoir of nutrients in cells under stressful environmental conditions. *Selective autophagy* is a group of misfolded protein- and organelle-specific autophagic processes that may dispose, *inter alia*, of sequence-specific proteins [18], mitochondria, viruses, and ribosomes [19]. In particular, *aggrephagy* [20,21] is a misfolded protein-specific process with strong implications on NDDs in general, and on tauopathies in particular. Aggrephagy is covered in detail in Chapter 5.

Non-selective autophagy deals with long-lived proteins and insoluble/UPS-resistant aggregates [22]. UPS and autophagy are interconnected, as the impairment of either one impacts on the other [21,23,24]. UPS, non-selective and selective autophagy processes manage the turnover for the vast majority of human proteins—including tau [25].

Five major MA steps—and the putative targets implied—can be identified. *Initiation* of autophagy, *nucleation* to form phagophores/isolation membranes, *expansion* to form APs, and finally AP *maturation* and *fusion* to yield fully degradative autolysosomes [26–28]. The process is widely studied in yeast cells, where at least 37 autophagy-related (Atg) proteins are identified and characterized [29]. Information about mammalian MA as a process and the proteins involved in its progression are much less abundant.

Autophagy commences at the PAS [30]. Yeast cells start either constitutive (cytosol-to-vacuole transport, Cvt pathway [31], nutrient-rich conditions) or stress-induced autophagy (starvation [32]) at PAS. Mammalian cells lack constitutive autophagy but show an autophagy-promoting site for stress-induced MA, which is located either at the endoplasmic reticulum (ER)–mitochondrial junction [33] or at the ER–Golgi intermediate compartment (ERGIC [34]). Remarkably, in mammalian cells phagophores are formed and evolve to other autophagic organelles throughout the cytoplasm, through several local PAS [26]. Subsequently, those organelles are transported to the autophagy-promoting site location by motor proteins—see below.

In yeast cells, autophagy initiation proceeds through recruitment of the Atg1 complex (an Atg1-Atg13-Atg17-Atg29-Atg31 dimer [35]) to the PAS [36], recruitment of Atg9-containing small vesicles [37] (corresponding to mammalian mAtg9), vesicle clustering on the Atg1 complex [35], and soluble N-ethylmaleimide-sensitive factor (NSF) attachment protein receptor (SNARE [38])-mediated vesicle fusion [39]. Vesicle fusion originates a cup-shaped phagophore that grows further by using material from ER [40], mitochondria [41], and the plasma membrane [42].

The assembly of vesicles and cellular material into pre-APs in mammalian cells is promoted by several molecular complexes. Among them, the ULK complex is functionally related to the Atg1 complex in yeast [43]. The ULK complex is composed of uncoordinated UNC-51-like kinases 1 and 2 (ULK1/2 [44], corresponding to Atg1 in yeast), the Ser-Thr kinase ATG13 [45] (corresponding to Atg13 in yeast), the Atg13-binding protein Atg101 [46] (no counterparts in yeast), and the 200kDa focal adhesion kinase family-interacting protein (FIP200 [47], functionally related to Atg17 in yeast). The role of the ULK complex in the creation of a phagophore mirrors the functions of the yeast Atg1 complex, although significant differences exist between the components of each complex [48]. Its structure and main functions are shown in Figure 4.2.

In basal conditions, the autophagy-regulating mammalian target of rapamycin/mTOR signaling complex 1 (mTORC1) is bound to the inactive form of the ULK complex and prevents its activation through the establishment of a phosphorylation pattern on ULK1, ATG13, and FIP200 (Figure 4.2, top [45]). Autophagy induction by external stimuli (nutrient deprivation, hypoxia, etc., step 1a) causes the release of mTORC1 from the ULK complex (step 1b) [45]. The subsequent activation of the ULK complex is triggered by the phosphorylation of ULK1, by AMP-activated protein kinase (AMPK) (step 2b [49]), and entails the spatial rearrangement of the ULK complex, and the phosphorylation of ATG13, FIP200, and ULK1 itself (autophosphorylation) by activated ULK1 [50] (step 2a). Translocation of activated ULK from the cytosol to pre-autophagosomal isolation membranes creates a local PAS [51] (step 3), and phosphorylates a component of another key autophagy complex (step 4, Figure 4.2, see also Figure 4.3). The ULK complex is negatively regulated by insulin and growth factor signaling through class I phosphatidylinositol-3-kinase (PI3K)-Akt kinases [52].

Autophagy initiation is influenced by the concentration of phosphatidylinositol-3-phosphate (PI3P), a promoter of membrane nucleation [53]. Yeast cells have two PI3P-producing complexes named class III PI3K complex I and II [54]. They both contain the PI3K vacuolar protein sorting protein 34 (Vps34 [55]), the Vps34-binding adaptor protein Atg6/Vsp30 [56], and the Vps34 regulatory Ser-Thr kinase Vsp15 [57]. The complex I is autophagy-specific due to Atg14 [58], an Atg17-, Atg6/Vps30-, and Vps34-binding protein responsible for the localization of PI3K complex I at the yeast PAS [59]. A similar mammalian complex, named class III PI3K complex, relies on the Atg6/Vps30 ortholog Bcl-2-homology (BH)-3 domain only protein beclin 1 [53] as a molecular scaffold. It contains hVPS34/PI3K [60], the Atg14-like barkor protein [61], and the Vps15-like, Vps34-regulatory protein kinase p150/hVPS15 [62]. Its structure and main functions are shown in Figure 4.3.

The class III PI3K complex is bound to the microtubules (MTs) *via* the interaction between autophagy/beclin 1 regulator 1 (AMBRA1 [63]) and dynein motor proteins under physiological, non-autophagic conditions

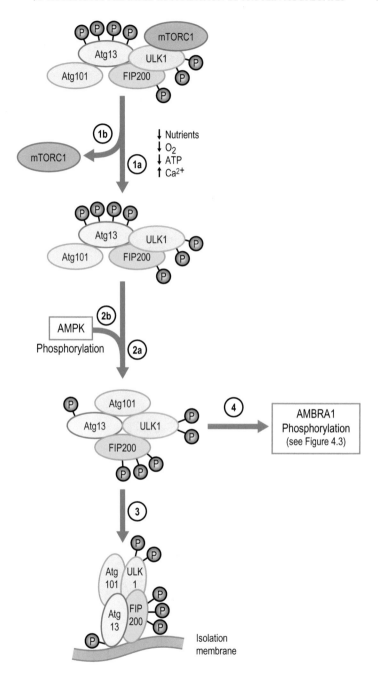

FIGURE 4.2   The ULK complex: structure and roles in autophagy initiation.

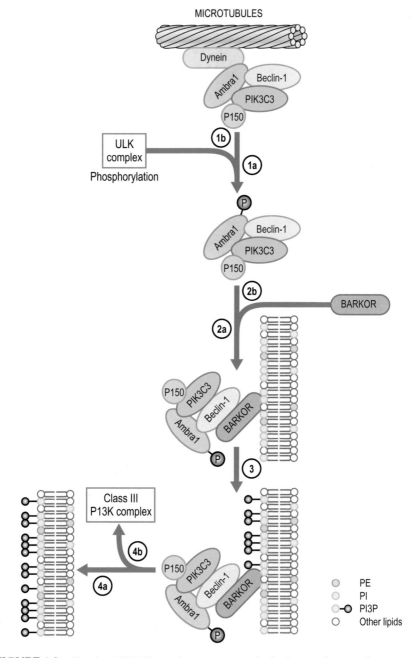

FIGURE 4.3   The class III PI3K complex: structure and roles in autophagy nucleation.

(Figure 4.3, top). Phosphorylation of AMBRA1 by ULK1 (step 1b) weakens its bond with the cytoskeleton, causes the release of the AMBRA1-class III PI3K complex from the cytoskeleton (step 1a), and its barkor-assisted relocation to the local PAS (steps 2a and 2b) [64]. Membrane-anchored PI3K now starts *nucleation* of the phagophore [53] by phosphorylation of phosphatidylinositol (PI, yellow circles), ensuring autophagy-promoting PI3P production on the isolation membrane (step 3). Eventually, the whole phagophore is highly PI3Pylated (step 4a), and the class III PI3K complex is lost from the membrane (step 4b, Figure 4.3).

Interestingly, AMBRA1 positively regulates ULK1 by binding to the E3 ubiquitin (UBQ) ligase tumor necrosis factor/TNF receptor associated factor 6 (TRAF6). Simultaneous ULK1 and TRAF6 binding by AMBRA1 promotes TRAF6-mediated K63-ubiquitination and activation of ULK1 *via* its self-association [65]. The positive feedback loop between the ULK1 and class III PI3K complexes, thus, accelerates autophagy progression once it is induced by external stimuli.

Local PI3P levels, and consequently various autophagy processes, are modulated by beclin 1-driven assembly of PI3K enzyme-containing complexes, but also by PI3P phosphatases [66,67]. Replacement of ATG14L with the beclin 1-binding partner UV irradiation resistance-associated gene (UVRAG [68]) leads to an autophagy-promoting complex that produces a PI3P pool recognized by different effectors [10]. Namely, while ATG14L-containing class III PI3K complexes are nucleation-specific, UVRAG-containing complexes interact with endothelin B1 (SH3GLB1, BIF-1 [69]), acquire the ability to expand phagophore and AP membranes and to induce their curvature [10] through the N-BAR domain of BIF-1 [70]. The replacement of BIF-1 with the RUN domain and cysteine-rich domain-containing, beclin 1-interacting protein rubicon [71] in the UVRAG-containing complex inhibits PI3K activity and autophagy progression [72].

The BH3 domain of beclin 1 promotes its interaction with anti-apoptotic Bcl-2 family members Bcl-2 and Bcl$_{XL}$ [73]. These heterodimers are negative autophagy regulators, as they prevent beclin 1 binding with UVRAG or barkor. The formation of Bcl-2/Bcl$_{XL}$-beclin 1 heterodimers can be prevented by displacement of beclin 1 by pro-apoptotic Bcl-2 family members (e.g., Bad and BNIP3 [74]), or by post-translational modification (PTM) of beclin 1 (e.g., phosphorylation on the BH3 domain of beclin 1 by death-associated protein kinase (DAPK) [75], K63-ubiquitination by TRAF6 [76]). Conversely, phosphorylation on the autophagy-relevant evolutionary conserved domain (ECD) of beclin 1 by the protein kinase B (PKB, Akt) negatively regulates autophagy [77]. Beclin 1 interacts also with high mobility group box 1 (HMGB1 [78]) and PTEN-induced putative kinase 1 (PINK1 [79]), which induce autophagy, and with the inositol-1,4,5-triphosphate receptor (IP$_3$R [80]), which inhibits autophagy. Positive regulation-induction of autophagy by mTORC1 will be treated in detail later.

After initiation and nucleation, *expansion* adds membrane material—lipids and proteins—to the pre-AP. As previously mentioned, the transport of "building material" happens through the action of *trans*-membrane protein carrier mAtg9-containing vesicles [81], which shuttle between the *trans*-Golgi network, late ESs and pre-APs in a largely unclarified process [82]. Mammalian ATG9, as its yeast counterpart, coordinates the transport and incorporation of membrane material into the phagophore [83]. In particular, vesicular mammalian ATG9 displays ULK complex-dependent cycling between *trans*-Golgi network and the ES [84].

Elongation mostly happens through the action of two UBQ-like (UBL) conjugating systems [85], which are extremely conserved in yeast and mammals. The first entails the activation of the UBL protein Atg12 [86] through the E1/UBQ-activating enzyme-like Atg7 that, after homodimerization (step 1a, Figure 4.4) [87], establishes a thioester bond between a Cys residue in the Atg7 active site and the Atg12 C-terminal Gly186 (step 1b).

A *trans*-thiolation reaction involving the E2/UBQ-conjugating enzyme-like Atg10 (step 2a) causes the release of homodimeric Atg7 (step 2b) and the transfer of C-terminal bound Atg12 to a Cys group on Atg10 [88] (step 2c). Breakage of the high-energy thioester bond between Atg10 and Atg12 releases Atg10 (step 3b) and provides the energy to connect the incoming E3/UBQ ligase autophagy protein Atg5 (step 3a) [88] with Atg12 *via* an isopeptide bond between the C-terminal Gly186 of Atg12 and the ε-amine function of the Lys149 residue in Atg5 [89] (step 3c).

Atg5 contains a membrane-binding domain, but its constitutive covalent conjugation with Atg12 prevents membrane association and provides a regulation mechanism for *in vivo* Atg5-membrane binding [90]. Atg12–Atg5 conjugates then interact with either Atg16 in yeast or Atg16L1 in humans (from now on "Atg16" to indicate Atg16 proteins from yeast and mammals). After "Atg16" dimerization (step 4a), a non-covalent interaction takes place between the N-terminal domain of "Atg16" and the opposite site of Atg5 with respect to Atg12 [91] (step 4b). The Atg5–Atg12/"Atg16" complex has a 2:2:2/dimeric stoichiometry [92], which may even grow further to multimeric, ≈800kDa species [93]. Atg5–Atg12/"Atg16" conjugation and its subsequent multimerization enables the ternary complex to specifically associate to the pre-autophagosomal membrane (step 5, shown for a dimer Atg5–Atg12/"Atg16" complex, Figure 4.4) [90]. In fact, Atg5–Atg12/"Atg16" complexes have a preference for PI3P-containing membranes [54], positioning themselves on the outer surface of the isolation membrane/phagophore [88]. Interestingly, phagophore clustering around di- or multimeric Atg5–Atg12/"Atg16" complexes seems to be important for autophagy [90].

The essential role of Atg5–Atg12/"Atg16" conjugates in AP formation is to be attributed, at least in part, to three UBL domains (the whole Atg12

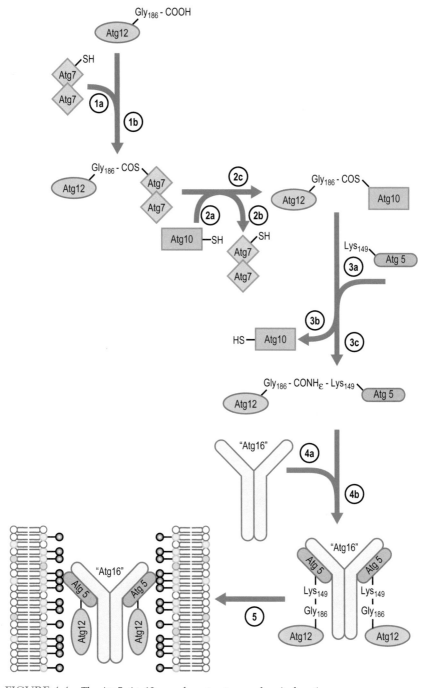

FIGURE 4.4   The Atg5–Atg12 complex: structure and main functions.

and two UBL domains on Atg5) that dynamically recruit and activate factors needed for phagophore expansion and closure [94].

Atg5–Atg12/"Atg16" conjugates are connected to a second UBL-conjugating system, centered onto the UBL Atg8 protein [95]. The single yeast Atg8 protein is replaced in mammalian cells by two subfamilies of Atg8-like UBL proteins (mAtg8s) named microtubule-associated protein 1 light chain 3 (LC3) proteins, and γ-aminobutyric acid (GABA) receptor-associated proteins (GABARAPs) [96]. From now on "Atg8" indicates Atg8 proteins from yeast and mammals.

Figures 4.5 and 4.6 illustrate the structure and roles of "Atg8"-based UBL-conjugating systems, and their interactions with Atg5–Atg12/ "Atg16" conjugates.

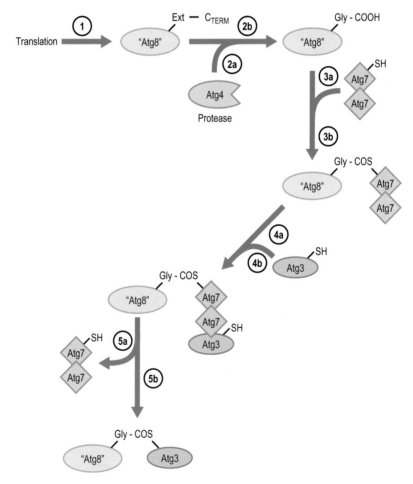

FIGURE 4.5   The "Atg8"–PE conjugate: the path to its synthesis—part 1.

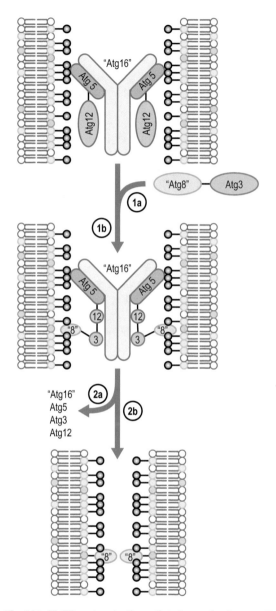

FIGURE 4.6   The "Atg8"–PE conjugate: the path to its synthesis—part 2.

"Atg8" is translated as a C-terminus extended peptide (step 1, Figure 4.5) that is cleaved by the cysteine protease Atg4 [97] (steps 2a, 2b). The exposed C-terminal Gly residue of "Atg8" is then recognized and activated by previously mentioned, homodimeric, E1/UBQ-activating enzyme-like Atg7 (step 3a) [98]. A high-energy thioester bond is established between an Atg7 molecule in the Atg7 homodimer and the C-terminus of

"Atg8" (step 3b). A non-covalent interaction then takes place between the E2/UBQ-conjugating enzyme-like Atg3 (step 4a) with the non-"Atg8"-binding Atg7 monomer (step 4b). The ternary complex eventually promotes the *trans*-thiolation reaction leading to Atg7 release (step 5a) and to "Atg8"–Atg3 connection *via* a thioester bond (step 5b, Figure 4.5) [99]. Similar processes lead to the conjugation of mammalian Atg3 with any mammalian Atg8 isoform.

The high-energy "Atg8"–Atg3 thioester connection promotes the formation of a covalent amide bound between "Atg8" and the membrane phospholipid phosphatidylethanolamine (PE) [98] (see below). "Atg8"–PE conjugation causes the compact "Atg8" proteins to assume an extended conformation [100] that better supports intramolecular interactions [101]. "Atg8"–PE conjugates are key elements in phagophore expansion and AP formation *in vitro* [102] and *in vivo* [103].

The Atg5–Atg12/"Atg16" conjugate influences the processing of "Atg8"–PE precursors, and the location of "Atg8"–PE conjugates (Figure 4.6).

Once bound to the isolation membrane, the Atg5–Atg12/"Atg16" conjugate acts as an E3-like enzyme complex, promoting "Atg8" recruitment and conjugation with PE [104]. In detail, an interaction takes place between the Atg7-binding region of the E2/UBQ-conjugating enzyme-like Atg3 (step 1a) and a region of Atg12 opposite to the Atg5–Atg12 bond (step 1b) [105]. Thus, "Atg8" nears Atg12 and translocates to the phagophore, where "Atg8" is conjugated with PE (green circles) *via* the RING E3-like activity of a structural region between Atg5 and Atg12 (step 1b) [106]. Membrane-bound "Atg8"–PE is thus retained (step 2b), while Atg3, Atg5, Atg12 and "Atg16" proteins are released from the membrane (step 2a, Figure 4.6). It is suggested that "Atg16", in addition to restoring the isolation membrane binding of Atg5, influences the PE lipidation site of "Atg8" [92].

Some mammalian Atg8-like proteins are called LC3 before their cleavage by Atg4. After cleavage they are called LC3-I, while LC3-II are the mammalian counterparts of Atg8–PE yeast conjugates. "Atg8" proteins recognize an "Atg8"-interacting motif (AIM [107], named LC3-interacting region (LIR) [108] for mAtg8s) and act as cargo adaptors, recruiting AIM/LIR-containing proteins to be degraded [109] once they are solidly anchored onto the autophagy membrane. The AIM entails several acidic residues followed by a W/Y-x-x-L/I/V sequence [110]. At least 67 AIM/LIR-containing proteins are known, with limited *in vitro* selectivity among LC3 and GABARAP mAtg8 paralog subfamilies [108]. Surprisingly, both yeast and human Atg12 proteins contain an AIM sequence [111].

A careful estimation of Atg5–Atg12/"Atg16" and "Atg8"–PE/LC3-II levels on the growing phagophore membrane shows that a limited Atg5–Atg12/"Atg16" concentration is sufficient to stimulate efficient "Atg8"–PE/LC3-II conjugation on the membrane. The majority of Atg5–Atg12/"Atg16" is recruited after "Atg8"–PE/LC3-II conjugation, likely due to an interaction between "Atg8" and the Atg12 AIM domain [111]. "Atg16", then, appears not only to promote Atg5–Atg12/"Atg16" dimerization, but also to provide a more structured arrangement to Atg5–Atg12/"Atg16"–"Atg8"–PE/LC3-II networks on the membrane. Namely, "Atg16" promotes the oligomerization of Atg5–Atg12/"Atg16" conjugates that contribute to "Atg8"–PE/LC3-II immobilization through multiple Atg12 AIM domain–"Atg8" interactions [111]. The resulting Atg5–Atg12/"Atg16"–"Atg8"–PE/LC3-II network is arranged into a structured membrane scaffold that covers the convex/outwards surface of the membrane. Such scaffold is flexible enough to allow isolation membrane growth, to modify its curvature and to capture cargo proteins in a progressing autophagy scenario [111].

The concentration of "Atg8"–PE/LC3-II conjugates is higher than the concentration of Atg5–Atg12/"Atg16" conjugates [112], possibly reflecting the double role as structured membrane scaffold component and cargo adaptor of the former conjugates. "Atg8" is the only Atg protein family that remains anchored onto the mature AP membrane [91]. Atg5–Atg12/"Atg16" conjugates are found only on the convex/outwards surface of the isolation membrane/phagophore, and dissociate from the membrane once the mature AP is formed [88,93]. Conversely, "Atg8"–PE/LC3-II conjugates are found on both sides of the membrane [111], as components of the membrane scaffold (convex/outwards surface) [113] and as cargo adaptors (concave/inwards surface) [107]. The former conjugates are released from mature APs, while the latter ones remain anchored onto the inner surface of mature APs.

The AIM domain of Atg12 (responsible for the insertion of "Atg8"–PE/LC3-II in the membrane scaffold), and the AIM domains of cargo proteins (binding to "Atg8"–PE/LC3-II adaptors for autophagic degradation) compete for "Atg8" binding, and the equilibrium among those interactions ensures that both key functions in AP formation and cargo recruitment are carried out by "Atg8"–PE conjugates [111].

Notwithstanding the overall similarity between yeast and mammalian UBL-conjugating systems, a single yeast Atg8 *versus* three subfamilies of mAtg8 paralogs indicate an additional layer of complexity in mammals. As to their structure, a basic N-terminus region is observed in the LC3 family (LC3A, LC3B, LC3B2, LC3C) while the corresponding region is acidic in the GABARAP family (GABARAP, GABARAPL1 GABARAPL2/GATE16, GABARAPL3) [105]. The N-terminus regions of both LC3B and GABARAPL2 promote fusion with the isolation membrane, but the former interaction is mediated by charged amino acids, while the latter stems from hydrophobic interactions [114]. As to the interactome of mAtg8s, they all interact aspecifically *in vitro* with most AIM-containing proteins [108]. *In vivo* subfamily selectivity may be provided by, *inter alia*, tissue expression and intracellular localization [89]. As to the biogenesis of the human AP, LC3 (and LC3B in particular) and GABARAP proteins (and GABARAPL2/GATE16 in particular) are essential for AP maturation [115]. Autophagy impairment results from both small interfering RNA (siRNA)-mediated knockdown of GABARAP proteins in cells stably expressing LC3B, and siRNA-mediated knockdown of LC3 proteins in cells stably expressing GABARAPL2/GATE16. Namely, GABARAP and LC3 knockdown cause respectively an ≈40% and 60% reduction in bulk protein degradation in starvation/autophagy-inducing conditions [115]. Their effect is downstream of the recruitment of mAtg8- and Atg12-based complexes onto the isolation membrane, but involves different steps leading to AP maturation. In fact, LC3 proteins seem to be essential for the elongation of the isolation membrane/phagophore, while GABARAP proteins act downstream of elongation, possibly by sealing the AP structure [115].

Atg12-specific hydrolases for the covalent Atg5–Atg12 bond are unknown [89]. The non-covalent interaction between "Atg8" and PE is regulated by the cysteine protease Atg4, which dynamically recycles the "Atg8"–PE pool during AP biogenesis (Figure 4.7) [116].

Atg4-defective mutants show impaired early recruiting of "Atg8"–PE on the isolation membrane, due to unproductive "Atg8"–PE tethering onto other membranes [117]. Atg4 dynamically increases the percentage of "good-phagophore" tethered "Atg8"–PE and lowers "bad-other endomembrane" tethering [117]. The larger membrane surface area of the latter organelles induces frequent bad "Atg8"–PE tethering events and requires a proofreading function to be performed by Atg4. Namely, Atg4 (step 1a) selectively cleaves endomembrane-bound "Atg8" (step 1c). Released "Atg8" (step 1b) is targeted to PE-rich phagophores (step 2a), where it forms "Atg8"–PE conjugates (step 2b). Previously described clustering of PI3P- and PE-rich isolation membranes around Atg5–Atg12/"Atg16" conjugates (see Figures 4.5 and 4.6) causes the tethering of "Atg8" on PE molecules on both sides of the membranes. Once mature APs are formed (step 3), Atg4 (step 4a) is crucial to release "Atg8" from their outer surface (step 4c), making them ready to fuse with LSs and at the same time recycling the "Atg8" pool for further use (step 4b, Figure 4.7) [118]. As already mentioned, inner surface-bound "Atg8"–PE conjugates remain anchored onto mature APs and act as cargo adaptors.

The structured membrane scaffold based on an Atg5–Atg12/"Atg16"–"Atg8"–PE network is an Atg4 substrate [111]. Atg4 first disrupts the scaffold structure, releasing the Atg5–Atg12/"Atg16" conjugate from the "Atg8"–PE conjugate. "Atg8"–PE, then, becomes accessible for deconjugation by Atg4. Deconjugation of mAtg8s–PE/LC3-II is more complex in humans, due to multiple mAtg8 and mAtg4 paralogs. The single yeast Atg4 protein is replaced in humans by four Atg4 paralogs, with varying specificities. Atg4B (autophagin-1) is the most promiscuous paralog, which cleaves any "Atg8-PE"/LC3-II bond [119,120]. Atg4C (autophagin-3) is the most abundant human isoform, but its role in autophagy is questionable and its

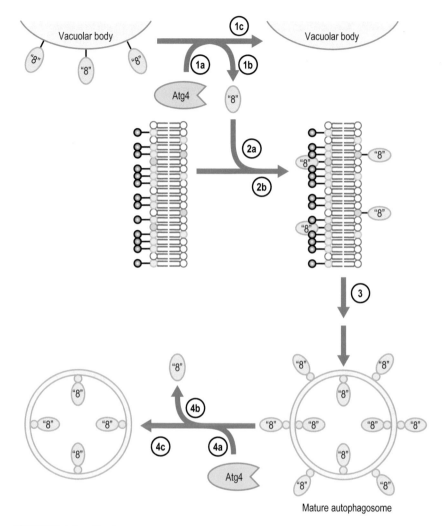

FIGURE 4.7   The "Atg8"–PE conjugate: Atg4-mediated recycling of "Atg8" from non-AP membranes and from outer AP membranes.

specificity for any mAtg8 paralog is unknown [121]. AtgA (autophagin-2) and AtgD (autophagin-4) are both GABARAP-selective proteases, with the former cleaving all GABARAP subfamily members [115] and the latter being GABARAPL1- (preferred) and GABARAPL2/GATE16-selective [122]. The impact of mAtg8s' and mAtg4s' networks, and of their differential expression, localization, and regulation (i.e., inactivation of a catalytic Cys in the active site of Atg4A and Atg4B by reactive oxygen species (ROS) close to the mitochondria [123]) in autophagy requires further studies in order to be fully understood.

AP growth, expansion, and closure will not be discussed further here. Its relevance in selective autophagy/aggrephagy, and the role played in mammals by the LC3-II complex, deserves further comment, which can be found in Chapter 5.

Maturation of functional APs involves the removal of LC3-II from their surfaces. LC3-II exposed onto the cytoplasmic/outward face of AP membranes may then be recycled *via* previously described Atg4-promoted delipidation [119], while AP-internalized LC3-II is degraded at a later stage [113]. Further maturation entails a multistep *fusion* process that starts merging an AP—also called initial autophagic vacuole (iAV)—with one or more endosomal vescicles, to form an amphisome [124]. Amphisomes contain ES-derived proton pumps and lysosomal enzymes, and show a more acidic pH than APs [125]. The final fusion step in MA entails merging of amphisomes with dense LSs, and the formation of fully degradative autolysosomes [126]. Amphisomes and autolysosomes are capable of autophagolytic degradation of their content by lysosomal enzymes, and are called degradative autophagic vacuoles (dAVs) [127].

Yeast cells grow mature APs at the PAS in close vicinity to vacuoles, facilitating fusion and degradation of the resulting autolysosomes [26]. Conversely, mammalian APs in most cell types are formed randomly throughout the cytoplasm, and—otherwise than immobile phagophores— they move bidirectionally along MTs [128]. Centripetal/perinuclear region-directed AP transport is carried out by MT *minus*-end-directed dynein motor proteins [129, 130], while centrifugal AP transport is probably mediated by MT *plus*-end-directed kinesin motor proteins [131,132]. AP initiation, elongation, and closure in primary neurons invariably happen distally from the cell body (Figure 4.8) [133].

MT-bound dynein and kinesin motors are associated with cargo-loaded, almost mature APs located at the distal axonal end. Both motors actively transport APs, likely to secure their movement in the distal compartment, to complete their biogenesis and cargo loading (Figure 4.8, left). In order to fuse with degradative organelles and complete the autophagic process, mature APs must move towards LS-enriched areas close to the microtubule-organizing center (MTOC), located at the neuronal cell soma (Figure 4.8, right) [128]. An unknown event causes the down-regulation/inactivation of anterograde, kinesin-mediated transport and the exit of mature APs from the distal compartment. They migrate towards the proximal axonal end through retrograde, dynein-mediated axonal movement (Figure 4.8, middle) [133,134]. During their migration, APs fuse with ESs and/or LSs, and complete the autophagic process in the vicinity of the cell soma— possibly to optimize the recycling efficiency of amino acidic and lipidic building blocks close to the primary protein synthesis site. AP fusion, conversely, does not happen on the distal tip of MTs, preventing the fusion of immature APs with ESs or LSs [133]. Once fusion with LSs is accomplished

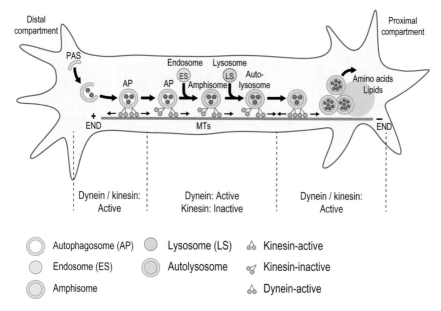

FIGURE 4.8   Transportation of autophagic organelles from the distal to the proximal compartment of axons: from isolation membranes to degradation-competent autolysosomes.

and the MTOC is reached, the acidified, degradation-competent autolysosomes reactivate their kinesin motors and are capable of bidirectional MT movements in the proximal compartment (Figure 4.8, right).

Although examples exist where AP fusion is MT-independent (i.e., starvation-induced autophagy [135]), the significant distance between APs and LSs in neurons [136] makes AP trafficking a crucial step in neuronal autophagy. The retrograde axonal transport of APs may be impaired either by destabilization of the MT structure [137] or by inhibition of dynein-dependent transport [138].

A number of proteins and protein complexes are involved in the regulation of AP maturation and fusion [139], affecting also AP transportation. Beclin 1 regulates maturation of APs and endocytic trafficking *via* complexes containing previously mentioned hVps34 and either UVRAG (upregulation) [135] or rubicon (down-regulation) [140]. The UVRAG-containing complex recruits the fusion machinery on the APs (see below) [141]. Rubicon inhibits autophagy through direct binding to, and sequestration of, UVRAG and other key fusion-promoting proteins [142]. Monomeric Ras-related in brain (Rabs) GTPases [143] are membrane organizer-vesicular transporters. They are either validated (Rab7 [144], Rab11 [145]) or suspected (Rab22 [146], Rab24 [147]) players in late autophagy. In particular, Rab7 is involved in several molecular mechanisms contributing to late-stage autophagy (Figure 4.9) [142].

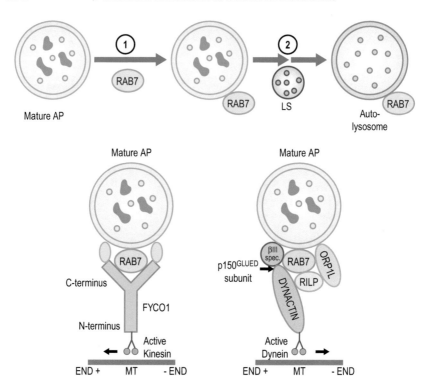

FIGURE 4.9   Rab7-promoted late-stage autophagy.

The attachment of Rab7 to late APs (top, step 1, Figure 4.9) predisposes them to transformation into autolysosomes through several fusion steps with ESs and/or LSs (step 2) [142]. Additionally, Rab7 regulates the AP transport activity on MTs by dynein and kinesin motors. As to anterograde/MT *plus*-end-directed transport, Rab7 and other conjugates on the AP outer surface bind to the C-terminal FYVE and Golgi dynamic domains of the large effector protein FYCO1 (FYVE (Fab1-YotB-Vac1p-EEA1) and coiled-coil domain containing 1 [132]), which is connected to kinesin motor proteins with its N-terminal coiled-coil domain (bottom left, Figure 4.9) [148]. Rab7 also regulates the retrograde/MT *minus*-end-directed transport through the proteins RILP (Rab7-interacting lysosomal protein [149]) and ORPL1 (oxysterol-binding protein-related protein 1 L [150]). The latter protein interacts with AP membrane-bound cholesterol, and in so doing it promotes the interaction between the RILP–Rab7 complex and the dynactin subunit p150[Glued] that stimulates the translocation-essential binding of dynactin with βIII spectrin receptors [151] and the recruitment of dynein motor proteins to APs (bottom right, Figure 4.9) [152].

Rab7 is also involved in the fusion of autophagic bodies with LSs, where the leading role is played by SNAREs [153]. SNAREs are ≈60 evolutionary conserved mediators of membrane fusion in a wealth of processes, including autophagy [154]. They are characterized by the presence of three highly conserved glutamine residues (respectively Qa-, Qb-, and Qc-SNARE subfamilies), or by one highly conserved arginine residue (R-SNARE subfamily) [155]. Among them, the Qb SNARE Vt1b (vesicle transport through interaction with t-SNAREs homolog 1B) and the R-SNARE VAMP8 (vesicle-associated membrane protein 8) are involved in the fusion between LSs and APs [156], or between LSs and structurally related xenophagosomes [157]. The R-SNAREs VAMP3 and VAMP7 are respectively involved in the earlier fusion between APs and ESs to generate amphisomes (VAMP3), and in the later fusion between APs or amphisomes with LSs to generate autolysosomes (VAMP7) [158]. Vt1b, VAMP3, -7, and -8 promote fusion in autophagy through their positioning either on ESs (Vt1b, VAMP3) or on LSs (VAMP7, -8). The Qa SNARE syntaxin 17 (Stx17 [159]) is the only characterized autophagosomal SNARE, which is recruited onto mature APs with an unknown, LC3-II-independent mechanism [160]. Stx17 specifically binds to mature APs, and does not interact with immature APs, providing a checkpoint to avoid premature fusion with LSs [160]. AP-bound Stx17 interacts with the lysosomal SNARE VAMP8 [156], and eventually a ternary complex, including the Qb/Qc SNARE SNAP-29 (synaptosomal-associated protein 29 [161]), is established. AP-LS fusion is mediated by the ternary complex, and causes the degradation of the material enclosed by the autolysosome [160].

The family of four endosomal sorting complexes required for transport 0–III (ESCRT0–III) participates in lysosomal destruction of ubiquitinated proteins [162]. Their knockdown leads to the accumulation of APs in *Drosophila* [163], and of amphisomes in HeLa cells [164]. ESCRT-III, and in particular its component named CHMPB2 (charged multivesicular body protein B2 [165]), contributes to the autophagic fusion processes leading respectively to amphisomes [164] and autolysosomes [163]. CHMPB2 interacts with Rab7 through the HOPS (homotypic vacuole fusion and vacuole protein sorting [166]) complex, and contributes to the recruitment of SNAREs [167]. ESCRT0, and in particular its hepatocyte growth factor-regulated tyrosine kinase substrate protein (Hrs) component, impairs AP–ES and AP–LS fusion, and contributes to AP maturation in a PI3K-dependent manner [168].

The FYVE finger-containing 1-phosphatidylinositol-3-phosphate 5-kinase (PIKfyve) is another endosomal target in late autophagy regulation, possibly connected to reduced fusion between APs and LSs [169]. The multispanning transmembrane lysosomal damage-regulated autophagy modulator (DRAM) both controls AP formation and influences late-stage autophagy progression [170]. Atg22 is a lysosomal amino acid transporter that regulates autophagy through recycling of amino acidic nutrients produced by lysosomal degradation [171].

Ablation of the transmembrane lysosomal-associated membrane proteins 1 and 2 (LAMP-1 and LAMP-2) shows a cell type-specific blockade of autophagy [172–174] due to impairment of AP–LS fusion. The protein isoform LAMP-2A acts as a receptor in chaperone-mediated autophagy (CMA, see also Chapter 5, Figure 5.1) [18], a process specific for proteins bearing a motif recognized by heat shock cognate 70 (Hsc70). The Hsc70–cargo protein complex binds to LAMP-2A, unfolds and crosses the lysosomal membrane to be degraded [175].

Vacuolar ATPases (v-ATPases) provide the acidic pH needed for lysosomal proteases to degrade the autolysosome-trapped proteic material [176]—proton pumping inhibition by v-ATPase inhibitors impairs the autophagic flux [177]. ATPases associated with various cellular activities (AAA ATPases) are intracellular membrane fusion proteins. Among them, NSF (N-ethylmaleimide-sensitive fusion protein [178]) may assist the AP–LS fusion process, as this process is impaired in yeast when the NSF homolog Sec18 is ablated [179], and is slowed in starvation-induced autophagy where NSF activity is attenuated [128]. Suppressor of K^+ transport growth defect 1 (SKD1-Vps4) is required for AP maturation in mammalian [180] and *Drosophila* cells [163], and in AP fusion in yeast [181].

The therapeutical relevance of many mentioned, putative targets in AP maturation and fusion will be more reliably determined when further basic information on these processes is acquired.

4.2 AUTOPHAGY-MEDIATED DEGRADATION OF PROTEIN AGGREGATES IN NDDs

As described in Chapter 3, the failure in disposing of pathologically misfolded/post-translationally misdecorated proteins through UPS [182] causes an increase in their cellular concentration, which eventually leads to their assembly into insoluble aggregates. The persistence of insoluble aggregates—although possibly not the main cause of proteinopathies, often caused by soluble oligomeric forms—in cellular environments, due to the impairment of specific autophagy processes [20], plays a "co-star" role in misfolding diseases.

NDDs are often related to the impairment of aggregate elimination [183,184], and autophagy-related pathologies are well-known contributors to NDDs [185–187]. Inherited forms of *Parkinson's disease (PD)* and parkinsonism are caused by mutations in genes involved in the autophagy/lysosomal pathway [188]. MA accepts both wild-type (WT) and mutated α-synuclein as degradation substrates [189], while CMA accepts only WT α-synuclein [190,191]. Rather, mutant α-synuclein inhibits CMA [191], that is impaired in brains from PD patients [192], and causes a compensatory increase in MA. Depending on the activation pathway, an increase of MA activity in PD may lead to toxic, MA-induced neuronal cell death (mTOR-dependent MA activation [193], see Section 4.3.1) or to neuro-protective/neuronal survival effects (mTOR-independent MA activation [194]). WT α-synuclein inhibits MA [189] *via* inhibition of the Ras-related protein Rab1a, which causes mislocalization of the autophagy protein Atg9 [195]. α-Synuclein-mediated inhibition of PD is strongly linked to mitophagy impairment [196] and mitochondrial dysfunctions [197].

mHtt, the mutant protein responsible for *Huntington disease (HD)*, can be degraded by both MA [198] and CMA [199]. Its degradation is en-hanced (and its aggregation is reduced) following PTMs [200]. A mutation in the Atg7 gene is associated with an earlier disease onset (≈4 years) in HD patient populations [201]. Early-stage HD shows impaired MA and constitutively active CMA [202]. Conflicting observations are reported in terms of autophagy modulation in HD. Impaired cargo recognition—in-cluding mHtt—is claimed to lead to mHtt accumulation and neurotoxicity in mouse HD models and in human cells from HD patients [203], while the formation and degradation of APs is unaffected. Conversely, autophagy inducers provide beneficial effects in HD models [204], including reduced accumulation of huntingtin and mortality in TG mice [205].

Spinal neurons from animal models [206] and *postmortem* samples from *amyotrophic lateral sclerosis (ALS)* patients [207] show increased au-tophagy coupled with UPS impairment. Mitochondrial dysfunctions caused by aggregates of Cu-Zn superoxide dismutase (SOD1) [208] and transactive response DNA binding protein 43 (TDP-43) [209] contribute to ALS, and may induce autophagy [206,210]. TG mice with a pathological mutation in SOD1, recapitulating ALS, show an increased level of MA [211]. Both SOD1 (either WT or mutated) [212] and TDP-43 [213] are MA substrates. Conflicting observations regarding the role of autophagy modulation in ALS models include the decrease of protein aggregation and the increase of neuronal survival caused by autophagy inducers [214,215], and the lack of MA dysfunctions in cells from ALS patients [216]. It is rea-sonable to assume a disease stage-dependent role of autophagy in ALS. In fact, MA stimulation provides beneficial effects in early SOD1-dependent HD mice models [217], while it promotes motor neuron degeneration at a later disease stage [218].

NDDs showing autophagy-dependent impairment in proteostasis include *prion disease*—disease progression increases MA activity [219]; *Lafora disease*—mTOR up-regulation-dependent decrease of mature AP formation [220]; *Machado–Joseph disease*—mitophagy impairment through ataxin-3 mutation (see also Chapter 3 [221]); *Paget disease*—mutated va-losin-containing protein (VCP/p97)-dependent reduction in AP matura-tion and mitophagy [222]; *Perry's syndrome*—AP transport blockage due to dynactin mutations [223]; *spinal and bulbar muscular atrophy*—MT-based transport impairment due to mutations in the dynein motor proteins [224]; and *frontotemporal dementia*—impaired amphisome formation due to mutations in the CHMPB2 gene [165].

The major autophagy markers—Atg5, Atg12, and LC3-II—are associated with both beta amyloid (Aβ) plaques and tau tangles in *Alzheimer's disease* (*AD*) [225], the most common tauopathy. Autophagy deregulation in AD includes accumulation of iAVs and dAVs [226] and dystrophic neurons, containing a large burden of waste proteins [227]. The characterization of *postmortem* AD brains reflects the therapeutical relevance of autophagy-related molecular targets in AD and, more in general, in tauopathies. The autophagy machinery in AD seems to be "stalling" at the level of intermediate AV structures—i.e., APs, autolysosomes—which contain a growing burden of proteins and organelles to be disposed of [228]. Accumulating AVs are resistant to clearance by fusion with LSs, and represent an intracellular reservoir of Aβ [229].

As to Aβ, its clearance by MA is proven by the co-localization of amyloid precursor peptides (APPs) and Aβ peptides with LC3-II-positive APs in APP-overexpressing AD mouse models [230]. MA enhances the translocation of γ-secretase complexes (responsible for the γ-cleavage of APPs yielding aggregation-prone Aβ peptides) to "stalling" APs [231] in a neurotoxic process.

As to tau, hyperphosphorylation (HP) and, in general, any nonphysiological PTM pattern on tau leads to an increased susceptibility to autophagy [232]. A study in cultured primary neurons claims MA/autophagy to be the main processing mechanism for endogenous tau, while UPS inhibition—and subsequent autophagy activation—leads to the reduction of tau levels [233]. Tau aggregate elimination in cultured primary neurons is promoted by chemical inducers of autophagy [233]. Insoluble tau aggregates cannot physically access the site of UPS degradation due to their dimensions [187]. Ubiquitination of a larger tau fragment resulting from caspase cleavage at Asp421 is an early pathological event in neurofibrillary tangle (NFT) formation, and promotes aggregation with full length (FL) tau to form NFTs [234]. Asp421-truncated tau is preferentially degraded through MA/autophagy, when compared with FL tau—preferentially processed by UBQ-independent UPS [235].

A genome-wide analysis highlights a number of dysregulated autophagy mechanisms in AD with respect to control brains [236]. Several mutated and/or dysregulated proteins involved in autophagy are observed in AD. The expression of pro-autophagic protein Atg6/beclin 1 decreases in mice carrying a heterozygous deletion of beclin 1, in aging human brains and in neuronal tissues from AD patients [237]. Beclin 1 depletion impairs MA, disrupts AP/LS-mediated degradation and alters APP metabolism [238]. Mutated presenilin-1 (PS1) leads to early AD onset and to an impaired AP/LS machinery in fibroblasts from AD patients [239]. The UPS- and autophagy-connected E3 ligase parkin shows reduced solubility and co-localizes with HP-tau and Aβ in AVs from AD brains [240].

4.3 MACROAUTOPHAGY—TARGETS

MA is the most relevant form of non-selective autophagy with regard to NDDs, and tauopathies in particular. Most of its complex machinery is used by cells to specifically dispose of protein aggregates (specific autophagy/aggrephagy [20]), and will be detailed in Chapter 5.

Any enzyme or receptor involved in the MA process may be considered a suitable target to modulate autophagic activity—almost invariably inducing it, and causing degradation of insoluble aggregates of misfolded proteins—in an impaired cellular environment. The delicate balance between the interactions among proteins in each complex, and their impact on catalytic and binding activities, has not been targeted yet in drug discovery projects. More information is needed before rationally targeting a single component of an extremely complex biochemical construct. Care should be taken to further elucidate the details of MA; to better understand the therapeutic potential of each involved target—selectivity and relevance; and to avoid impacting on essential mechanisms of neuronal development and activity through aggressive tackling of autophagy-related targets—side effects and toxicity.

Autophagy regulation is a therapeutic option in NDDs [241,242], as long as autophagy is stimulated at the proper stage. Further increasing the number of "stalling" iAVs and dAVs with early autophagy induction may be detrimental, while acting downstream of AP formation and closure, at the AP/ES and/or AP/LS fusion stage, should be often desirable [243]. Conversely, neurodegeneration induced by autophagy up-regulation is sometimes reported [244,245]. These observations are explained by autophagy being a neuron- [243] and NDD-dependent process [246]. Autophagy implications in healthy and diseased tissue, and in particular in neuronal samples, must be further elucidated to properly weigh the benefits of autophagy-modulating small molecules in specific NDDs. Only then, a more solid target selection for drug discovery projects will be granted—possibly through the identification of targets providing therapeutic effects through synchronized UPS inhibition and autophagy induction. We cover in detail induction of autophagy by the mammalian target of rapamycin/mTOR complex 1 (see the next section), i.e., the single, fully validated molecular target in autophagy.

4.3.1 mTORC1

The *mammalian target of rapamycin (mTOR)* [247] is a large, multidomain 289 kDa Ser/Thr kinase, initially isolated in yeast while characterizing the mode of action of naturally occurring rapamycin [248]. It belongs to the phosphoinositide kinase-related kinase (PIKK) family, and is highly conserved in eukaryotes. Its N-terminus contains up to 20 tandem repeats of the protein–protein interaction domain Huntingtin, Elongation factor-3, protein phosphatase 2A, Tor1 (HEAT). HEAT repeats are followed by a FRAP, ATM, TRRAP (FAT) domain. Its C-terminal portion contains the kinase domain, the FKBP–rapamycin binding (FRB) domain, and a FRAP, ATM, TRRAP C-terminal (FATC) domain that is paired with the FAT domain to modulate the kinase activity of mTOR in all PIKKs [249].

mTOR associates with multiple proteins and is post-translationally modified/phosphorylated to act as a key cellular pathway regulator [250]. It is ubiquitously expressed in the human body [251], with cellular levels and activity dependent on the concentration of signaling factors [252], and on metabolic changes [253]. mTOR is the catalytic core of two regulatory complexes, mTOR complex 1 (*mTORC1*) and 2 (*mTORC2*) [254].

The former is built onto the rapamycin-sensitive scaffolding protein of mTOR (*raptor*) [255], a docking platform for mTORC1 constituents and substrates, that regulates the strength of the mTOR–raptor interaction [256]. mTORC1 includes the proline-rich Akt substrate p40 (*PRAS40*) [257], which may act—when overexpressed—as an mTORC1 inhibitor through binding competition with its main substrates [258]; the 36 kDa, mTORC1-stabilizer mammalian lethal with Sec13 protein 8 (*mLST8*) [259], which is essential for rapamycin–mTORC1 binding [260]; and the DEP domain-containing mTOR interacting protein (*deptor*) [261], which negatively regulates mTORC1 activity [262]. The expression and activity of mTORC1 determines the switch between cellular growth when conditions are favorable—availability of growth factors, nutrients, or energy—and catabolic processes when conditions are unfavorable—starvation, hypoxia, stress [250].

mTORC2 shares mTOR, mLST8, and deptor with mTORC1, but is built onto the rapamycin-insensitive companion of mTOR (*rictor*) [263], which shows acetylation-dependent positive regulation of mTORC2 [264]. mTORC2 includes the mammalian stress-activated protein kinase interacting protein (*mSIN1*) [265], which is necessary for mTORC2 to be functional [266]; the protein observed with rictor-1 (*protor*) [267], which does not show an influence on the main mTORC2 substrates [268]; and the proapoptotic proline-rich protein 5 (*PRR5*) [269], which acts also downstream of mTORC2 [270]. mTORC2 regulates cytoskeletal organization through its PKCα substrate [271], cell-to-cell contact through Rho signaling [272], and cell migration through Akt-mediated pathways [273]. As mTORC2 has no direct relationship with autophagy, it will not be discussed in detail here.

mTORC1 activity is regulated by each of its constituents in a complex relationship [250], but is also controlled by upstream modulators. The tuberous sclerosis complex (TSC1/2) [274], composed of TSC1-hamartin and TSC2-tuberin, is a phosphorylation-dependent, negative mTORC1 regulator through the small G-protein Ras homolog enriched in brain (Rheb)

[275]. The Rheb-GTP active form interacts with raptor [276] and with some mTORC1 substrates [277]. It controls the binding of mTORC1 to the endogenous mTORC1 inhibitor FKBP38 [278]. The GTP-activating protein TSC2 converts active Rheb-GTP to inactive Rheb-GDP [279]. mTORC1 regulation of TSC1/2 is controlled by cellular energy levels—activation of AMPK, TSC2 phosphorylation, and mTORC1 inhibition in low-energy conditions [280]; by growth factors—Akt-mediated phosphorylation of TSC2 on multiple sites, TSC2 binding to protein 14-3-3, disruption of negative regulation of Rheb, mTORC1 activation in growth signals-cell growth conditions [281]; by nutrients—Ras-related small GTP binding protein (Rag)-dependent lysosomal recruitment and activation by Rheb of mTORC1 in the presence of amino acids [282]; and by oxygen levels—AMPK-mediated expression of transcriptional regulation of DNA damage response 1 protein (REDD1), release of TSC2 from the complex with protein 14-3-3, mTORC1 inhibition in hypoxia conditions [283].

AMPK and Akt show TSC-independent effects on mTORC1. AMPK phosphorylates raptor in energy stress conditions, and creates a docking site for protein 14-3-3 [253], leading to mTORC1 inhibition. Akt phosphorylates PRAS40 [284], preventing its negative regulation of mTORC1 and leading to mTORC1 activation-cell growth.

The eukaryotic initiation factor 4E-binding protein 1 (4EBP1) [285] and p70 ribosome S6 kinase (p70 S6K) [286] are the main mTORC1 substrates. Both substrates bind to raptor, and their binding can be antagonized by activation of PRAS40 [258]. 4EBP1 phosphorylation by mTORC1 has a stimulating effect on mTORC1-mediated p70 S6K phosphorylation [287]. Phosphorylated 4EBP1 dissociates from the eukaryotic translation initiation factor 4 epsilon (eIF4E), and eventually promotes mRNA translation [288]. Phosphorylation of p70 S6K controls, *inter alia*, food intake [289], cell cycle progression [290], ribosome biogenesis [291], neuronal synaptic signaling [292], and stem cell differentiation [293].

In metazoans, mTORC1 regulates autophagy through interactions with the ULK complex [294], composed by ULK1/2, Atg13, Atg101, and FIP200 (see also Section 4.1). In basal conditions, mTORC1 and the ULK complex associate through binding between raptor and ULK1/2 [295]. Once bound, mTOR phosphorylates several epitopes on ULK1/2 and Atg13—including, but not limited to, Ser638 and Ser758 on human ULK1 [296]. As a result, the kinase activity of Ulk1/2 is suppressed. Conversely, if mTORC1 activity is down-regulated by external signals, or is pharmacologically inhibited, dephosphorylation of mTOR kinase epitopes on ULK1/2 and Atg13 leads to ULK1/2 activation [45]. Activated ULK1/2 undergoes autophosphorylation, and phosphorylates both Atg13 and Fip200 [297] on multiple epitopes, in a largely unclarified process [298]. Eventually, the entire ULK complex is translocated to the pre-AP, and autophagy is induced [297,298].

The equation "mTORC1 inhibition = autophagy induction" is somewhat simplistic. In starvation conditions, mTORC1 inhibition leads to extensive autophagy, and provides cells with fresh building blocks. Once a defined autophagy threshold is reached, mTORC1 is reactivated by unknown mechanisms to stop autophagy and to restore the lysosomal reservoir in a process named autophagic lysosomal reformation (ALR) [299]. Another negative-feedback process entails the mTORC1 substrate death associated protein 1 (DAP1), an autophagy repressor inactivated by mTOR phosphorylation [300]. How sustained pharmacological inhibition of mTORC1 would affect ALR or DAP1 is unclear, but the risk of an extreme scenario of autophagic cell death [301] cannot be discounted *a priori*.

A naturally occurring mTORC1 inhibitor, and some of its semisynthetic analogs, are approved treatments for transplant rejection, restenosis, and some cancers [302]. Therapeutic modulation of mTORC1 in pathological conditions has different outcomes—sometimes even in animal models of the same human disease. mTORC1 inhibition leads to beneficial effects on cardiac and kidney functions [303], on obesity [304], on autoimmune disorders [305], and as immunosuppressant therapy [306]. As to CNS, beneficial effects are observed in anxiety and post-traumatic stress disorders [307], epilepsy [308], and depression [309]. Stimulating [310] or inhibiting [311] mTOR kinase activity in cancer patients may be therapeutically relevant in different settings. Similarly, mTOR kinase activation [312] and mTOR kinase inhibition [313] are both reported to provide neuroprotection and increased cell survival in different animal models of stroke. mTOR regulation is important in CNS tissue repair and regeneration following traumatic CNS injuries [314]. Importantly, mTORC1 inhibition shows generalized anti-aging effects [302,315] and increases the lifespan of several model organisms [316–320]. In addition to previously mentioned effects, age-related inflammatory diseases [321], age-associated obesity [322], and age-associated macular degeneration [323] benefit from mTORC1 inhibition. Therapeutic effects of mTORC1 inhibition on NDDs often increase the lifespan in model organisms [302], although toxic effects are sometimes observed. As to *PD*, autophagy induction leads to the death of oxidatively stressed dopamine neurons [324], but the mTORC1 inhibitor rapamycin is neuroprotective in PD models [325]. mTORC1 inhibition prevents accumulation of toxic huntingtin aggregates in *HD* [326], and delays disease progression in *frontotemporal lobar dementia* (*FTLD*) models [327]. mTOR signaling is down-regulated in models of *AD* [328,329]. Reduced mTOR signaling is connected to the accumulation of APs, while increased Aβ levels lead to MA "stalling"-decrease, observed in AD [330,331]. Thus, mTORC1 inhibition provides therapeutic benefits in terms of Aβ pathology [332] in AD, although loss of mTOR kinase activity is detrimental in Aβ-based AD models [333].

mTOR acts on *tau*, both *in vitro* and *in vivo*. Tau phosphorylation is regulated through mTORC1 activation or inhibition either directly or indirectly through other kinases. mTOR immunoreactivity is increased in tangle-bearing neurons in AD, and mTOR itself phosphorylates tau *in vitro* [334]. Glycogen synthase kinase (GSK-3β) is among the most relevant tau kinases [335], while protein phosphatase 2A (PP2A) is the major tau phosphatase [336]. This tau kinase/phosphatase couple acts in a concerted manner to keep a steady/physiological phosphorylation level on tau, and is regulated by PI3K and mTOR signaling [337]. mTORC1 inhibition increases tau phosphorylation levels in an Akt- and GSK-3β-independent manner, although Aβ-induced tau HP is mTORC1- and GSK-3β-dependent [338]. Conversely, the levels of tau pSer214—a site which primes tau for pathological HP—are reduced by mTORC1 inhibition, possibly by regulation of the kinase activity of cAMP-dependent protein kinase (PKA) [339]. In general, mTOR imbalances tau homeostasis, through the regulation of several tau kinases, and is a key player in the onset and progression of tauopathies [334].

Dysregulation of insulin signaling contributes to the pathogenesis of AD, and specifically to tau HP, in an mTOR- and GSK-3β-dependent manner *in vitro* [340] and *in vivo* [341]. Interestingly, diabetic AD rats show increased mTOR and HP-tau levels in the hippocampus if compared with their "AD only" controls [342]. An overactivation of mTOR in the hippocampus by impairing insulin signaling is likely to increase the extent of tau HP and to promote the occurrence of AD-like neurodegeneration [342].

mTORC1 influences protein translation in general, and tau translation in particular, through regulation of its main substrates. Aberrant changes in the protein translation machinery in AD are connected with higher levels of p70 S6K [343] and p-eIF4E [344]. Higher levels of these p-isoforms, and of total and phosphorylated levels of mTOR, 4E-BP1, eEF2, and eEF2K, are coupled with higher tau and p-tau levels in AD brains [345]. mTOR-driven p70 SK6 controls axon formation through increased axonal translation of tau and other neuronal polarity proteins [346]. mTOR overactivation in somatodendritic compartments may lead to tau mis-sorting, an early pathological event in AD [347].

Autophagy regulation by mTORC1 has a strong impact on tauopathies [348]. mTORC1 inhibition increases autophagy and reduces the extent of tau and Aβ pathology in an AD animal model [349]. Similar observations are made on a tau-mTOR animal model, establishing a direct/non-Aβ-mediated tau-mTOR link that controls tau phosphorylation, translocation, and degradation in AD and other tauopathies [350]. Thus, mTORC1 is a meaningful target to find putative, autophagy-directed treatments against tauopathies. Further efforts are needed to better understand how to rationally design and develop disease-selective, mTORC1-directed autophagy modulators/anti-aging treatments against NDDs.

4.4 DISEASE-MODIFYING COMPOUNDS

This chapter deals with mid-late steps leading to neuropathological alterations related to protein misfolding and aggregation in general, and to tau and/or tau-connected events in particular. A single molecular target is chosen and discussed in detail. Tens of other targets are just mentioned here, mostly for reasons of space. Thirty-seven compounds/scaffolds acting on mTORC1 are diffusely covered in a chemistry-oriented companion book [351] devoted to disease-modifying compounds, and are briefly summarized below in Table 4.1. Each compound class is numbered as in the chemistry oriented companion book, and its chemical core is structurally defined; its mechanism of action and molecular target are mentioned; the public or private laboratory that develops the compound is listed; and the development status—according to publicly available information—is provided.

TABLE 4.1   Compounds 4.1–4.29: Chemical Class, Target, Developing Organization, Development Status

Number	Chemical cpd./class	Target	Organization	Dev. status
4.1a	Rapamycin-sirolimus-rapamune™	mTOR inhibition	Pfizer	MKTD
4.1b	Everolimus-RAD-001-afinitor™	mTOR inhibition	Novartis	MKTD
4.1c	Temsirolimus-CCI-779-torisel™	mTOR inhibition	Wyeth	MKTD
4.1d	Ridaforolimus-AP23573-deforolimus	mTOR inhibition	Merck/ARIAD	Ph III
4.1e	Umirolimus-TRM-986-biolimus A9	mTOR inhibition	Biosensors Intl.	MKTD
4.1f	Zotarolimus-ABT-578	mTOR inhibition	Medtronics	MKTD
4.2	Tricyclic pyrimidine-morpholines, PI-103	PI3K/mTOR dual inhibitors	Piramed Pharma	LO
4.3a,b	Tricyclic pyrimidine-morpholines PI-540 (a), PI-620 (b)	PI3K/mTOR dual inhibitors	Piramed Pharma	PE

TABLE 4.1 Compounds 4.1–4.29: Chemical Class, Target, Developing Organization, Development Status *(cont.)*

Number	Chemical cpd./class	Target	Organization	Dev. status
4.4	Tricyclic pyrimidine-morpholine GDC-0980	PI3K/mTOR dual inhibitors	Genentech	Ph II
4.5	Imidazo[4,5-c] quinolones, NVP-BEZ235	PI3K/mTOR dual inhibitors	Novartis	Ph II
4.6	Imidazo[4,5-c] quinolones, NVP-BGT226	PI3K/mTOR dual inhibitors	Novartis	PhI/II
4.7	Quinoxalines, XL765-SAR245408	PI3K/mTOR dual inhibitors	Sanofi	Ph I/II
4.8	Quinolones, GSK2126458	PI3K/mTOR dual inhibitors	GSK	Ph I
4.9	Triazolopyrimidines, PKI-402	PI3K/mTOR— dual inhibitors	Wyeth	PE
4.10	Morpholinotriazines, PKI-587/PF-05212384	PI3K/mTOR dual inhibitors	Pfizer	Ph II
4.11, 4.12	Morpholylbenzopyra-nones, LY294002 (4.11), prodrug (SF1129, 4.12)	PI3K/mTOR dual inhibitors	Eli Lilly	Ph I
4.13	Pyrazolo[3,4-d] pyrimidines, PP242	Pan-mTOR inhibition	University of San Francisco	PE
4.14	Pyrazolo[3,4-d] pyrimidines, MLN0128-INK128	Pan-mTOR inhibition	Millennium	Ph I/II
4.15	Pyrazolo[3,4-d] pyrimidines, WAY600	Pan-mTOR inhibition	Wyeth	LO
4.16	Pyrazolo[3,4-d] pyrimidines, WYE-132-WYE125132	Pan-mTOR inhibition	Wyeth	PE
4.17a	Pyrido[2,3-d] pyrimidines, Ku0063794	Pan-mTOR inhibition	AstraZeneca	LO
4.17b	Pyrido[2,3-d] pyrimidines, AZD8055	Pan-mTOR inhibition	AstraZeneca	Ph I
4.17c	Pyrido[2,3-d] pyrimidines, AZD2014	Pan-mTOR inhibition	AstraZeneca	Ph II
4.18	Imidazo[1,5-f] [1,2,4]triazines, OSI-027	Pan-mTOR inhibition	Astellas	Ph I
4.19	Benzochromenones, palomid 529	Pan-mTOR inhibition	Paloma Pharmaceu-ticals	Ph I

(Continued)

TABLE 4.1   Compounds 4.1–4.29: Chemical Class, Target, Developing Organization, Development Status *(cont.)*

Number	Chemical cpd./class	Target	Organization	Dev. status
4.20	Benzonaphthiridinones, torin1	Pan-mTOR inhibition	Dana Farber Cancer Inst., Boston, USA	LO
4.21	Benzonaphthiridinones, torin2	Pan-mTOR inhibition	Dana Farber Cancer Inst., Boston, USA	PE
4.22	Caffeine	mTORC1 regulation	Juntendo University, Tokio	FS
4.23	Phosphatidic acid	mTORC1 regulation	University of Leicester, UK	FS
4.24	Dexamethasone	mTORC1 regulation	Case Western Reserve Univ., Cleveland, USA	MKTD
4.25	Metformin	mTORC1 regulation	Several	MKTD
4.26	Aspirin	mTORC1 regulation	Several	MKTD
4.27	Δ^9-Tetrahydrocannabinol	mTORC1 regulation	Alcala Univ., Madrid, Spain	MKTD
4.28, 4.29	Resveratrol (4.28), resveratrol analogs (4.29)	mTORC1 regulation	Feinstein Institute, New York, USA	LO

Not progressed, NP; lead optimization, LO; preclinical evaluation, PE; clinical Phase I-II-III, Ph I–Ph III; marketed, MKTD; food supplement, FS.

References

1. Hershko, A.; Ciechanover, A. The ubiquitin system. *Annu. Rev. Biochem.* **1998**, *67*, 425–479.
2. Weissman, A. M.; Shabek, N.; Ciechanover, A. The predator becomes the prey: regulating the ubiquitin system by ubiquitylation and degradation. *Nat. Rev. Mol. Cell Biol.* **2011**, *12*, 605–620.
3. Wong, E.; Cuervo, A. M. Integration of clearance mechanisms: the proteasome and autophagy. *Cold Spring Harbor Perspect. Biol.* **2010**, *2*, a006734.
4. Ericsson, J. L. E. Mechanisms of cellular autophagy. *Lysosomes Biol. Pathol.* **1969**, *2*, 345–394.
5. Codogno, P.; Mehrpour, M.; Proikas-Cezanne, T. Canonical and non-canonical autophagy: variations on a common theme of self-eating? *Nat. Rev. Mol. Cell Biol.* **2012**, *13*, 7–12.
6. Korolchuk, V. I.; Rubinsztein, D. C. On signals controlling autophagy: It's time to eat yourself healthy. *Biochemist* **2012**, *34*, 8–13.

7. He, C.; Klionsky, D. J. Regulation mechanisms and signaling pathways of autophagy. *Annu. Rev. Genet.* **2009**, *43*, 67–93.

8. Kroemer, G.; Marino, G.; Levine, B. Autophagy and the integrated stress response. *Mol. Cell* **2010**, *40*, 280–293.

9. Kraft, C.; Martens, S. Mechanisms and regulation of autophagosome formation. *Curr. Opin. Cell Biol.* **2012**, *24*, 496–501.

10. Simonsen, A.; Tooze, S. A. Coordination of membrane events during autophagy by multiple class III PI3-kinase complexes. *J. Cell Biol.* **2009**, *186*, 773–782.

11. Rubinsztein, D. C.; Codogno, P.; Levine, B. Autophagy modulation as a potential therapeutic target for diverse diseases. *Nat. Rev. Drug Discov.* **2012**, *11*, 709–730.

12. Li, W.-w.; Li, J.; Bao, J.-k. Microautophagy: lesser-known self-eating. *Cell. Mol. Life Sci.* **2012**, *69*, 1125–1136.

13. Mijaljica, D.; Prescott, M.; Devenish, R. J. Microautophagy in mammalian cells: revisiting a 40-year-old conundrum. *Autophagy* **2011**, *7*, 673–682.

14. Mueller, O.; Sattler, T.; Floetenmeyer, M.; Schwarz, H.; Plattner, H.; Mayer, A. Autophagic tubes: vacuolar invaginations involved in lateral membrane sorting and inverse vesicle budding. *J. Cell Biol.* **2000**, *151*, 519–528.

15. Sattler, T.; Mayer, A. Cell-free reconstitution of microautophagic vacuole invagination and vesicle formation. *J. Cell Biol.* **2000**, *151*, 529–538.

16. Uttenweiler, A.; Mayer, A. Microautophagy in the yeast *Saccharomyces cerevisiae. Methods Mol. Biol.* **2008**, *445*, 245–259.

17. Epple, U. D.; Suriapranata, I.; Eskelinen, E. L.; Thumm, M. Aut5/Cvt17p, a putative lipase essential for disintegration of autophagic bodies inside the vacuole. *J. Bacteriol.* **2001**, *183*, 5942–5955.

18. Li, W.; Yang, Q.; Mao, Z. Chaperone-mediated autophagy: machinery, regulation and biological consequences. *Cell. Mol. Life Sci.* **2011**, *68*, 749–763.

19. Komatsu, M.; Ichimura, Y. Selective autophagy regulates various cellular functions. *Genes to Cells* **2010**, *15*, 923–933.

20. Lamark, T.; Johansen, T. Aggrephagy: selective disposal of protein aggregates by macroautophagy. *Int. J. Cell Biol.* **2012**, 736905.

21. Tyedmers, J.; Moegk, A.; Bukau, B. Cellular strategies for controlling protein aggregation. *Nat. Rev. Mol. Cell Biol.* **2010**, *11*, 777–788.

22. Mizushima, N.; Levine, B.; Cuervo, A. M.; Klionsky, D. J. Autophagy fights disease through cellular self-digestion. *Nature* **2008**, *451*, 1069–1075.

23. Korolchuk, V. I.; Mansilla, A.; Menzies, F. M.; Rubinsztein, D. C. Autophagy inhibition compromises degradation of ubiquitin-proteasome pathway substrates. *Mol. Cell* **2009**, *33*, 517–527.

24. Ding, W. X.; Ni, H. M.; Gao, W.; Yoshimori, T.; Stolz, D. B.; Ron, D.; Yin, X. M. Linking of autophagy to ubiquitin–proteasome system is important for the regulation of endoplasmic reticulum stress and cell viability. *Am. J. Pathol.* **2007**, *171*, 513–524.

25. Wang, Y.; Mandelkow, E. Degradation of tau protein by autophagy and proteasomal pathways. *Bioch. Soc. Trans.* **2012**, *40*, 644–652.

26. Ravikumar, B.; Sarkar, S.; Davies, J. E.; Futter, M.; Garcia-Arencibia, M.; Green-Thompson, Z. W., et al. Regulation of mammalian autophagy in physiology and pathophysiology. *Physiol. Rev.* **2010**, *90*, 1383–1435.

27. Puyal, J.; Ginet, V.; Grishchuk, Y.; Truttmann, A. C.; Clarke, P. G. H. Neuronal autophagy as a mediator of life and death: contrasting roles in chronic neurodegenerative and acute neural disorders. *The Neuroscientist* **2012**, *18*, 224–236.

28. Choi, A. M. K.; Ryter, S. W.; Levine, B. Autophagy in human health and disease. *N. Engl. J. Med.* **2013**, *368*, 651–662.

29. Lamb, C. A.; Yoshimori, T.; Tooze, S. A. The autophagosome: origins unknown, biogenesis complex. *Nat. Rev. Mol. Cell. Biol.* **2013**, *14*, 759–774.

30. Suzuki, K.; Ohsumi, Y. Current knowledge of the pre-autophagosomal structure (Pas). *FEBS Lett.* **2010**, *584*, 1280–1286.

31. Lynch-Day, M. A.; Klionsky, D. J. The Cvt pathway as a model for selective autophagy. *FEBS Lett.* **2010**, *584*, 1359–1366.

32. Song, Q.; Kumar, A. An overview of autophagy and yeast pseudohyphal growth: integration of signaling pathways during nitrogen stress. *Cells* **2012**, *1*, 263–283.

33. Hamasaki, M.; Furuta, N.; Matsuda, A.; Nezu, A.; Yamamoto, A.; Fujita, N., et al. Autophagosomes form at ER–mitochondria contact sites. *Nature* **2012**, *495*, 389–393.

34. Ge, L.; Melville, D.; Zhang, M.; Schekman, R. The ER-Golgi intermediate compartment is a key membrane source for the LC3 lipidation step of autophagosome biogenesis. *Elife* **2013**, *2*, e00947.

35. Ragusa, M. J.; Stanley, R. E.; Hurley, J. H. Architecture of the Atg17 complex as a scaffold for autophagosome biogenesis. *Cell* **2012**, *151*, 1501–1512.

36. Kabeya, Y.; Noda, N. N.; Fujioka, Y.; Suzuki, K.; Inagaki, F.; Ohsumi, Y. Characterization of the Atg17–Atg29–Atg31 complex specifically required for starvation-induced autophagy in *Saccharomyces cerevisiae*. *Biochem. Biophys. Res. Commun.* **2009**, *389*, 612–615.

37. Mari, M.; Griffith, J.; Rieter, E.; Krishnappa, L.; Klionsky, D. J.; Reggiori, F. An Atg9-containing compartment that functions in the early steps of autophagosome biogenesis. *J. Cell Biol.* **2010**, *190*, 1005–1022.

38. Moreau, K.; Renna, M.; Rubinsztein, D. C. Connections between SNAREs and autophagy. *Trends Biochem. Sci.* **2013**, *38*, 57–63.

39. Sudhof, T. C.; Rothman, J. E. Membrane fusion: grappling with SNARE and SM proteins. *Science* **2009**, *323*, 474–477.

40. Hayashi-Nishino, M.; Fujita, N.; Noda, T.; Yamaguchi, A.; Yoshimori, T.; Yamamoto, A. A subdomain of the endoplasmic reticulum forms a cradle for autophagosome formation. *Nat. Cell. Biol.* **2009**, *11*, 1433–1437.

41. Hailey, D. W.; Ranbold, A. S.; Satpute-Krishran, P.; Mitra, K.; Sougrat, R.; Kim, P. K.; Lippincott-Schwartz, J. Mitochondria supply membranes for autophagosome biogenesis during starvation. *Cell* **2010**, *141*, 656–667.

42. Moreau, K.; Rubinsztein, D. C. The plasma membrane as a control center for autophagy. *Autophagy* **2012**, *8*, 861–863.

43. Mizushima, N. The role of the Atg1/ULK1 complex in autophagy regulation. *Curr. Opin. Cell Biol.* **2010**, *22*, 132–139.

44. Chan, E. Y.; Kir, S.; Tooze, S. A. siRNA screening of the kinome identifies ULK1 as a multidomain modulator of autophagy. *J. Biol. Chem.* **2007**, *282*, 25464–25474.

45. Ganley, I. G.; Lam, du H.; Wang, J.; Ding, X.; Chen, S.; Jiang, X. ULK1. ATG13. FIP200 complex mediates mTOR signaling and is essential for autophagy. *J. Biol. Chem.* **2009**, *284*, 12297–12305.

46. Hosokawa, N.; Sasaki, T.; Iemura, S. -I.; Natsume, T.; Hara, T.; Mizushima, N. Atg101, a novel mammalian autophagy protein interacting with Atg13. *Autophagy* **2009**, *5*, 973–979.

47. Hara, T.; Takamura, A.; Kishi, C.; Iemura, S.; Natsume, T.; Guan, J. L.; Mizushima, N. FIP200, a ULK-interacting protein, is required for autophagosome formation in mammalian cells. *J. Cell. Biol.* **2008**, *181*, 497–510.

48. Stanley, R. E.; Ragusa, M. J.; Hurley, J. H. The beginning of the end: how scaffolds nucleate autophagosome biogenesis. *Trends Cell Biol.* **2014**, *24*, 73–81.

49. Meijer, A. J.; Codogno, P. AMP-activated protein kinase and autophagy. *Autophagy* **2007**, *3*, 238–240.

50. Wirawan, E.; Vanden Berghe, T.; Lippens, S.; Agostinis, P.; Vandenabeele, P. Autophagy: for better or for worse. *Cell Res.* **2012**, *22*, 43–61.

51. Itakura, E.; Mizushima, N. Characterization of autophagosome formation site by a hierarchical analysis of mammalian Atg proteins. *Autophagy* **2010**, *6*, 764–776.

52. Arico, S.; Petiot, A.; Bauvy, C.; Dubbelhuis, P. F.; Meijer, A. J.; Codogno, P.; Ogier-Denis, E. The tumor suppressor PTEN positively regulates macroautophagy by inhibiting the phosphatidylinositol 3-kinase/protein kinase B pathway. *J. Biol. Chem.* **2001**, *276*, 35243–35246.

53. He, C.; Levine, B. The Beclin 1 interactome. *Curr. Opin. Cell Biol.* **2010**, *22*, 140–149.
54. Kihara, A.; Noda, T.; Ishihara, N.; Ohsumi, Y. Two distinct Vps34 phosphatidylinositol 3-kinase complexes function in autophagy and carboxypeptidase Y sorting in *Saccharomyces cerevisiae. J. Cell Biol.* **2001**, *152*, 519–530.
55. Stack, J. H.; Emr, S. D. Vps34p required for yeast vacuolar protein sorting is a multiple specificity kinase that exhibits both protein kinase and phosphatidylinositol-specific PI 3-kinase activities. *J. Biol. Chem.* **1994**, *269*, 31552–31562.
56. Kametaka, S.; Okano, T.; Ohsumi, M.; Ohsumi, Y. Apg14p and Apg6/Vps30p form a protein complex essential for autophagy in the yeast, *Saccharomyces cerevisiae. J. Biol. Chem.* **1998**, *273*, 22284–22291.
57. Herman, P. K.; Stack, J. H.; DeModena, J. A.; Emr, S. D. A novel protein kinase homolog essential for protein sorting to the yeast lysosome-like vacuole. *Cell* **1991**, *64*, 425–437.
58. Obara, K.; Sekito, T.; Ohsumi, Y. Assortment of phosphatidylinositol 3-kinase complexes—Atg14p directs association of complex I to the pre-autophagosomal structure in *Saccharomyces cerevisiae. Mol. Biol. Cell* **2006**, *17*, 1527–1539.
59. Obara, K.; Ohsumi, Y. Atg14: A key player in orchestrating autophagy. *Int. J. Cell Biol.* **2011**, 713435.
60. Backer, J. M. The regulation and function of Class III PI3Ks: novel roles for Vps34. *Biochem. J.* **2008**, *410*, 1–17.
61. Sun, Q.; Fan, W.; Chen, K.; Ding, X.; Chen, S.; Zhong, Q. Identification of Barkor as a mammalian autophagy-specific factor for Beclin 1 and class III phosphatidylinositol 3-kinase. *Proc. Natl. Acad. Sci. U.S.A.* **2008**, *105*, 19211–19216.
62. Stein, M. P.; Cao, C.; Tessema, M.; Feng, Y.; Romero, E.; Welford, A.; Wandinger-Ness, A. Interaction and functional analyses of human VPS34/p150 phosphatidylinositol 3-kinase complex with Rab7. *Methods Enzymol.* **2005**, *403*, 628–649.
63. Fimia, G. M.; Stoykova, A.; Romagnoli, A.; Giunta, L.; Di Bartolomeo, S.; Nardacci, R., et al. Ambra1 regulates autophagy and development of the nervous system. *Nature* **2007**, *447*, 1121–1125.
64. Di Bartolomeo, S.; Corazzari, M.; Nazio, F.; Oliverio, S.; Lisi, G.; Antonioli, M., et al. The dynamic interaction of AMBRA1 with the dynein motor complex regulates mammalian autophagy. *J. Cell Biol.* **2010**, *191*, 155–168.
65. Nazio, F.; Strappazzon, F.; Antonioli, M.; Bielli, P.; Cianfanelli, V.; Bordi, M., et al. mTOR inhibits autophagy by controlling ULK1 ubiquitylation, self-association and function through AMBRA1 and TRAF6. *Nat. Cell. Biol.* **2013**, *15*, 406–416.
66. Vergne, I.; Roberts, E.; Elmaoued, R. A.; Tosch, V.; Delgado, M. A.; Proikas-Cezanne, T., et al. Control of autophagy initiation by phosphoinositide 3-phosphatase Jumpy. *EMBO J.* **2009**, *28*, 2244–2258.
67. Taguchi-Atarashi, N.; Hamasaki, M.; Matsunaga, K.; Omori, H.; Ktistakis, N. T. Modulation of local PtdIns3P levels by the PI phosphatase MTMR3 regulates constitutive autophagy. *Traffic* **2010**, *11*, 468–478.
68. Liang, C.; Sir, D.; Lee, S.; Ou, J. H.; Jung, J. U. Beyond autophagy: the role of UVRAG in membrane trafficking. *Autophagy* **2008**, *4*, 817–820.
69. Takahashi, Y.; Meyerkord, C. L.; Wang, H. G. Bif-1/endophilin B1: a candidate for crescent driving force in autophagy. *Cell Death Differ.* **2009**, *16*, 947–955.
70. Itoh, T.; De Camilli, P. BAR, F-BAR (EFC) and ENTH/ANTH domains in the regulation of membrane-cytosol interfaces and membrane curvature. *Biochim. Biophys. Acta 1761*, **2006**, 897–912.
71. Zhong, Y.; Wang, Q. J.; Li, X.; Yan, Y.; Backer, J. M.; Chait, B. T., et al. Distinct regulation of autophagic activity by Atg14L and Rubicon associated with Beclin 1-phosphatidylinositol-3-kinase complex. *Nat. Cell. Biol.* **2009**, *11*, 468–476.
72. Kang, R.; Zeh, H. J.; Lotze, M. T.; Tang, D. The Beclin 1 network regulates autophagy and apoptosis. *Cell Death Dis.* **2011**, *18*, 571–580.

73. Pattingre, S.; Tassa, A.; Qu, X.; Garuti, R.; Liang, X. H.; Mizushima, N., et al. Bcl-2 anti-apoptotic proteins inhibit Beclin 1-dependent autophagy. *Cell* 2005, *122*, 927–939.

74. Sinha, S.; Levine, B. The autophagy effector Beclin 1: a novel BH3-only protein. *Oncogene* 2008, *27*, S137–S148.

75. Zalckvar, E.; Berissi, H.; Mizrachy, L.; Idelchuk, Y.; Koren, I.; Eisenstein, M., et al. DAP-kinase mediated phosphorylation on the BH3 domain of beclin 1 promotes dissociation of beclin 1 from Bcl-XL and induction of autophagy. *EMBO Rep.* 2009, *10*, 285–292.

76. Shi, C. S.; Kehrl, J. H. TRAF6 and A20 regulate lysine 63-linked ubiquitination of Beclin-1 to control TLR4-induced autophagy. *Sci. Signal.* 2010, *3*, ra42.

77. Wang, R. C.; Wei, Y.; An, Z.; Zou, Z.; Xiao, G.; Bhagat, G., et al. Akt-mediated regulation of autophagy and tumorigenesis through Beclin 1 phosphorylation. *Science* 2012, *338*, 956–959.

78. Kang, R.; Livesey, K. M.; Zeh, H. J.; Loze, M. T.; Tang, D. HMGB1: a novel Beclin 1-binding protein active in autophagy. *Autophagy* 2010, *6*, 1209–1211.

79. Michiorri, S.; Gelmetti, V.; Giarda, E.; Lombardi, F.; Romano, F.; Marongiu, R., et al. The Parkinson associated protein PINK1 interacts with Beclin 1 and promotes autophagy. *Cell Death Differ.* 2010, *17*, 962–974.

80. Vicencio, J. M.; Ortiz, C.; Criollo, A.; Jones, A. W.; Kepp, O.; Galluzzi, L., et al. The inositol 1,4,5-trisphosphate receptor regulates autophagy through its interaction with Beclin 1. *Cell Death Differ* 2009, *16*, 1006–1017.

81. Webber, J. L.; Young, A. R.; Tooze, S. A. Atg9 trafficking in mammalian cells. *Autophagy* 2007, *3*, 54–56.

82. Yamamoto, H.; Kakuta, S.; Watanabe, T. M.; Kitamura, A.; Sekito, T.; Kondo-Kakuta, C., et al. Atg9 vesicles are an important membrane source during early steps of autophagosome formation. *J. Cell Biol.* 2012, *198*, 219–233.

83. Orsi, A.; Razi, M.; Dooley, H. C.; Robinson, D.; Weston, A. E.; Collinson, L. M.; Tooze, S. A. Dynamic and transient interactions of Atg9 with autophagosomes, but not membrane integration, are required for autophagy. *Mol. Biol. Cell* 2012, *23*, 1860–1873.

84. Young, A. R. J.; Chan, E. Y. W.; Hu, X. W.; Köchl, R.; Crawshaw, S. G.; High, S., et al. Starvation and ULK1-dependent cycling of mammalian Atg9 between the TGN and endosomes. *J. Cell Sci.* 2006, *119*, 3888–3900.

85. Esclatine, A.; Chaumorcel, M.; Codogno, P. Macroautophagy signaling and regulation. *Curr. Top. Microbiol. Immunol.* 2009, *335*, 33–70.

86. Mizushima, N.; Sugita, H.; Yoshimori, T.; Ohsumi, Y. A new protein conjugation system in human. The counterpart of the yeast Apg12p conjugation system essential for autophagy. *J. Biol. Chem.* 1998, *273*, 33889–33892.

87. Yamaguchi, M.; Matoba, K.; Sawada, R.; Fujioka, Y.; Nakatogawa, H.; Yamamoto, H., et al. Noncanonical recognition and UBL loading of distinct E2s by autophagy-essential Atg7. *Nature Struct. Mol. Biol.* 2012, *19*, 1250–1256.

88. Mizushima, N.; Yamamoto, A.; Hatano, M.; Kobayashi, Y.; Kabeya, Y.; Suzuki, K., et al. Dissection of autophagosome formation using Apg5-deficient mouse embryonic stem cells. *J. Cell. Biol.* 2001, *152*, 657–668.

89. Van der Veen, A. G.; Ploegh, H. L. Ubiquitin-like proteins. *Annu. Rev. Biochem.* 2012, *81*, 323–357.

90. Romanov, J.; Walczak, M.; Ibiricu, I.; Schuechner, S.; Ogris, E.; Kraft, C.; Martens, S. Mechanism and functions of membrane binding by the Atg5-Atg12/Atg16 complex during autophagosome formation. *EMBO J.* 2012, *31*, 4304–4317.

91. Shpilka, T.; Mizushima, N.; Elazar, Z. Ubiquitin-like proteins and autophagy at a glance. *J. Cell Sci.* 2012, *125*, 2343–2348.

92. Fujita, N.; Itoh, T.; Omori, H.; Fukuda, M.; Noda, T.; Yoshimori, T. The Atg16L complex specifies the site of LC3 lipidation for membrane biogenesis in autophagy. *Mol. Biol. Cell* 2008, *19*, 2092–2100.

93. Mizushima, N.; Kuma, A.; Kobayashi, Y.; Yamamoto, A.; Matsubae, M.; Takao, T., et al. Mouse Apg16L, a novel WD-repeat protein, targets to the autophagic isolation membrane with the Apg12-Apg5 conjugate. *J. Cell. Sci.* **2003**, *116*, 1679–1688.

94. Rogov, V.; Doetsch, V.; Johansen, T.; Kirkin, V. Interactions between autophagy receptors and ubiquitin-like proteins form the molecular basis for selective autophagy. *Mol. Cell* **2014**, *53*, 167–178.

95. Geng, J.; Klionsky, D. J. The Atg8 and Atg12 ubiquitin-like conjugation systems in macroautophagy. *EMBO Rep.* **2008**, *9*, 859–864.

96. Shpilka, T.; Weidberg, H.; Pietrokovski, S.; Elazar, Z. Atg8: an autophagy-related ubiquitin-like protein family. *Genome Biol.* **2011**, *12*, 226.

97. Kirisako, T.; Ichimura, Y.; Okada, H.; Kabeya, Y.; Mizushima, N.; Yoshimori, T., et al. The reversible modification regulates the membrane-binding state of Apg8/Aut7 essential for autophagy and the cytoplasm to vacuole targeting pathway. *J. Cell Biol.* **2000**, *151*, 263–276.

98. Ichimura, Y.; Kirisako, T.; Takao, T.; Satomi, Y.; Shimonishi, Y.; Ishihara, N., et al. A ubiquitin like system mediates protein lipidation. *Nature* **2000**, *408*, 488–492.

99. Taherbhoy, A. M.; Tait, S. W.; Kaiser, S. E.; Williams, A. H.; Deng, A.; Nourse, A., et al. Atg8 transfer from Atg7 to Atg3: a distinctive E1-E2 architecture and mechanism in the autophagy pathway. *Mol. Cell* **2011**, *44*, 451–461.

100. Coyle, J. E.; Qamar, S.; Rajashankar, K. R.; Nikolov, D. B. Structure of GABARAP in two conformations: implications for GABA(A) receptor localization and tubulin binding. *Neuron* **2002**, *33*, 63–74.

101. Ichimura, Y.; Imamura, Y.; Emoto, K.; Umeda, M.; Noda, T.; Ohsumi, Y. In vivo and in vitro reconstitution of Atg8 conjugation essential for autophagy. *J. Biol. Chem.* **2004**, *279*, 40584–40592.

102. Nakatogawa, H.; Ichimura, Y.; Ohsumi, Y. Atg8, a ubiquitin-like protein required for autophagosome formation, mediates membrane tethering and hemifusion. *Cell* **2007**, *130*, 165–178.

103. Weidberg, H.; Shvets, E.; Elazar, Z. Biogenesis and cargo selectivity of autophagosomes. *Annu. Rev. Biochem.* **2011**, *80*, 125–156.

104. Hanada, T.; Noda, N. N.; Satomi, Y.; Ichimura, Y.; Fujioka, Y.; Takao, T., et al. The Atg12-Atg5 conjugate has a novel E3-like activity for protein lipidation in autophagy. *J. Biol. Chem.* **2007**, *282*, 37298–37302.

105. Sugawara, K.; Suzuki, N. N.; Fujioka, Y.; Mizushima, N.; Ohsumi, Y.; Inagaki, F. The crystal structure of microtubule-associated protein light chain 3, a mammalian homologue of *Saccharomyces cerevisiae* Atg8. *Genes Cells* **2004**, *9*, 611–618.

106. Otomo, C.; Metlagel, Z.; Takaesu, G.; Otomo, T. Structure of the human ATG12~ATG5 conjugate required for LC3 lipidation in autophagy. *Nat. Struct. Mol. Biol.* **2013**, *20*, 59–66.

107. Kondo-Okamoto, N.; Noda, N. N.; Suzuki, S. W.; Nakatogawa, H.; Takahashi, I.; Matsunami, M., et al. Autophagy-related protein 32 acts as autophagic degron and directly initiates mitophagy. *J. Biol. Chem.* **2012**, *287*, 10631–10638.

108. Behrends, C.; Sowa, M. E.; Gygi, S. P.; Harper, J. W. Network organization of the human autophagy system. *Nature* **2010**, *466*, 68–76.

109. Huang, W. P.; Scott, S. V.; Kim, J.; Klionsky, D. J. The itinerary of a vesicle component, Aut7p/Cvt5p, terminates in the yeast vacuole via the autophagy/Cvt pathways. *J. Biol. Chem.* **2000**, *275*, 5845–5851.

110. Noda, N. N.; Ohsumi, Y.; Inagaki, F. Atg8-family interacting motif crucial for selective autophagy. *FEBS Lett.* **2010**, *584*, 1379–1385.

111. Kaufmann, A.; Beier, V.; Franquelim, H. G.; Wollert, T. Molecular mechanism of autophagic membrane-scaffold assembly and disassembly. *Cell* **2014**, *156*, 469–481.

112. Geng, J.; Baba, M.; Nair, U.; Klionsky, D. J. Quantitative analysis of autophagy-related protein stoichiometry by fluorescence microscopy. *J. Cell Biol.* **2008**, *182*, 129–140.

113. Xie, Z.; Nair, U.; Klionsky, D. J. Atg8 controls phagophore expansion during autophagosome formation. *Mol. Biol. Cell* **2008**, *19*, 3290–3298.
114. Weidberg, H.; Shpilka, T.; Shvets, E.; Abada, A.; Shimron, F.; Elazar, Z. LC3 and GATE-16 N termini mediate membrane fusion processes required for autophagosome biogenesis. *Dev. Cell* **2011**, *20*, 444–454.
115. Weidberg, H.; Shvets, E.; Shpilka, T.; Shimron, F.; Shinder, V.; Elazar, Z. LC3 and GATE-16/GABARAP subfamilies are both essential yet act differently in autophagosome biogenesis. *EMBO J.* **2010**, *29*, 1792–1802.
116. Ohsumi, Y. Molecular dissection of autophagy: two ubiquitin-like systems. *Nat. Rev. Mol. Cell Biol.* **2001**, *2*, 211–216.
117. Nair, U.; Yen, W. L.; Mari, M.; Cao, Y.; Xie, Z.; Baba, M., et al. A role for Atg8-PE deconjugation in autophagosome biogenesis. *Autophagy* **2012**, *8*, 780–793.
118. Yu, Z. Q.; Ni, T.; Hong, B.; Wang, H. Y.; Jiang, F. J.; Zou, S., et al. Dual roles of Atg8-PE deconjugation by Atg4 in autophagy. *Autophagy* **2012**, *8*, 883–892.
119. Tanida, I.; Sou, Y. S.; Ezaki, J.; Minematsu-Ikeguchi, N.; Ueno, T.; Kominami, E. HsAtg4B/HsApg4B/autophagin-1 cleaves the carboxyl termini of three human Atg8 homologues and delipidates microtubule-associated protein light chain 3- and GABAA receptor associated protein-phospholipid conjugates. *J. Biol. Chem.* **2004**, *279*, 36268–36276.
120. Hemelaar, J.; Lelyveld, V. S.; Kessler, B. M.; Ploegh, H. L. A single protease, Apg4B, is specific for the autophagy-related ubiquitin-like proteins GATE-16, MAP1-LC3, GABARAP, and Apg8L. *J. Biol.Chem.* **2003**, *278*, 51841–51850.
121. Marino, G.; Salvador-Montoliu, N.; Fueyo, A.; Knecht, E.; Mizushima, N.; Lopez-Otin, C. Tissue-specific autophagy alterations and increased tumorigenesis in mice deficient in Atg4C/autophagin-3. *J. Biol. Chem.* **2007**, *282*, 18573–18583.
122. Betin, V. M.; Lane, J. D. Caspase cleavage of Atg4D stimulates GABARAP-L1 processing and triggers mitochondrial targeting and apoptosis. *J. Cell Sci.* **2009**, *122*, 2554–2566.
123. Scherz-Shouval, R.; Shvets, E.; Fass, E.; Shorer, H.; Gil, L.; Elazar, Z. Reactive oxygen species are essential for autophagy and specifically regulate the activity of Atg4. *EMBO J.* **2007**, *26*, 1749–1760.
124. Longatti, A.; Tooze, S. A. Vesicular trafficking and autophagosome formation. *Cell Death Differ.* **2009**, *16*, 956–965.
125. Dunn, W. A., Jr. Studies on the mechanism of autophagy: maturation of the autophagic vacuole. *J. Cell. Biol.* **1997**, *136*, 61–70.
126. Fengsrud, M.; Lunde Sneve, M.; Øverbye, A.; Seglen, P. O. Structural aspects of mammalian autophagy. In *Autophagy*; Klionsky, D. J., Ed.; Landes Bioscience: Georgetown, TX, 2009; pp. 11–25.
127. Eskelinen, E. L. Maturation of autophagic vacuoles in mammalian cells. *Autophagy* **2005**, *1*, 1–10.
128. Fass, E.; Shvets, E.; Degani, I.; Hirschberg, K.; Elazar, Z. Microtubules support production of starvation-induced autophagosomes but not their targeting and fusion with lysosomes. *J. Biol. Chem.* **2006**, *281*, 36303–36316.
129. Jahreiss, L.; Menzies, F. M.; Rubinsztein, D. C. The itinerary of autophagosomes: from peripheral formation to kiss-and-run fusion with lysosomes. *Traffic* **2008**, *9*, 574–587.
130. Kimura, S.; Noda, T.; Yoshimori, T. Dynein-dependent movement of autophagosomes mediates efficient encounters with lysosomes. *Cell Struct. Funct.* **2008**, *33*, 109–122.
131. Cardoso, C. M.; Groth-Pedersen, L.; Høyer-Hansen, M.; Kirkegaard, T.; Corcelle, E.; Andersen, J. S., et al. Depletion of kinesin 5B affects lysosomal distribution and stability and induces peri-nuclear accumulation of autophagosomes in cancer cells. *PLoS One* **2009**, *4*, e4424.
132. Pankiv, S.; Alemu, E. A.; Brech, A.; Bruun, J. A.; Lamark, T.; Overvatn, A., et al. FYCO1 is a Rab7 effector that binds to LC3 and PI3P to mediate microtubule plus end-directed vesicle transport. *J. Cell Biol.* **2010**, *188*, 253–269.

133. Maday, S.; Wallace, K. E.; Holzbaur, E. L. F. Autophagosomes initiate distally and mature during transport toward the cell soma in primary neurons. *J. Cell Sci.* **2012**, *196*, 407–417.

134. Lee, S.; Sato, Y.; Nixon, R. A. Lysosomal proteolysis inhibition selectively disrupts axonal transport of degradative organelles and causes an Alzheimer's-like axonal dystrophy. *J. Neurosci.* **2011**, *31*, 7817–7830.

135. Liang, C.; Feng, P.; Ku, B.; Dotan, I.; Canaani, D.; Oh, B. H.; Jung, J. U. Autophagic and tumour suppressor activity of a novel Beclin1-binding protein UVRAG. *Nat. Cell. Biol.* **2006**, *8*, 688–699.

136. Wang, Q. J.; Ding, Y.; Kohtz, S.; Mizushima, N.; Cristea, I. M.; Rout, M. P., et al. Induction of autophagy in axonal dystrophy and degeneration. *J. Neurosci.* **2006**, *26*, 8057–8068.

137. Webb, J. L.; Ravikumar, B.; Rubinsztein, D. C. Microtubule disruption inhibits autophagosome-lysosome fusion: implications for studying the roles of aggresomes in polyglutamine diseases. *Int. J. Biochem. Cell. Biol.* **2004**, *36*, 2541–2550.

138. Ravikumar, B.; Acevedo-Arozena, A.; Imarisio, S.; Berger, Z.; Vacher, C.; O'Kane, C. J., et al. Dynein mutations impair autophagic clearance of aggregate-prone proteins. *Nat. Genet.* **2005**, *37*, 771–776.

139. Mehrpour, M.; Esclatine, A.; Beau, I.; Codogno, P. Overview of macroautophagy regulation in mammalian cells. *Cell Res.* **2010**, *20*, 748–762.

140. Matsunaga, K.; Saitoh, T.; Tabata, K.; Omori, H.; Satoh, T.; Kurotori, N., et al. Two Beclin 1-binding proteins, Atg14L and Rubicon, reciprocally regulate autophagy at different stages. *Nat. Cell. Biol.* **2009**, *11*, 385–396.

141. Liang, C.; Lee, J. S.; Inn, K. S.; Gack, M. U.; Li, Q.; Roberts, E. A., et al. Beclin1-binding UVRAG targets the class C Vps complex to coordinate autophagosome maturation and endocytic trafficking. *Nat. Cell Biol.* **2008**, *10*, 776–787.

142. Hyttinen, J. M. T.; Niittykoski, M.; Salminen, A.; Kaarniranta, K. Maturation of autophagosomes and endosomes: a key role for Rab7. *Biochim. Biophys. Acta* **1833**, *2013*, 503–510.

143. Zerial, M.; McBride, H. Rab proteins as membrane organizers. *Nat. Rev. Mol. Cell. Biol.* **2001**, *2*, 107–117.

144. Gutierrez, M. G.; Munafo, D. B.; Beron, W.; Colombo, M. I. Rab7 is required for the normal progression of the autophagic pathway in mammalian cells. *J. Cell. Sci.* **2004**, *117*, 2687–2697.

145. Fader, C. M.; Sanchez, D.; Furlan, M.; Colombo, M. I. Induction of autophagy promotes fusion of multivesicular bodies with autophagic vacuoles in k562 cells. *Traffic* **2008**, *9*, 230–250.

146. Mesa, R.; Salomon, C.; Roggero, M.; Stahl, P. D.; Mayorga, L. S. Rab22a affects the morphology and function of the endocytic pathway. *J. Cell. Sci.* **2001**, *114*, 4041–4049.

147. Egami, Y.; Kiryu-Seo, S.; Yoshimori, T.; Kiyama, H. Induced expressions of Rab24 GTPase and LC3 in nerve-injured motor neurons. *Biochem. Biophys. Res. Commun.* **2005**, *337*, 1206–1213.

148. Wang, T.; Ming, Z.; Xiaochun, W.; Hong, W. Rab7: role of its protein interaction cascades in endo-lysosomal traffic. *Cell. Signal.* **2011**, *23*, 516–521.

149. Harrison, R. E.; Bucci, C.; Vieira, O. V.; Schroer, T. A.; Grinstein, S. Phagosomes fuse with late endosomes and/or lysosomes by extension of membrane protrusions along microtubules: role of Rab7 and RILP. *Mol. Cell. Biol.* **2003**, *23*, 6494–6506.

150. Johansson, M.; Lehto, M.; Tanhuanpää, K.; Cover, T. L.; Olkkonen, V. M. The oxysterol-binding protein homologue ORP1L interacts with Rab7 and alters functional properties of late endocytic compartments. *Mol. Biol. Cell* **2005**, *16*, 5480–5492.

151. Johansson, M.; Rocha, N.; Zwart, W.; Jordens, I.; Janssen, L.; Kuijl, C., et al. Activation of endosomal dynein motors by stepwise assembly of Rab7-RILP-p150Glued, ORP1L, and the receptor betaIII spectrin. *J. Cell Biol.* **2007**, *176*, 459–471.

152. Rocha, N.; Kuijl, C.; van der Kant, R.; Janssen, L.; Houben, D.; Janssen, H., et al. Cholesterol sensor ORP1L contracts the ER protein VAP to control Rab7-RILP-p150 Glued and late endosome positioning. *J. Cell Biol.* **2009**, *185*, 1209–1225.
153. Gurkan, C.; Koulov, A. V.; Balch, W. E. An evolutionary perspective on eukaryotic membrane trafficking. *Adv. Exp. Med. Biol.* **2007**, *607*, 73–83.
154. Jahn, R.; Scheller, R. H. SNAREs—engines for membrane fusion. *Nat. Rev. Mol. Cell Biol.* **2006**, 631–643.
155. Fasshauer, D.; Sutton, R. B.; Brunger, A. T.; Jahn, R. Conserved structural features of the synaptic fusion complex: SNARE proteins reclassified as Q- and R-SNAREs. *Proc. Natl Acad. Sci. U.S.A.* **1998**, *95*, 15781–15786.
156. Furuta, N.; Fujita, N.; Noda, T.; Yoshimori, T.; Amano, A. Combinational soluble N-ethylmaleimide-sensitive factor attachment protein receptor proteins VAMP8 and Vti1b mediate fusion of antimicrobial and canonical autophagosomes with lysosomes. *Mol. Biol. Cell* **2010**, *21*, 1001–1010.
157. Atlashkin, V.; Kreykenbohm, V.; Eskelinen, E. L.; Wenzel, D.; Fayyazi, A.; Fischer von Mollard, G. Deletion of the SNARE vti1b in mice results in the loss of a single SNARE partner, syntaxin 8. *Mol. Cell. Biol.* **2003**, *23*, 5198–5207.
158. Fader, C. M.; Sanchez, D. G.; Mestre, M. B.; Colombo, M. I. TIVAMP/VAMP7 and VAMP3/cellubrevin: two v-SNARE proteins involved in specific steps of the autophagy/multivesicular body pathways. *Biochim. Biophys. Acta* **1793**, *2009*, 1901–1916.
159. Hong, W. SNAREs and traffic. *Biochim. Biophys. Acta* **1744**, *2005*, 120–144.
160. Itakura, E.; Kishi-Itakura, C.; Mizushima, N. The hairpin-type tail-anchored SNARE syntaxin 17 targets to autophagosomes for fusion with endosomes/lysosomes. *Cell* **2012**, *151*, 1256–1269.
161. Weng, N.; Thomas, D. D.; Groblewski, G. E. Pancreatic acinar cells express vesicle-associated membrane protein 2- and 8-specific populations of zymogen granules with distinct and overlapping roles in secretion. *J. Biol. Chem.* **2007**, *282*, 9635–9645.
162. Raiborg, C.; Stenmark, H. The ESCRT machinery in endosomal sorting of ubiquitylated membrane proteins. *Nature* **2009**, *458*, 445–452.
163. Rusten, T. E.; Vaccari, T.; Lindmo, K.; Rodahl, L. M.; Nezis, I. P.; Sem-Jacobsen, C., et al. ESCRTs and Fab1 regulate distinct steps of autophagy. *Curr. Biol.* **2007**, *17*, 1817–1825.
164. Filimonenko, M.; Stuffers, S.; Raiborg, C.; Yamamoto, A.; Malerød, L.; Fisher, E. M. C., et al. Functional multivesicular bodies are required for autophagic clearance of protein aggregates associated with neurodegenerative disease. *J. Cell. Biol.* **2007**, *179*, 485–500.
165. Skibinski, G.; Parkinson, N. J.; Brown, J. M.; Chakrabarti, L.; Lloyd, S. L.; Hummerich, H., et al. Mutations in the endosomal ESCRTIII-complex subunit CHMP2B in frontotemporal dementia. *Nat. Genet.* **2005**, *37*, 806–808.
166. Seals, D. F.; Eitzen, G.; Margolis, N.; Wickner, W. T.; Price, A. A Ypt/Rab effector complex containing the Sec1 homolog Vps33p is required for homotypic vacuole fusion. *Proc. Natl. Acad. Sci. U.S.A.* **2000**, *97*, 9402–9407.
167. Luzio, J. P.; Pryor, P. R.; Bright, N. A. Lysosomes: fusion and function. *Nat. Rev. Mol. Cell Biol.* **2007**, *8*, 622–632.
168. Tamai, K.; Tanaka, N.; Nara, A.; Yamamoto, A.; Nakagawa, I.; Yoshimori, T., et al. Role of Hrs in maturation of autophagosomes in mammalian cells. *Biochem. Biophys. Res. Commun.* **2007**, *360*, 721–727.
169. de Lartigue, J.; Polson, H.; Feldman, M.; Shokat, K.; Tooze, S. A.; Urbe, S.; Clague, M. J. PIKfyve regulation of endosome-linked pathways. *Traffic* **2009**, *10*, 883–893.
170. Crighton, D.; Wilkinson, S.; O'Prey, J.; Syed, N.; Smith, P.; Harrison, P. R., et al. DRAM, a p53-induced modulator of autophagy, is critical for apoptosis. *Cell* **2006**, *126*, 121–134.
171. Yang, Z.; Huang, J.; Geng, J.; Nair, U.; Klionsky, D. J. Atg22 recycles amino acids to link the degradative and recycling functions of autophagy. *Mol. Biol. Cell* **2006**, *17*, 5094–5104.

172. Tanaka, Y.; Guhde, G.; Suter, A.; Eskelinen, E. L.; Hartmann, D.; Lullmann-Rauch, R., et al. Accumulation of autophagic vacuoles and cardiomyopathy in LAMP-2-deficient mice. *Nature* **2000**, *406*, 902–906.

173. Eskelinen, E. L.; Schmidt, C. K.; Neu, S.; Willenborg, M.; Fuertes, G.; Salvador, N., et al. Disturbed cholesterol traffic but normal proteolytic function in LAMP-1/LAMP-2 double-deficient fibroblasts. *Mol. Biol. Cell.* **2004**, *15*, 3132–3145.

174. Huynh, K. K.; Eskelinen, E. L.; Scott, C. C.; Malevanets, A.; Saftig, P.; Grinstein, S. LAMP proteins are required for fusion of lysosomes with phagosomes. *EMBO J.* **2007**, *26*, 313–324.

175. Saftig, P.; Schroeder, B.; Blanz, J. Lysosomal membrane proteins: life between acid and neutral conditions. *Biochem. Soc. Trans.* **2010**, *38*, 1420–1423.

176. Ramachandran, N.; Munteanu, I.; Wang, P.; Aubourg, P.; Rilstone, J. J.; Israelian, N., et al. VMA21 deficiency causes an autophagic myopathy by compromising V-ATPase activity and lysosomal acidification. *Cell* **2009**, *137*, 235–246.

177. Mousavi, S. A.; Kjeken, R.; Berg, T. O.; Seglen, P. O.; Berg, T.; Brech, A. Effects of inhibitors of the vacuolar proton pump on hepatic heterophagy and autophagy. *Biochim. Biophys. Acta* **1510**, *2001*, 243–257.

178. Wang, C. W.; Klionsky, D. J. The molecular mechanism of autophagy. *Mol. Med.* **2003**, *9*, 65–76.

179. Ishihara, N.; Hamasaki, M.; Yokota, S.; Suzuki, K.; Kamada, Y.; Kihara, A., et al. Autophagosome requires specific early Sec proteins for its formation and NSF/ SNARE for vacuolar fusion. *Mol. Biol. Cell.* **2001**, *12*, 3690–3702.

180. Nara, A.; Mizushima, N.; Yamamoto, A.; Kabeya, Y.; Ohsumi, Y.; Yoshimori, T. SKD1 AAA ATPase-dependent endosomal transport is involved in autolysosome formation. *Cell. Struct. Funct.* **2002**, *27*, 29–37.

181. Shirahama, K.; Noda, T.; Ohsumi, Y. Mutational analysis of Csc1/Vps4p: involvement of endosome in regulation of autophagy in yeast. *Cell. Struct. Funct.* **1997**, *22*, 501–509.

182. Bedford, L.; Lowe, J.; Dick, L. R.; Mayer, R. J.; Brownell, J. E. Ubiquitin-like protein conjugation and the ubiquitin-proteasome system as drug targets. *Nat. Rev. Drug Discov.* **2011**, *10*, 29–46.

183. Harris, H.; Rubinsztein, D. C. Control of autophagy as a therapy against neurodegenerative diseases. *Nat. Rev. Neurol.* **2012**, *8*, 108–117.

184. Son, J. H.; Shim, J. H.; Kim, K.-H.; Ha, J. Y.; Han, J.-Y. Neuronal autophagy and neurodegenerative diseases. *Exp. Mol. Med.* **2012**, *44*, 89–98.

185. Liberski, P. P.; Budka, H.; Yanagihara, R.; Gajdusek, D. C. Neuroaxonal dystrophy in experimental Creutzfeldt–Jakob disease: electron microscopical and immunohistochemical demonstration of neurofilament accumulations within affected neurites. *J. Comp. Pathol.* **1995**, *112*, 243–255.

186. Yue, Z.; Horton, A.; Bravin, M.; DeJager, P. L.; Selimi, F.; Heintz, N. A novel protein complex linking the delta 2 glutamate receptor and autophagy: implications for neurodegeneration in lurcher mice. *Neuron* **2002**, *35*, 921–933.

187. Rubinsztein, D. C. The roles of intracellular protein-degradation pathways in neurodegeneration. *Nature* **2006**, *443*, 780–786.

188. Manzoni, C.; Lewis, P. A. Dysfunction of the autophagy/lysosomal degradation pathway is a shared feature of the genetic synucleinopathies. *FASEB J.* **2013**, *27*, 3424–3429.

189. Vogiatzi, T.; Xilouri, M.; Vekrellis, K.; Stefanis, L. Wild type alpha-synuclein is degraded by chaperone-mediated autophagy and macroautophagy in neuronal cells. *J. Biol. Chem.* **2008**, *283*, 23542–23556.

190. Cuervo, A. M.; Stefanis, L.; Fredenburg, R.; Lansbury, P. T.; Sulzer, D. Impaired degradation of mutant alpha-synuclein by chaperone-mediated autophagy. *Science* **2004**, *305*, 1292–1295.

191. Alvarez-Erviti, L.; Rodriguez-Oroz, M. C.; Cooper, J. M.; Caballero, C.; Ferrer, I.; Obeso, J. A.; Schapira, A. H. V. Chaperone-mediated autophagy markers in Parkinson disease brains. *Arch. Neurol.* **2010**, *67*, 1464–1472.

192. Anglade, P.; Vyas, S.; Javoy-Agid, F.; Herrero, M. T.; Michel, P. P.; Marquez, J., et al. Apoptosis and autophagy in nigral neurons of patients with Parkinson's disease. *Histol. Histopathol.* **1997**, *12*, 25–31.

193. Xilouri, M.; Vogiatzi, T.; Vekrellis, K.; Park, D.; Stefanis, L. Aberrant α-synuclein confers toxicity to neurons in part through inhibition of chaperone-mediated autophagy. *PLoS One* **2009**, *4*, e5515.

194. Spencer, B.; Potkar, R.; Trejo, M.; Rockenstein, E.; Patrick, C.; Gindi, R., et al. Beclin1 gene transfer activates autophagy and ameliorates the neurodegenerative pathology in α-synuclein models of Parkinson's and Lewy body diseases. *J. Neurosci.* **2009**, *29*, 13578–13588.

195. Winslow, A. R.; Chen, C. W.; Corrochano, S.; Acevedo-Arozena, A.; Gordon, D. E.; Peden, A. A., et al. α-Synuclein impairs macroautophagy: implications for Parkinson's disease. *J. Cell Biol.* **2010**, *190*, 1023–1037.

196. Narendra, D.; Tanaka, A.; Suen, D. F.; Youle, R. J. Parkin is recruited selectively to impaired mitochondria and promotes their autophagy. *J. Cell Biol.* **2008**, *183*, 795–803.

197. Arduino, D. M.; Esteves, A. R.; Cardoso, S. M. Mitochondria drive autophagy pathology via microtubule disassembly: a new hypothesis for Parkinson disease. *Autophagy* **2013**, *9*, 112–114.

198. Carra, S.; Seguin, S. J.; Lambert, H.; Landry, J. HspB8 chaperone activity toward poly(Q)-containing proteins depends on its association with Bag3, a stimulator of macroautophagy. *J. Biol. Chem.* **2008**, *283*, 1437–1444.

199. Bauer, P. O.; Goswami, A.; Wong, H. K.; Okuno, M.; Kurosawa, M.; Yamada, M., et al. Harnessing chaperone-mediated autophagy for the selective degradation of mutant huntingtin protein. *Nat. Biotechnol.* **2010**, *28*, 256–263.

200. Gu, X.; Greiner, E. R.; Mishra, R.; Kodali, R.; Osmand, A.; Finkbeiner, S., et al. Serines 13 and 16 are critical determinants of full-length human mutant huntingtin induced disease pathogenesis in HD mice. *Neuron* **2009**, *64*, 828–840.

201. Metzger, S.; Saukko, M.; Van Che, H.; Tong, L.; Puder, Y.; Riess, O.; Nguyen, H. P. Age at onset in Huntington's disease is modified by the autophagy pathway: implication of the V471A polymorphism in Atg7. *Hum Genet.* **2010**, *128*, 453–459.

202. Koga, H.; Martinez-Vicente, M.; Arias, E.; Kaushik, S.; Sulzer, D.; Cuervo, A. M. Constitutive upregulation of chaperone-mediated autophagy in Huntington's disease. *J. Neurosci.* **2011**, *31*, 18492–18505.

203. Martinez-Vicente, M.; Talloczy, Z.; Wong, E.; Tang, G. M.; Koga, H.; Kaushik, S., et al. Cargo recognition failure is responsible for inefficient autophagy in Huntington's disease. *Nat. Neurosci.* **2010**, *13*, 567–576.

204. Wang, P.; Li, B.; Zhou, L.; Fei, E.; Wang, G. The kdel receptor induces autophagy to promote the clearance of neurodegenerative disease-related proteins. *Neuroscience* **2011**, *190*, 43–55.

205. Ravikumar, B.; Vacher, C.; Berger, Z.; Davies, J. E.; Luo, S.; Oroz, L. G., et al. Inhibition of mTOR induces autophagy and reduces toxicity of polyglutamine expansions in fly and mouse models of Huntington disease. *Nat. Genet.* **2004**, *36*, 585–595.

206. Li, L.; Zhang, X.; Le, W. Altered macroautophagy in the spinal cord of SOD1 mutant mice. *Autophagy* **2008**, *4*, 290–293.

207. Sasaki, S. Autophagy in spinal cord motor neurons in sporadic amyotrophic lateral sclerosis. *J. Neuropathol. Exp. Neurol.* **2011**, *70*, 349–359.

208. Shi, P.; Wei, Y. M.; Zhang, J. Y.; Gal, J.; Zhu, H. N. Mitochondrial dysfunction is a converging point of multiple pathological pathways in amyotrophic lateral sclerosis. *J. Alzheimers Dis.* **2010**, *20*, S311–S324.

209. Neumann, M.; Sampathu, D. M.; Kwong, L. K.; Truax, A. C.; Micsenyi, M. C.; Chou, T. T., et al. Ubiquitinated TDP-43 in frontotemporal lobar degeneration and amyotrophic lateral sclerosis. *Science* **2006**, *314*, 130–133.

210. Crippa, V.; Sau, D.; Rusmini, P.; Boncoraglio, A.; Onesto, E.; Bolzoni, E., et al. The small heat shock protein B8 (Hspb8) promotes autophagic removal of misfolded proteins involved in amyotrophic lateral sclerosis (ALS). *Hum. Mol. Genet.* **2010**, *19*, 3440–3456.
211. Morimoto, N.; Nagai, M.; Ohta, Y.; Miyazaki, K.; Kurata, T.; Morimoto, M., et al. Increased autophagy in transgenic mice with a G93A mutant SOD1 gene. *Brain Res.* **2007**, *1167*, 112–117.
212. Kabuta, T.; Suzuki, Y.; Wada, K. Degradation of amyotrophic lateral sclerosis-linked mutant Cu, Zn-superoxide dismutase proteins by macroautophagy and the proteasome. *J. Biol. Chem.* **2006**, *281*, 30524–30533.
213. Johnson, J. O.; Mandrioli, J.; Benatar, M.; Abramzon, Y.; Van Deerlin, V. M.; Trojanowski, J. Q., et al. Exome sequencing reveals VCP mutations as a cause of familial ALS. *Neuron* **2010**, *68*, 857–864.
214. Gomes, C.; Escrevente, C.; Costa, J. Mutant superoxide dismutase 1 overexpression in NSC-34 cells: effect of trehalose on aggregation, TDP-43 localization and levels of co-expressed glycoproteins. *Neurosci. Lett.* **2010**, *475*, 145–149.
215. Kim, D.; Nguyen, M. D.; Dobbin, M. M.; Fischer, A.; Sananbenesi, F.; Rodgers, J. T., et al. SIRT1 deacetylase protects against neurodegeneration in models for Alzheimer's disease and amyotrophic lateral sclerosis. *EMBO J.* **2007**, *26*, 3169–3179.
216. Sala, G.; Tremolizzo, L.; Melchionda, L.; Stefanoni, G.; Derosa, M.; Susani, E., et al. A panel of macroautophagy markers in lymphomonocytes of patients with amyotrophic lateral sclerosis. *Amyotroph. Lateral Scler.* **2012**, *13*, 119–124.
217. Zhang, K.; Shi, P.; An, T.; Wang, Q.; Wang, J.; Li, Z., et al. Food restriction-induced autophagy modulates degradation of mutant SOD1 in an amyotrophic lateral sclerosis mouse model. *Brain Res.* **1519**, *2013*, 112–119.
218. Zhang, X.; Li, L.; Chen, S.; Yang, D.; Wang, Y.; Zhang, X., et al. Rapamycin treatment augments motor neuron degeneration in SOD1G93A mouse model of amyotrophic lateral sclerosis. *Autophagy* **2011**, *7*, 412–425.
219. Xu, Y.; Tian, C.; Wang, S.-B.; Xie, W.-L.; Guo, Y.; Zhang, J., et al. Activation of the macroautophagic system in scrapie-infected experimental animals and human genetic prion diseases. *Autophagy* **2012**, *8*, 1604–1620.
220. Aguado, C.; Sarkar, S.; Korolchuk, V. I.; Criado, O.; Vernia, S.; Boya, P., et al. Laforin, the most common protein mutated in Lafora disease, regulates autophagy. *Hum. Mol. Genet.* **2010**, *19*, 2867–2876.
221. Durcan, T. M.; Kontogiannea, M.; Bedard, N.; Wing, S. S.; Fon, E. A. Ataxin-3 deubiquitination is coupled to Parkin ubiquitination via E2 ubiquitin-conjugating enzyme. *J. Biol. Chem.* **2012**, *287*, 531–541.
222. Tresse, E.; Salomons, F. A.; Vesa, J.; Bott, L. C.; Kimonis, V.; Yao, T.-P., et al. VCP/p97 is essential for maturation of ubiquitin-containing autophagosomes and this function is impaired by mutations that cause IBMPFD. *Autophagy* **2010**, *6*, 217–227.
223. Farrer, M. J.; Hulihan, M. M.; Kachergus, J. M.; Dächsel, J.; Stoessl, A. J.; Grantier, L. L., et al. DCTN1 mutations in Perry syndrome. *Nat. Genet.* **2009**, *41*, 163–165.
224. Ferrucci, M.; Fulceri, F.; Toti, L.; Soldani, P.; Siciliano, G.; Paparelli, A.; Fornai, F. Protein clearing pathways in ALS. *Arch. Ital. Biol.* **2011**, *149*, 121–149.
225. Ma, J. F.; Huang, Y.; Chen, S. D.; Halliday, G. Immunohistochemical evidence for macroautophagy in neurones and endothelial cells in Alzheimer's disease. *Neuropathol. Appl. Neurobiol.* **2010**, *36*, 312–319.
226. Nixon, R. A.; Cataldo, A. M. Lysosomal system pathways: genes to neurodegeneration in Alzheimer's disease. *J. Alzheimers Dis.* **2006**, *9*, 277–289.
227. Schmidt, M. L.; DiDario, A. G.; Lee, V. M.-Y.; Trojanowski, J. Q. An extensive network of PHF tau-rich dystrophic neuritis permeates neocortex and nearly all neuritic and diffuse amyloid plaques in Alzheimer disease. *FEBS Lett.* **1994**, *344*, 69–73.

228. Nixon, R. A.; Wegiel, J.; Kumar, A.; Yu, W. H.; Peterhoff, C.; Cataldo, A.; Cuervo, A. M. Extensive involvement of autophagy in Alzheimer disease: an immuno-electron microscopy study. *J. Neuropathol. Exp. Neurol.* **2005**, *64*, 113–122.

229. Boland, B.; Kumar, A.; Lee, S.; Platt, F. M.; Wegiel, J.; Yu, W. H.; Nixon, R. A. Autophagy induction and autophagosome clearance in neurons: relationship to autophagic pathology in Alzheimer's disease. *J. Neurosci.* **2008**, *28*, 6926–6937.

230. Lunemann, J. D.; Schmidt, J.; Schmid, D.; Barthel, K.; Wrede, A.; Dalakas, M. C.; Munz, C. Beta-amyloid is a substrate of autophagy in sporadic inclusion body myositis. *Ann. Neurol.* **2007**, *61*, 476–483.

231. Yu, W. H.; Cuervo, A. M.; Kumar, A.; Peterhoff, C. M.; Schmidt, S. D.; Lee, J. H., et al. Macroautophagy—a novel beta-amyloid peptide-generating pathway activated in Alzheimer's disease. *J. Cell Biol.* **2005**, *171*, 87–98.

232. Wang, Y.; Martinez-Vicente, M.; Kruger, U.; Kaushik, S.; Wong, E.; Mandelkow, E. M., et al. Tau fragmentation, aggregation and clearance: the dual role of lysosomal processing. *Hum. Mol. Genet.* **2009**, *18*, 4153–4170.

233. Kruger, U.; Wang, Y.; Kumar, S.; Mandelkow, E. M. Autophagic degradation of tau in primary neurons and its enhancement by trehalose. *Neurobiol. Aging* **2012**, *33*, 2291–2305.

234. García-Sierra, F.; Jarero-Basulto, J. J.; Kristofikova, Z.; Majer, E.; Binder, L. I.; Ripova, D. Ubiquitin is associated with early truncation of tau protein at aspartic acid[421] during the maturation of neurofibrillary tangles in Alzheimer's disease. *Brain Pathol.* **2012**, *22*, 240–250.

235. Dolan, P. J.; Johnson, G. V. A caspase cleaved form of tau is preferentially degraded through the autophagy pathway. *J. Biol. Chem.* **2010**, *285*, 21978–21987.

236. Lipinski, M. M.; Zheng, B.; Lu, T.; Yan, Z.; Py, B. F.; Ng, A., et al. Genome-wide analysis reveals mechanisms modulating autophagy in normal brain aging and in Alzheimer's disease. *Proc. Natl. Acad. Sci. U.S.A.* **2010**, *107*, 14164–14169.

237. Pickford, F.; Masliah, E.; Britschgi, M.; Lucin, K.; Narasimhan, R.; Jaeger, P. A., et al. The autophagy-related protein beclin 1 shows reduced expression in early Alzheimer disease and regulates amyloid beta accumulation in mice. *J. Clin. Invest.* **2008**, *118*, 2190–2199.

238. Jaeger, P. A.; Pickford, F.; Sun, C. H.; Lucin, K. M.; Masliah, E.; Wyss-Coray, T. Regulation of amyloid precursor protein processing by the Beclin 1 complex. *PLoS One* **2010**, *5*, e11102.

239. Lee, J. H.; Yu, W. H.; Kumar, A.; Lee, S.; Mohan, P. S.; Peterhoff, C. M., et al. Lysosomal proteolysis and autophagy require presenilin 1 and are disrupted by Alzheimer-related PS1 mutations. *Cell* **2010**, *141*, 1146–1158.

240. Lonskaya, I.; Shekoyan, A. R.; Hebron, M. L.; Desforges, N.; Algarzae, N. K.; Moussa, C. E.-H. Diminished parkin solubility and co-localization with intraneuronal amyloid-β are associated with autophagic defects in Alzheimer's disease. *J. Alzheimer's Dis.* **2013**, *33*, 231–247.

241. Ghavami, S.; Shojaei, S.; Yeganeh, B.; Ande, S. R.; Jangamreddy, J. R.; Mehrpour, M., et al. Autophagy and apoptosis dysfunction in neurodegenerative disorders. *Progr. Neurobiol.* **2014**, *112*, 24–49.

242. Nixon, R. A. The role of autophagy in neurodegenerative disease. *Nat. Med.* **2013**, *19*, 983–997.

243. Nixon, R. A.; Yang, D. S. Autophagy failure in Alzheimer's disease—locating the primary defect. *Neurobiol. Dis.* **2011**, *43*, 38–45.

244. Cataldo, A. M.; Mathews, P. M.; Boiteau, A. B.; Hassinger, L. C.; Peterhoff, C. M.; Jiang, Y., et al. Down syndrome fibroblast model of Alzheimer-related endosome pathology. Accelerated endocytosis promotes late endocytic defects. *Am. J. Pathol.* **2008**, *173*, 370–384.

245. Zheng, S.; Clabough, E. B. D.; Sarkar, S.; Futter, M.; Rubinsztein, D. C.; Zeitlin, S. O. Deletion of the huntingtin polyglutamine stretch enhances neuronal autophagy and longevity in mice. *PLoS Genetics* **2010**, *6*, e10008.

246. Wong, E.; Cuervo, A. M. Autophagy gone awry in neurodegenerative diseases. *Nat. Neurosci. Rev.* **2010**, *13*, 805–811.
247. Weber, J. D.; Gutmann, D. H. Deconvoluting mTOR biology. *Cell Cycle* **2012**, *11*, 236–248.
248. Heitman, J.; Movva, N. R.; Hall, M. N. Targets for cell cycle arrest by the immunosuppressant rapamycin in yeast. *Science* **1991**, *253*, 905–909.
249. Zoncu, R.; Efeyan, A.; Sabatini, D. M. mTOR: from growth signal integration to cancer, diabetes and ageing. *Nature Rev. Mol. Cell Biol.* **2011**, *12*, 21–35.
250. Zhong, C. Z.; Shang, Y. C.; Wang, S.; Maiese, K. Shedding new light on neurodegenerative diseases through the mammalian target of rapamycin. *Progr. Neurobiol.* **2012**, *99*, 128–148.
251. Murakami, M.; Ichisaka, T.; Maeda, M.; Oshiro, N.; Hara, K.; Edenhofer, F., et al. mTOR is essential for growth and proliferation in early mouse embryos and embryonic stem cells. *Mol. Cell. Biol.* **2004**, *24*, 6710–6718.
252. Mounier, C.; Dumas, V.; Posner, B. I. Regulation of hepatic insulin-like growth factor-binding protein-1 gene expression by insulin: central role for mammalian target of rapamycin independent of forkhead box O proteins. *Endocrinology* **2006**, *147*, 2383–2391.
253. Gwinn, D. M.; Shackelford, D. B.; Egan, D. F.; Mihaylova, M. M.; Mery, A.; Vasquez, D. S., et al. AMPK phosphorylation of raptor mediates a metabolic checkpoint. *Mol. Cell* **2008**, *30*, 214–226.
254. Loewith, R.; Jacinto, E.; Wullschleger, S.; Lorberg, A.; Crespo, J. L.; Bonenfant, D., et al. Two TOR complexes, only one of which is rapamycin sensitive, have distinct roles in cell growth control. *Mol. Cell* **2002**, *10*, 457–468.
255. Kim, D. H.; Sarbassov, D. D.; Ali, S. M.; King, J. E.; Latek, R. R.; Erdjument-Bromage, H., et al. mTOR interacts with raptor to form a nutrient-sensitive complex that signals to the cell growth machinery. *Cell* **2002**, *110*, 163–175.
256. Carriere, A.; Cargnello, M.; Julien, L. A.; Gao, H.; Bonneil, E.; Thibault, P.; Roux, P. P. Oncogenic MAPK signaling stimulates mTORC1 activity by promoting RSK-mediated raptor phosphorylation. *Curr. Biol.* **2008**, *18*, 1269–1277.
257. Vander Haar, E.; Lee, S. I.; Bandhakavi, S.; Griffin, T. J.; Kim, D. H. Insulin signalling to mTOR mediated by the Akt/PKB substrate PRAS40. *Nat. Cell. Biol.* **2007**, *9*, 316–323.
258. Wang, L.; Harris, T. E.; Roth, R. A.; Lawrence, J. C., Jr. PRAS40 regulates mTORC1 kinase activity by functioning as a direct inhibitor of substrate binding. *J. Biol. Chem.* **2007**, *282*, 20036–20044.
259. Guertin, D. A.; Stevens, D. M.; Thoreen, C. C.; Burds, A. A.; Kalaany, N. Y.; Moffat, J., et al. Ablation in mice of the mTORC components raptor, rictor, or mLST8 reveals that mTORC2 is required for signaling to Akt-FOXO and PKCalpha, but not S6K1. *Dev. Cell* **2006**, *11*, 859–871.
260. Chen, E. J.; Kaiser, C. A. LST8 negatively regulates amino acid biosynthesis as a component of the TOR pathway. *J. Cell Biol.* **2003**, *161*, 333–347.
261. Peterson, T. R.; Laplante, M.; Thoreen, C. C.; Sancak, Y.; Kang, S. A.; Kuehl, W. M., et al. DEPTOR is an mTOR inhibitor frequently overexpressed in multiple myeloma cells and required for their survival. *Cell* **2009**, *137*, 873–886.
262. Zhao, Y.; Xiong, X.; Sun, Y. DEPTOR, an mTOR inhibitor, is a physiological substrate of SCF(betaTrCP) E3 ubiquitin ligase and regulates survival and autophagy. *Mol. Cell* **2011**, *44*, 304–316.
263. Masri, J.; Bernath, A.; Martin, J.; Jo, O. D.; Vartanian, R.; Funk, A.; Gera, J. mTORC2 activity is elevated in gliomas and promotes growth and cell motility via overexpression of rictor. *Cancer Res.* **2007**, *67*, 11712–11720.
264. Glidden, E. J.; Gray, L. G.; Vemuru, S.; Li, D.; Harris, T. E.; Mayo, M. W. Multiple site acetylation of Rictor stimulates mammalian target of rapamycin complex 2 (mTORC2)-dependent phosphorylation of Akt protein. *J. Biol. Chem.* **2012**, *287*, 581–588.
265. Yang, Q.; Inoki, K.; Ikenoue, T.; Guan, K. L. Identification of Sin1 as an essential TORC2 component required for complex formation and kinase activity. *Genes Dev.* **2006**, *20*, 2820–2832.

266. Frias, M. A.; Thoreen, C. C.; Jaffe, J. D.; Schroder, W.; Sculley, T.; Carr, S. A.; Sabatini, D. M. mSin1 is necessary for Akt/PKB phosphorylation, and its isoforms define three distinct mTORC2s. *Curr. Biol.* **2006**, *16*, 1865–1870.
267. Pearce, L. R.; Huang, X.; Boudeau, J.; Pawlowski, R.; Wullschleger, S.; Deak, M., et al. Identification of Protor as a novel Rictor-binding component of mTOR complex-2. *Biochem. J.* **2007**, *405*, 513–522.
268. Pearce, L. R.; Sommer, E. M.; Sakamoto, K.; Wullschleger, S.; Alessi, D. R. Protor-1 is required for efficient mTORC2-mediated activation of SGK1 in the kidney. *Biochem. J.* **2011**, *436*, 169–179.
269. Thedieck, K.; Polak, P.; Kim, M. L.; Molle, K. D.; Cohen, A.; Jeno, P., et al. PRAS40 and PRR5-like protein are new mTOR interactors that regulate apoptosis. *PLoS One* **2007**, *2*, e1217.
270. Woo, S. Y.; Kim, D. H.; Jun, C. B.; Kim, Y. M.; Haar, E. V.; Lee, S. I., et al. PRR5, a novel component of mTOR complex 2, regulates platelet-derived growth factor receptor beta expression and signaling. *J. Biol. Chem.* **2007**, *282*, 25604–25612.
271. Sarbassov, D. D.; Ali, S. M.; Kim, D. H.; Guertin, D. A.; Latek, R. R.; Erdjument-Bromage, H., et al. Rictor, a novel binding partner of mTOR, defines a rapamycin-insensitive and raptor-independent pathway that regulates the cytoskeleton. *Curr. Biol.* **2004**, *14*, 1296–1302.
272. Gulhati, P.; Bowen, K. A.; Liu, J.; Stevens, P. D.; Rychahou, P. G.; Chen, M., et al. mTORC1 and mTORC2 regulate EMT, motility, and metastasis of colorectal cancer via RhoA and Rac1 signaling pathways. *Cancer Res.* **2011**, *71*, 3246–3256.
273. Hernandez-Negrete, I.; Carretero-Ortega, J.; Rosenfeldt, H.; Hernandez-Garcia, R.; Calderon-Salinas, J. V.; Reyes-Cruz, G., et al. PRex1 links mammalian target of rapamycin signaling to Rac activation and cell migration. *J. Biol. Chem.* **2007**, *282*, 23708–23715.
274. Huang, J.; Manning, B. D. The TSC1-TSC2 complex: a molecular switchboard controlling cell growth. *Biochem. J.* **2008**, *412*, 179–190.
275. Long, X.; Lin, Y.; Ortiz-Vega, S.; Yonezawa, K.; Avruch, J. Rheb binds and regulates the mTOR kinase. *Curr. Biol.* **2005**, *15*, 702–713.
276. Sancak, Y.; Peterson, T. R.; Shaul, Y. D.; Lindquist, R. A.; Thoreen, C. C.; Bar-Peled, L.; Sabatini, D. M. The Rag GTPases bind raptor and mediate amino acid signaling to mTORC1. *Science* **2008**, *320*, 1496–1501.
277. Sato, T.; Nakashima, A.; Guo, L.; Tamanoi, F. Specific activation of mTORC1 by Rheb G-protein in vitro involves enhanced recruitment of its substrate protein. *J. Biol. Chem.* **2009**, *284*, 12783–12791.
278. Bai, X.; Ma, D.; Liu, A.; Shen, X.; Wang, Q. J.; Liu, Y.; Jiang, Y. Rheb activates mTOR by antagonizing its endogenous inhibitor FKBP38. *Science* **2007**, *318*, 977–980.
279. Inoki, K.; Li, Y.; Zhu, T.; Wu, J.; Guan, K. L. TSC2 is phosphorylated and inhibited by Akt and suppresses mTOR signalling. *Nat. Cell Biol.* **2002**, *4*, 648–657.
280. Inoki, K.; Zhu, T.; Guan, K. L. TSC2 mediates cellular energy response to control cell growth and survival. *Cell* **2003**, *115*, 577–590.
281. Manning, B. D.; Tee, A. R.; Logsdon, M. N.; Blenis, J.; Cantley, L. C. Identification of the tuberous sclerosis complex-2 tumor suppressor gene product tuberin as a target of the phosphoinositide 3-kinase/akt pathway. *Mol. Cell* **2002**, *10*, 151–162.
282. Sancak, Y.; Bar-Peled, L.; Zoncu, R.; Markhard, A. L.; Nada, S.; Sabatini, D. M. Ragulator-Rag complex targets mTORC1 to the lysosomal surface and is necessary for its activation by amino acids. *Cell* **2010**, *141*, 290–303.
283. Brugarolas, J.; Lei, K.; Hurley, R. L.; Manning, B. D.; Reiling, J. H.; Hafen, E., et al. Regulation of mTOR function in response to hypoxia by REDD1 and the TSC1/TSC2 tumor suppressor complex. *Genes Dev.* **2004**, *18*, 2893–2904.
284. Oshiro, N.; Takahashi, R.; Yoshino, K.; Tanimura, K.; Nakashima, A.; Eguchi, S., et al. The proline-rich Akt substrate of 40 kDa (PRAS40) is a physiological substrate of mammalian target of rapamycin complex 1. *J. Biol. Chem.* **2007**, *282*, 20329–20339.

285. Wang, L.; Rhodes, C. J.; Lawrence, J. C. Activation of mammalian target of rapamycin (mTOR) by insulin is associated with stimulation of 4EBP1 binding to dimeric mTOR complex 1. *J. Biol. Chem.* **2006**, *281*, 24293–24303.

286. Fenton, I. R.; Gout, I. T. Functions and regulation of the 70kDa ribosomal S6 kinases. *Int. J. Biochem. Cell Biol.* **2011**, *43*, 47–59.

287. Hara, K.; Maruki, Y.; Long, X.; Yoshino, K.; Oshiro, N.; Hidayat, S., et al. Raptor, a binding partner of target of rapamycin (TOR), mediates TOR action. *Cell* **2002**, *110*, 177–189.

288. Mader, S.; Lee, H.; Pause, A.; Sonenberg, N. The translation initiation factor eIF-4E binds to a common motif shared by the translation factor eIF-4 gamma and the translational repressors 4E-binding proteins. *Mol. Cell. Biol.* **1995**, *15*, 4990–4997.

289. Cota, D.; Proulx, K.; Blake Smith, K. A.; Kozma, S. C.; Thomas, G.; Woods, S. C.; Seeley, R. J. Hypothalamic mTOR signaling regulates food intake. *Science* **2006**, *312*, 927–930.

290. Fingar, D. C.; Richardson, C. J.; Tee, A. R.; Cheatham, L.; Tsou, C.; Blenis, J. mTOR controls cell cycle progression through its cell growth effectors S6K1 and 4EBP1/eukaryotic translation initiation factor 4E. *Mol. Cell. Biol.* **2004**, *24*, 200–216.

291. Jastrzebski, K.; Hannan, K. M.; Tchoubrieva, E. B.; Hannan, R. D.; Pearson, R. B. Coordinate regulation of ribosome biogenesis and function by the ribosomal protein S6 kinase, a key mediator of mTOR function. *Growth Factors* **2007**, *25*, 209–226.

292. Lenz, G.; Avruch, J. Glutamatergic regulation of the p70S6 kinase in primary mouse neurons. *J. Biol. Chem.* **2005**, *280*, 38121–38124.

293. Easley, C. A.; Ben-Yehudah, A.; Redinger, C. J.; Oliver, S. L.; Varum, S. T.; Eisinger, V. M., et al. mTOR-mediated activation of p70 S6K induces differentiation of pluripotent human embryonic stem cells. *Cell. Reprogr.* **2011**, *12*, 263–273.

294. Alers, S.; Loeffler, A. S.; Wesselborg, S.; Stork, B. Role of AMPK-mTOR-Ulk1/2 in the regulation of autophagy: cross-talk, shortcuts, and feedbacks. *Mol. Cell. Biol.* **2012**, *32*, 2–11.

295. Hosokawa, N.; Hara, T.; Kaizuka, T.; Kishi, C.; Takamura, A.; Miura, Y., et al. Nutrient-dependent mTORC1 association with the ULK1-Atg13-FIP200 complex required for autophagy. *Mol. Biol. Cell* **2009**, *20*, 1981–1991.

296. Shang, L.; Chen, S.; Du, F.; Li, S.; Zhao, L.; Wang, X. Nutrient starvation elicits an acute autophagic response mediated by Ulk1 dephosphorylation and its subsequent dissociation from AMPK. *Proc. Natl. Acad. Sci. U.S.A.* **2011**, *108*, 4788–4793.

297. Jung, C. H.; Jun, C. B.; Ro, S. H.; Kim, Y. M.; Otto, N. M.; Cao, J., et al. ULK-Atg13-FIP200 complexes mediate mTOR signaling to the autophagy machinery. *Mol. Biol. Cell* **2009**, *20*, 1992–2003.

298. Shang, L.; Wang, X. AMPK and mTOR coordinate the regulation of Ulk1 and mammalian autophagy initiation. *Autophagy* **2011**, *7*, 924–926.

299. Yu, L.; McPhee, C. K.; Zheng, L.; Mardones, G. A.; Rong, Y.; Peng, J., et al. Termination of autophagy and reformation of lysosomes regulated by mTOR. *Nature* **2010**, *465*, 942–946.

300. Koren, I.; Reem, E.; Kimchi, A. DAP1, a novel substrate of mTOR, negatively regulates autophagy. *Curr. Biol.* **2010**, *20*, 1093–1098.

301. Le, X. F.; Mao, W.; Lu, Z.; Carter, B. Z.; Bast, R. C., Jr. Dasatinib induces autophagic cell death in human ovarian cancer. *Cancer* **2010**, *116*, 4980–4990.

302. Johnson, S. C.; Rabinovitch, P. S.; Kaeberlein, M. mTOR is a key modulator of ageing and age-related disease. *Nature* **2013**, *493*, 338–345.

303. Jaber, N.; Dou, Z.; Chen, J. S.; Catanzaro, J.; Jiang, Y. P.; Ballou, L. M., et al. Class III PI3K Vps34 plays an essential role in autophagy and in heart and liver function. *Proc. Natl. Acad. Sci. U.S.A.* **2012**, *109*, 2003–2008.

304. Um, S. H.; Frigerio, F.; Watanabe, M.; Picard, F.; Joaquin, M.; Sticker, M., et al. Absence of S6K1 protects against age- and diet-induced obesity while enhancing insulin sensitivity. *Nature* **2004**, *431*, 200–205.

305. Pierdominici, M.; Vacirca, D.; Delunardo, F.; Ortona, E. mTOR signaling and metabolic regulation of T cells: new potential therapeutic targets in autoimmune diseases. *Curr. Pharm. Des.* **2011**, *17*, 3888–3897.
306. Amiel, E.; Everts, B.; Freitas, T. C.; King, I. L.; Curtis, J. D.; Pearce, E. L.; Pearce, E. J. Inhibition of mechanistic target of rapamycin promotes dendritic cell activation and enhances therapeutic autologous vaccination in mice. *J. Immunol.* **2012**, *189*, 2151–2158.
307. Sui, L.; Wang, J.; Li, B. M. Role of the phosphoinositide 3-kinase-Akt-mammalian target of the rapamycin signaling pathway in long-term potentiation and trace fear conditioning memory in rat medial prefrontal cortex. *Learning & Memory* **2008**, *15*, 762–776.
308. Ryther, R. C.; Wong, M. Mammalian target of rapamycin (mTOR) inhibition: potential for antiseizure, antiepileptogenic, and epileptostatic therapy. *Curr. Neurol. Neurosci. Rep.* **2012**, *12*, 410–418.
309. Halloran, J.; Hussong, S. A.; Burbank, R.; Podlutskaya, N.; Fischer, K. E.; Sloane, L. B., et al. Chronic inhibition of mammalian target of rapamycin by rapamycin modulates cognitive and non-cognitive components of behavior throughout lifespan in mice. *Neuroscience* **2012**, *223*, 102–113.
310. Brech, A.; Ahlquist, T.; Lothe, R. A.; Stenmark, H. Autophagy in tumour suppression and promotion. *Mol. Oncol.* **2009**, *3*, 366–375.
311. Dalby, K. N.; Tekedereli, I.; Lopez-Berestein, G.; Ozpolat, B. Targeting the prodeath and prosurvival functions of autophagy as novel therapeutic strategies in cancer. *Autophagy* **2010**, *6*, 322–329.
312. Koh, P. O. Melatonin prevents ischemic brain injury through activation of the mTOR/p70S6 kinase signaling pathway. *Neurosci. Lett.* **2008**, *444*, 74–78.
313. Chong, Z. Z.; Shang, Y. C.; Wang, S.; Maiese, K. Shedding new light on neurodegenerative diseases through the mammalian target of rapamycin. *Progr. Neurobiol.* **2012**, *99*, 128–148.
314. Arachchige Don, A. S.; Tsang, C. K.; Kazdoba, T. M.; D'Arcangelo, G.; Young, W.; Zheng, X. F. S. Targeting mTOR as a novel therapeutic strategy for traumatic CNS injuries. *Drug Discov. Today* **2012**, *17*, 861–868.
315. Lamming, D. W.; Ye, L.; Sabatini, D. M.; Baur, J. A. Rapalogs and mTOR inhibitors as anti-aging therapeutics. *J. Clin. Invest.* **2013**, *123*, 980–989.
316. Jia, K.; Chen, D.; Riddle, D. L. The TOR pathway interacts with the insulin signaling pathway to regulate *C. elegans* larval development, metabolism and life span. *Development* **2004**, *131*, 3897–3906.
317. Partridge, L.; Alic, N.; Bjedov, I.; Piper, M. D. W. Ageing in *Drosophila*: The role of the insulin/Igf and TOR signalling network. *Exp Gerontol.* **2011**, *46*, 376–381.
318. Medvedik, O.; Lamming, D. W.; Kim, K. D.; Sinclair, D. A. MSN2 and MSN4 link calorie restriction and TOR to sirtuin-mediated lifespan extension in *Saccharomyces cerevisiae*. *PLoS Biol.* **2007**, *5*, e261.
319. Harrison, D. E.; Strong, R.; Sharp, Z. D.; Nelson, J. F.; Astle, C. M.; Flurkey, K., et al. Rapamycin fed late in life extends lifespan in genetically heterogeneous mice. *Nature* **2009**, *460*, 392–395.
320. Kolosova, N. G.; Vitovtov, A. O.; Muraleva, N. A.; Akulov, A. E.; Stefanova, N. A.; Blagosklonny, M. V. Rapamycin suppresses brain aging in senescence-accelerated OXYS rats. *Aging* **2013**, *5*, 474–484.
321. Abdulrahman, B. A.; Khweek, A. A.; Akhter, A.; Caution, K.; Kotrange, S.; Abdelaziz, D. H., et al. Autophagy stimulation by rapamycin suppresses lung inflammation and infection by *Burkholderia cenocepacia* in a model of cystic fibrosis. *Autophagy* **2011**, *7*, 1359–1370.
322. Yang, S. B.; Tien, A.-C.; Boddupalli, G.; Xu, A. W.; Jan, Y. N.; Jan, L. Y. Rapamycin ameliorates age-dependent obesity associated with increased mTOR signaling in hypothalamic POMC neurons. *Neuron* **2012**, *75*, 425–436.

323. Nussenblatt, R. B.; Byrnes, G.; Sen, H. N.; Yeh, S.; Faia, L.; Meyerle, C., et al. A randomized pilot study of systemic immunosuppression in the treatment of age-related macular degeneration with choroidal neovascularization. *Retina* **2010**, *30*, 1579–1587.

324. Choi, K. C.; Kim, S. H.; Ha, J. Y.; Kim, S. T.; Son, J. H. A novel mTOR activating protein protects dopamine neurons against oxidative stress by repressing autophagy related cell death. *J. Neurochem.* **2010**, *112*, 366–376.

325. Malagelada, C.; Jin, Z. H.; Jackson-Lewis, V.; Przedborski, S.; Greene, L. A. Rapamycin protects against neuron death in in vitro and in vivo models of Parkinson's disease. *J. Neurosci.* **2010**, *30*, 1166–1175.

326. Floto, R. A.; Sarkar, S.; Perlstein, E. O.; Kampmann, B.; Schreiber, S. L.; Rubinsztein, D. C. Small molecule enhancers of rapamycin-induced TOR inhibition promote autophagy, reduce toxicity in Huntington's disease models and enhance killing of mycobacteria by macrophages. *Autophagy* **2007**, *3*, 620–622.

327. Wang, I. F.; Guo, B. S.; Liu, Y. C.; Wu, C. C.; Yang, C. H.; Tsai, K. J.; Shen, C. K. Autophagy activators rescue and alleviate pathogenesis of a mouse model with proteinopathies of the TAR DNA-binding protein 43. *Proc. Natl Acad. Sci. U.S.A.* **2012**, *109*, 15024–15029.

328. Yoon, S. Y.; Choi, J. E.; Kweon, H. S.; Choe, H.; Kim, S. W.; Hwang, O., et al. Okadaic acid increases autophagosomes in rat neurons: implications for Alzheimer's disease. *J. Neurosci. Res.* **2008**, *86*, 3230–3239.

329. Morel, M.; Couturier, J.; Pontcharraud, R.; Gil, R.; Fauconneau, B.; Paccalin, M.; Page, G. Evidence of molecular links between PKR and mTOR signalling pathways in Abeta neurotoxicity: role of p53, Redd1 and TSC2. *Neurobiol. Dis.* **2009**, *36*, 151–161.

330. Caccamo, A.; Maldonado, M. A.; Majumder, S.; Medina, D. X.; Holbein, W.; Magrí, A.; Oddo, S. Naturally secreted amyloid-β increases mammalian target of rapamycin (mTOR) activity via a PRAS40–mediated mechanism. *J. Biol. Chem.* **2011**, *286*, 8924–8932.

331. Shen, D.; Coleman, J.; Chan, E.; Nicholson, T. P.; Dai, L.; Sheppard, P. W.; Patton, W. F. Novel cell- and tissue-based assays for detecting misfolded and aggregated protein accumulation within aggresomes and inclusion bodies. *Cell. Biochem. Biophys.* **2011**, *60*, 173–185.

332. Spilman, P.; Podlutskaya, N.; Hart, M. J.; Debnath, J.; Gorostiza, O.; Bredesen, D., et al. Inhibition of mTOR by rapamycin abolishes cognitive deficits and reduces amyloid-beta levels in a mouse model of Alzheimer's disease. *PLoS One* **2010**, *5*, e9979.

333. Paccalin, M.; Pain-Barc, S.; Pluchon, C.; Paul, C.; Besson, M. N.; Carret-Rebillat, A. S., et al. Activated mTOR and PKR kinases in lymphocytes correlate with memory and cognitive decline in Alzheimer's disease. *Dem. Geriat. Cognit. Disord.* **2006**, *22*, 320–326.

334. Tang, Z.; Bereczki, E.; Zhang, H.; Wang, S.; Li, C.; Ji, X., et al. Mammalian target of rapamycin (mTor) mediates tau protein dyshomeostasis. *J. Biol. Chem.* **2013**, *288*, 25556–25570.

335. Hanger, D. P.; Noble, W. Functional implications of glycogen synthase kinase-3-mediated tau phosphorylation. *Int. J. Alzheim. Dis.* **2011**, 352805.

336. Tian, Q.; Wang, J. Role of serine/threonine protein phosphatase in Alzheimer's disease. *NeuroSignals* **2002**, *11*, 262–269.

337. Meske, V.; Albert, F.; Ohm, T. G. Coupling of mammalian target of rapamycin with phosphoinositide 3-kinase signaling pathway regulates protein phosphatase 2A- and glycogen synthase kinase-3-dependent phosphorylation of Tau. *J. Biol. Chem.* **2008**, *283*, 100–109.

338. Marwarha, G.; Dasari, B.; Prabhakara, J. P. R.; Schommer, J.; Ghribi, O. β-Amyloid regulates leptin expression and tau phosphorylation through the mTORC1 signaling pathway. *J. Neurochem.* **2010**, *115*, 373–384.

339. Liu, Y.; Su, Y.; Wang, J.; Sun, S.; Wang, T.; Qiao, X., et al. Rapamycin decreases tau phosphorylation at Ser214 through regulation of cAMP dependent kinase. *Neurochem. Int.* **2013**, *62*, 458–467.

340. Nemoto, T.; Satoh, S.; Maruta, T.; Kanai, T.; Yoshikawa, N.; Miyazaki, S., et al. Homologous posttranscriptional regulation of insulin-like growth factor-I receptor level via glycogen synthase kinase-3beta and mammalian target of rapamycin in adrenal chromaffin cells: effect on tau phosphorylation. *Neuropharmacology* **2010**, *58*, 1097–1108.
341. Deng, Y.; Li, B.; Liu, Y.; Iqbal, K.; Grundke-Iqbal, I.; Gong, C. X. Dysregulation of insulin signaling, glucose transporters, O-GlcNAcylation, and phosphorylation of tau and neurofilaments in the brain: implication for Alzheimer's disease. *Am. J. Pathol.* **2009**, *175*, 2089–2098.
342. Ma, Y. Q.; Wu, D. K.; Liu, J. K. mTOR and tau phosphorylated proteins in the hippocampal tissue of rats with type 2 diabetes and Alzheimer's disease. *Mol. Med. Rep.* **2013**, *7*, 623–627.
343. An, W. L.; Cowburn, R. F.; Li, L.; Braak, H.; Alafuzoff, I.; Iqbal, K., et al. Up-regulation of phosphorylated/activated p70 S6 kinase and its relationship to neurofibrillary pathology in Alzheimer's disease. *Am. J. Pathol.* **2003**, *163*, 591–607.
344. Li, X.; An, W. L.; Alafuzoff, I.; Soininen, H.; Winblad, B.; Pei, J. J. Phosphorylated eukaryotic translation factor 4E is elevated in Alzheimer brain. *NeuroReport* **2004**, *15*, 2237–2240.
345. Li, X.; Alafuzoff, I.; Soininen, H.; Winblad, B.; Pei, J. J. Levels of mTOR and its downstream targets 4E-BP1, eEF2, and eEF2 kinase in relationships with tau in Alzheimer's disease brain. *FEBS J.* **2005**, *272*, 4211–4220.
346. Morita, T.; Sobue, K. Specification of neuronal polarity regulated by local translation of CRMP2 and tau via the mTOR–p70S6K pathway. *J. Biol. Chem.* **2009**, *284*, 27734–27745.
347. O'Neill, C. PI3-kinase/Akt/mTOR signaling: impaired on/off switches in aging, cognitive decline and Alzheimer's disease. *Exp. Gerontol.* **2013**, *48*, 647–653.
348. Cai, Z.; Zhao, B.; Li, K.; Zhang, L.; Li, C.; Quazi, S. H.; Tan, Y. Mammalian target of rapamycin: a valid therapeutic target through the autophagy pathway for Alzheimer's disease? *J. Neurosci. Res.* **2012**, *90*, 1105–1118.
349. Caccamo, A.; Majumder, S.; Richardson, A.; Strong, R.; Oddo, S. Molecular interplay between mammalian target of rapamycin (mTOR), amyloid-beta, tau: effects on cognitive impairments. *J. Biol. Chem.* **2010**, *285*, 13107–13120.
350. Caccamo, A.; Magrí, A.; Medina, D. X.; Wisely, E. V.; Lopez-Aranda, M. F.; Silva, A. J.; Oddo, S. mTOR regulates tau phosphorylation and degradation: implications for Alzheimer's disease and other tauopathies. *Aging Cell* **2013**, *12*, 370–380.
351. Seneci, P. Chemical modulators of protein misfolding and neurodegenerative disease. Elsevier, accepted for publication, **2015**.

5

Selective Disposal of Insoluble Protein Aggregates
Pick, Transport and Remove to Cure

5.1 AGGREPHAGY-MEDIATED DEGRADATION OF PROTEIN AGGREGATES

Autophagy, and starvation-induced macroautophagy (MA, Chapter 4) [1,2] in particular, is often perceived as a non-specific, stimulus-initiated process to replenish nutrients—store energy in cells under stress conditions.

Selective autophagy processes are needed, either in particular cell types, or in specific environmental conditions. However, it may happen that a specific type of cell component must be disposed of *via* selective autophagy, while other cell contents do not experience autophagy through starvation or other stress conditions.

Chaperone-mediated autophagy (CMA, Figure 5.1) [3] is a selective, autophagosome (AP)-free process that involves translocation of soluble, KFERQ sequence-containing cytosolic proteins into the lysosome (LS) by interacting with the constitutive cytosolic Hsc70 (cyt-Hsc70) chaperone [4], and their subsequent degradation.

Translation (step 1, Figure 5.1) produces a large number of KFERQ-containing proteins. When properly folded (step 2a), they hide their CMA-specific sequence and perform their physiological functions (step 3), escaping recognition by cyt-Hsc70. Conversely, the KFERQ sequence becomes "visible" to the cyt-Hsc70/co-chaperone complex when the protein is misfolded/dysfunctional (step 2b [5]). The cyt-Hsc70/KFERQ-containing protein complexes are then formed (step 4) and transported to the LS [6], where they bind to the lysosomal membrane receptor glycoprotein Lgp96 (LAMP-2A) (step 5c [7]). LAMP-2A is translated (step 5a) and inserted into the LS membrane (step 5b), where its concentration increases in stress/autophagy-promoting conditions. Together with lysosomal

Molecular Targets in Protein Misfolding and Neurodegenerative Disease. DOI: 10.1016/B978-0-12-800186-8.00005-5

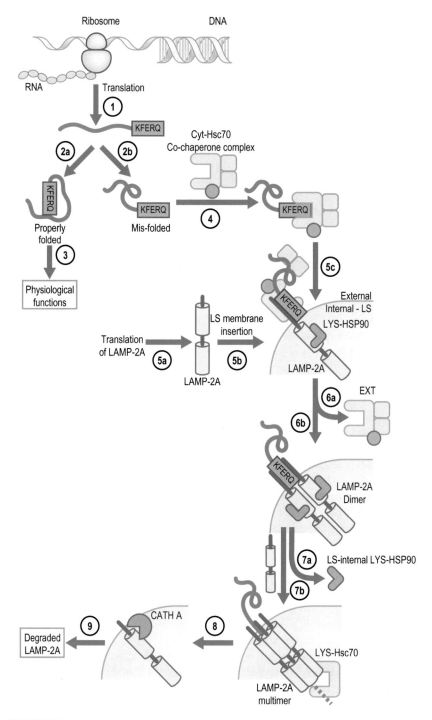

FIGURE 5.1   Chaperone-mediated autophagy (CMA): main features.

Hsp90 (LYS-Hsp90), LAMP-2A receptors assist the unfolding of KFERQ-containing proteins (step 6b [8]), while cyt-Hsc70 and its co-chaperones are released (step 6a) and re-enter the CMA cycle. Together with lysosomal Hsc70 (LYS-Hsc70), LAMP-2A receptors then promote the translocation of KFERQ-containing protein into the LS for their degradation (step 7b [9]), while LYS-Hsp90 is released (step 7a). The binding of LAMP-2A receptors to cyt-Hsc70/KFERQ-containing protein complexes stimulates their supramolecular assembly into multimeric LAMP-2A/cyt-Hsc70/KFERQ-containing protein complexes (steps 6b and 7b [3]). Once translocation and degradation of the substrate protein is completed, unstable multimeric LAMP-2A complexes are disassembled (step 8), and LAMP-2A monomers are degraded by cathepsin A (Cath A) in specific lysosomal compartments (step 9, Figure 5.1). LAMP-2A translation/membrane insertion and its lysosomal degradation provide an overall regulation of CMA [10].

Organelle-specific autophagic processes (Figure 5.2) "clean the house" when cells contain defective, foreign, or surplus insoluble material that threatens the cell survival. They take advantage of the AP structure, and of inner AP membrane-anchored selective autophagy receptors (SARs) that recognize and drive different cargos (CAR) to selective autophagy/degradation.

Mitophagy (removal of damaged mitochondria, 1–2 μm) [11] happens through the activation of MA and selective priming of damaged mitochondria either by the PTEN-induced putative kinase 1 (PINK1)–parkin signaling pathway or by the mitophagic receptors Bcl-2/E1B-19 kDa interacting protein 3 (Bnip3) and Nix/Bnip3L [12]. *Xenophagy* (bacteria or viruses, 1–5 μm) [13] is a specific autophagy mechanism activated by Toll-like receptors during infection, which implies the engulfment of ubiquitinated intracellular pathogens by APs after their recognition by autophagy receptors [14]. *Reticulophagy* (surplus endoplasmic reticulum (ER), 1–5 μm) [15] prevents unfolded protein response (UPR)-induced ER expansion by causing ER fragmentation and partial recycling to create ER-containing APs [16]. *Ribophagy* (ribosomes, 20–25 nm) [17] is a selective autophagy process leading to ribosomal turnover, i.e., a large supply of amino acids and a protein translation-down-regulating factor that ensures cell survival during nutrient limiting conditions [18]. *Lipophagy* (lipid droplets, 100–1000 nm) [19] contributes at least partially to the supply of free fatty acids for energy production through β-oxidation in response to fasting [20]. *Pexophagy* (peroxisomes, 200–500 nm) [21] regulates peroxisome homeostasis to adapt to external conditions through either micropexophagy (sequestration and degradation of a cluster of peroxisomes) or macropexophagy (individual sequestration and engulfment of a peroxisome in a pexophagosome) [22]. *Zymophagy* (zymogen granules, 100–1000 nm) [23] is a protective cell response in pancreatic acinar cells to degrade pancreatitis-activated zymogen granules that otherwise could self-digest the acinar

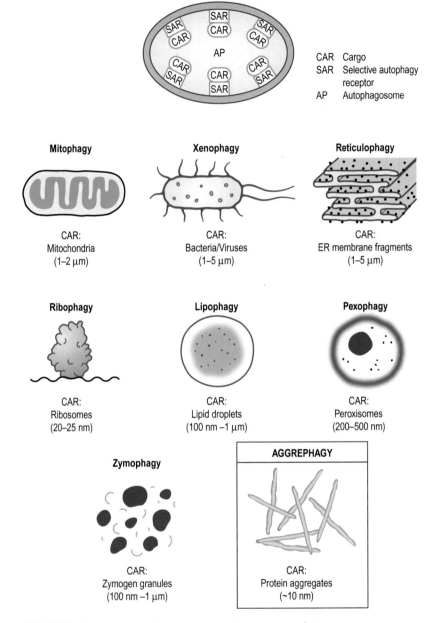

FIGURE 5.2   Features of the main organelle-selective autophagy processes.

cells and cause the progression of pancreatitis [24]. Finally, *aggrephagy* (protein aggregates, ≈10 nm, Figure 5.2) [25] is the main topic of this chapter.

Misfolded proteins are chaperone substrates that must be properly refolded to be fully functional. When the refolding chaperone system fails, soluble "damaged-beyond-repair" proteins may be degraded by complementary ubiquitin–proteasome system (UPS) and CMA processes. If UPS and CMA fail to clean up damaged proteins, their aggregation into soluble, toxic oligomeric species is a common event [26]. Truth be told, further aggregation of soluble oligomers into insoluble protein aggregates is a rescuing event, as the latter entities are less toxic and more amenable to autophagic degradation [27]. Soluble tau oligomers and insoluble tau fibrils exemplify the concept [28].

Nevertheless, insoluble tau, or any misfolded insoluble protein aggregate needs eventually to be disposed of in a cell. Neurons, in particular, are a privileged location for selective protein autophagy/aggrephagy [25,29]. They are post-mitotic cells, and cannot "dilute/partition" their proteic waste through cell division. Thus, when healthy, they need a significant induction and clearance of APs, in a dynamic equilibrium, to "clean the house" [30]. Rather than continuously engage in bulk-aspecific autophagy, they often need to deploy selective autophagic processes targeting only large protein aggregates, which become more and more abundant in aging neurons [31,32].

Non-selective autophagy is a complex, robust and flexible system, becoming activated or deactivated following a number of external and internal signals. It is also an ideal machinery to execute selective autophagy processes, as long as very few proteins are capable of switching it between substrate-selective and non-selective autophagy. The non-selective autophagy processes and organelles (APs, endosomes (ESs), LSs, and the molecular entities leading to their formation, maturation, fusion, and regulation), once activated by starvation or stress-related inputs, aspecifically confine mixed cellular contents in soon-to-be-degraded compartments.

Conversely, the conversion of freshly translated proteins into either misfolded soluble species (step 1, Figure 5.3) or insoluble aggregates (step 2) in otherwise viable cellular environments needs a more specific degradation pathway to avoid cellular toxicity (step 3). In addition to previously mentioned CMA for KFERQ-containing proteins, soluble misfolded proteins can be ubiquitinated with a K48 isopeptide bond (step 4a) that drives the K48-ubiquitinated proteins towards proteasomal degradation (step 4b, see Chapter 3).

K63 ubiquitination of soluble (step 5a) or insoluble species (step 5b) leads to small K63-ubiquitinated aggregates. Small K63-ubiquitinated aggregates bind to K63 UBQ-recognizing aggrephagy transporters (ATs, step 6, see below). Dynein motors (DYN) bind the resulting protein complexes

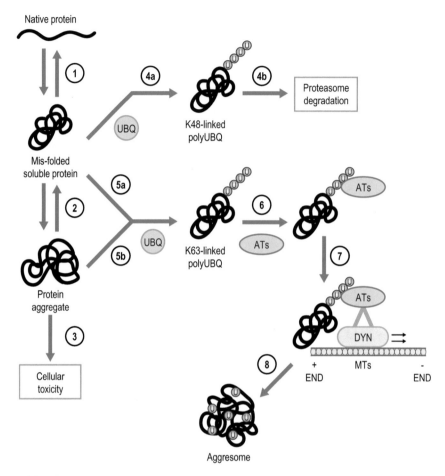

FIGURE 5.3   The aggrephagy process: from misfolded soluble proteins to aggresomes.

and transport them to the *minus* end of microtubules (MTs, step 7), similarly to AP transportation in macroautophagy (MA, see Chapter 4). Once there, K63-ubiquitinated aggregates merge into *aggresomes*, large insoluble inclusion bodies (step 8, Figure 5.3 [25]). Aggresomes are usually formed at the microtubule organizing center (MTOC) [33], close to the nuclear envelope, but aggresome-like structures may be found even in cell periphery [34]. Aggresomes mostly contain K63-ubiquitinated proteins [35] and are enclosed by filamentous proteins that protect their integrity [34]. Localized, large aggresomes are less likely to create toxic effects than a larger number of smaller protein aggregates throughout a cell. They may also more rapidly be degraded by MTOC-localized disposal machineries [36].

Neurons are different from other cells in terms of MTOCs and aggresomes. The former structure does not exist as a single entity in adult neurons, but rather a number of "local" MT nucleation points may either extend MTs [37] or, through them, may transport nuclear inclusion bodies to a limited number of decentralized spots. Multiple aggresome-like structures, thus, may be present in a single, adult neuron [38]. They contain neuron-specific constituents, such as neurofilaments (NFs) [39], but their nature and function—to concentrate toxic, mostly ubiquitinated protein aggregates and to speed up their degradation—defines them as atypical aggresomes.

The transport of protein aggregates to aggresome-forming locations is carried out by *aggrephagy transporters* (*ATs*). Histone deacetylase *HDAC6* [40] is the most relevant transporter, which will be discussed in detail later as a molecular target. Aggregate protein transport is regulated by several proteins. Among them, the nuclear exchange factor (NEF) co-chaperone Bcl2-associated athanogene 3 (BAG-3) [41], which is more abundant in an age-dependent equilibrium over its UPS-favoring congener BAG-1 [42]. BAG-3 requires the chaperone machinery to recruit non-ubiquitinated protein aggregates to aggresomes [43], and to promote chaperone-assisted selective autophagy (CASA) [44]. In detail, BAG-3 uses the specificity of Hsp70 chaperones for misfolded proteins to direct the release-loading of Hsp70-bound substrates onto the dynein motor complex, mediating aggresome-targeting and autophagic degradation of chaperoned substrates [43]. The AAA ATPase VCP/p97 [45] is also involved in later aggrephagy steps [46], being essential for maturation of ubiquitin-containing APs. p97 is associated, in its mutated form, with the neurodegenerative Paget's disease [47]. The chaperone ubiquilin-1 [48] is a CMA substrate that is localized in the APs of dystrophic neurites in lesions of a double transgenic (TG) presenilin (PS)/amyloid precursor protein (APP) mouse model of Alzheimer's disease (AD) [49]. Ubiquilin shows decreased levels in AD patients [50].

The recognition and loading of ubiquitinated protein aggregates to be transported to aggresomes (protein cargo function), and the subsequent recruitment of aggresomes onto AP structures (aggresome recognition), are carried out by *aggrephagy receptors* (AgRs). AgRs are often complexed with autophagy-relevant proteins through *aggrephagy scaffolds* (AgSs). AgR-rich aggresomes recruit nascent isolation membranes (IMs, step 1, Figure 5.4) and drive the formation of a functional, mature AP around the aggresome (step 2, Figure 5.4).

A small number of aggrephagy receptors is known. *p62-Sequestosome 1* (p62-SQSTM1) [51] is a well-characterized AgR, and its putative target role in neurodegeneration will be described in detail later in this chapter. The more recently identified neighbor of BCRA gene (NBR1) [52] shows similarities and differences in its AgR role when compared with p62 [53]. It is composed of 966 amino acids, i.e., it is twice as long as p62, containing the same p62-characterizing domains [35]. Additional domains, such as the four tryptophan domain (FW, NBR1 box [54]), are shared with

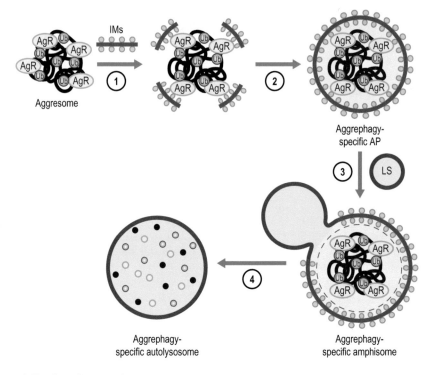

AgR = Aggrephagy receptors
IM = "Atg8"-bound membrane fragments
Ub = Ubiquitin
LS = Lysosome
AP = Autophagosome

FIGURE 5.4   The aggrephagy process: from aggresomes to autolysosomes.

non-autophagy-related proteins. NBR1 is more distributed throughout living organisms than metazoan-specific p62, as it is found in plants and fungi [53]. Optineurin (OPN [55]) possesses several coiled-coil domains that regulate its oligomerization, an LC3-interacting region (LIR) to anchor OPN onto APs, an UBAN (C-terminal ubiquitin binding in A20 binding inhibitor of nuclear factor kappa-light-chain-enhancer of activated B cells (NF-κB)/(ABIN) and NF-κB essential modulator/(NEMO) domain, and an UBZ (ubiquitin-binding zinc finger) domain. The last two domains allow the aggrephagy-targeted, UBQ-dependent and -independent interaction of OPN with misfolded proteins [56]. OPN colocalizes with protein inclusions in amyotrophic lateral sclerosis (ALS) [57], Huntington disease (HD) [58], AD, and Parkinson's disease (PD) [59]. Nuclear dot protein 52 (NDP52) [60] possesses a coiled-coil region/

oligomerization, a non-canonical LIR domain/AP interaction, and a C-terminal UBZ domain/UBQ-dependent misfolded protein interaction [61]. NDP52 is localized to APs, undergoes autophagic turnover [62], and mediates the degradation of various proteins [60].

The 395 kDa autophagy linked FYVE protein (ALFY) [63] is the only AgS characterized so far [64]. Its N-terminal two-thirds are not predicted to contain any functional domains, while its C-terminal region is involved in cell trafficking and selective autophagy [64]. Namely, it contains a p62-interacting, Pleckstrin homology (PH)-Beige and Chediak-Higashi (BEACH) domain [65], five ATG5-interacting WD40 repeats [64], and a PtdIns(3)P (PI3P)-binding FYVE domain [66]. As both NBR1 and ALFY contribute to the receptor activity of p62, and to its connections to neurodegenerative conditions, they are mentioned later. Similarly, components of the MA machinery with relevance for the activity of HDAC6 and p62 in selective autophagy are mentioned later. The fate of aggrephagy-specific APs, formed through steps 5–8 in Figure 5.3 and steps 1–2 in Figure 5.4, is identical to their MA counterparts. In fact, aggrephagy-specific APs merge with LSs (step 3 , Figure 5.4) and cause degradation of protein aggregates in the resulting autolysosomes (step 4, Figure 5.4).

5.2 SELECTIVE AUTOPHAGY-MEDIATED DEGRADATION OF PROTEIN AGGREGATES IN NDDs

Wild-type (WT) α-*synuclein* is a confirmed CMA substrate, but its pathogenic mutants are less susceptible—sometimes even inhibitory—to this autophagic process [67]. CMA appears also to have a role in *HD* [68].

Although tau contains two CMA-targeting motifs in its microtubule-binding repeats (MTBRs), full-length (FL) tau seems not to be a CMA substrate. A small, MTBR-containing tau fragment, conversely, is processed through an unconventional CMA process to yield smaller, highly amyloidogenic fragments [69]. FL, soluble tau may be polyubiquitinated and degraded by the proteasome [70], but may also be disposed of through MA [69] or selective autophagy [71]. Hyperphosphorylation and, in general, any non-physiological post-translational modification (PTM) pattern on tau leads to an increased susceptibility to selective autophagy [69].

MA is the most relevant form of non-selective autophagy with regard to neurodegeneration, and tauopathies in particular. Most of its complex machinery is used by cells also to specifically dispose of protein aggregates by aggrephagy [25]. The numerous and validated connections between aggrephagy targets and neurodegenerative diseases (NDDs) are discussed in detail in the next section.

5.3 SELECTIVE AUTOPHAGY—TARGETS

5.3.1 p62

p62 was originally discovered as an Lck kinase ligand [72]. Its ability to form aggregates was observed shortly thereafter, earning the alternative sequestosome-1 denomination [73]. It took around 10 years to firmly establish the p62–autophagy link [74], but since then the relevance of p62 in autophagy has been established. p62 is composed by 440 amino acids, and contains several functional domains (Figure 5.5).

p62 interacts with a large number of proteins, many among which are still unknown. Its domains include two nuclear localization systems (NLS1 and NLS2) and a nuclear export system (NES), involved in the nucleocytoplasmic shuttling of the scaffold protein ALFY [64] and of p62 [75]; two regions rich in proline, glutamate, serine, and threonine (PEST1 and 2), serving as sequences for proteolytic degradation [76]; a zinc finger motif (ZZ), which interacts either with the receptor-interacting protein 1 (RIP-1) contributing to the activation of cellular signaling cascades [77], and with the AMPA receptor enhancing synaptic plasticity through its membrane translocation [78]; a tumor necrosis factor (TNF) receptor associated factor 6 (TRAF6) binding site (TBS), acting as a scaffold protein

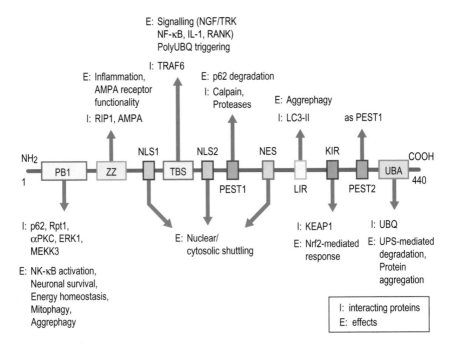

FIGURE 5.5   The aggrephagy receptor p62: interactions and functions of its main domains.

for the ubiquitin (UBQ) E3 ligase TRAF6, which regulates, *inter alia*, nerve growth factor (NGF)/tropomyosin receptor kinase (Trk) [79] and NF-κB [80] signaling; and a Keap1-interacting region (KIR), which binds to the kelch-like ECH-associated protein 1 (Keap1), an adaptor protein for UBQ E3 ligases, causing stabilization and activation of the NF-E2-related factor 2 (Nrf2) transcription factor [81] and leading, *inter alia*, to cellular adaptation to stress conditions. A recent addition is the interaction of p62 with the mTORC1 complex through raptor [82], which is mapped to the nameless p62 portion between the ZZ and TB domains and hints towards a role for p62 in MA-related nutrient sensing. The interactions with Keap1, TRAF6, and mTORC1 may have relevance for structure-based design of p62-driven autophagy modulators.

Three domains contribute to the role played by p62 in aggrephagy. We examine them following their "order of appearance" in the aggrephagy process. Chaperone-bound, misfolded proteins are at first ubiquitinated by UPS enzymes [45]. The recruitment of p62 by mono- or polyUBQ proteins happens *via* its C-terminal *ubiquitin-associated (UBA) domain*, leads to p62-promoted protein aggregation first, and eventually to aggrephagy [45,83]. UBA-mediated p62 recruitment is a general process that does not show preference for any misfolded protein family [84]. The UBA domain binds to both mono- and polyubiquitinated proteins [85], with a preference for K63-ubiquitinated proteins [86]. K63-ubiquitinated proteins are usually proteasome-resistant, and aggregation-prone/aggrephagy-directed in NDDs [87]. p62 is also involved in the regulation of K63 ubiquitination through interactions with K63-ubiquitinating E3 ligases (e.g., TRAF6 [88]) and with deubiquitinases (DUBs, e.g. ataxin-3 [36] and cylindromatosis (CYLD) [89]). p62 promotes selective autophagy of a few non-ubiquitinated proteins, such as superoxide dismutase (SOD1) [90] and signal transducer and activator of transcription 5A (STAT5A) [91].

UBQ/p62-containing protein aggregates continue their growth and form p62 bodies, that are transported to AP-forming locations. Both growth and transportation are assisted by the *N-terminal Phox and Bem 1p domain (PB1)*. The establishment of p62 bodies depends on p62 dimerization, followed by further oligomerization [74]. p62 dimerization is accomplished through the PB1 domain and is prevented in p62 variants lacking the PB1 domain, or bearing point mutations in the same domain [74]. In addition, the PB1 domain mediates p62 heterodimerization with the close congener NBR1 [92], which cannot homodimerize and whose presence in p62 bodies is important. The transportation of UBQ/p62-containing protein aggregates to AP-forming locations (see Section 5.3.2) is PB1- and p62 dimerization-dependent, although the detailed mechanism regulating it has not been elucidated yet [93]. A similar mechanism seems to regulate the transportation of NBR1 [93].

Finally, the *LC3-interacting region (LIR)* of p62 promotes the birth of selective APs—i.e., APs containing only p62 bodies. The LIR domain is a short oligomeric sequence shared by tens of proteins [94], among which p62 and a few others are functionally characterized. The DDDWTHL core LIR sequence in p62 binds to the interface of previously described LC3 proteins (see Section 4.1, Chapter 4) between the N-terminal arm (electrostatic interactions of the DDD LIR motif with basic residues) and the C-terminal UBQ-like (UBL) domain (electrostatic interactions, plus interactions of W and L side chains of the LIR motif with two hydrophobic pockets) [95]. p62–LC3 binding anchors p62 bodies onto the LC3-containing AP, which then continues its maturation and elongation as seen in MA [96]. Due to the selective p62–LC3 interaction, though, the evolving AP contains only protein aggregates and leads to aggrephagy, rather than non-selective MA.

The UBA and PB1 domains are involved in other processes in addition to aggrephagy. As to the former, single copies of UBQ/p62-coupled misfolded proteins—including tau [97]— undergo UPS-driven degradation [98]. p62 dimerization inhibits *in vitro* UBQ binding and possibly UPS degradation [99], and p62 interacts with the autophagy-promoting NEF co-chaperone BAG-3 [52]. These and other observations explain how aggrephagy, rather than proteasomal degradation, is the preferred fate for UBQ/p62-complexed proteins.

Aggrephagy-required homodimerization through the PB1 domain is a peculiar property of p62, as even its close congener NBR1 cannot homodimerize [92]. Rather, the conserved PB1 domain is often a heterodimerization motif that promotes the interaction among different PB1 domain-carrying proteins [100]. Namely, type I/type A proteins contain an acidic PB1 domain, which heterodimerizes with type II/type B basic domains from other PB1 proteins [101] through salt bridges. The PB1 domain of p62 contains both a type A/acidic and a type B/basic region, which dimerize in a head-to-tail manner to give p62 dimers [102], and eventually assist in the formation of p62 bodies. p62 may also heterodimerize with either type A or type B PB1-containing proteins, and its role in cellular signaling through PB1-mediated heterodimerization is known. NF-κB activation takes place through the formation of p62-mitogen-activated protein kinase kinase kinase 3 (MEKK3) [102] or p62-atypical protein kinase C (PKC) heterodimers [103]. The latter heterodimers induce neuronal survival through modulation of protein kinase B (Akt) activity [104] and modulation of voltage-gated potassium (K_v) channels [105]. It is known that levels of p62 bodies—i.e., p62 homodimerization—increase when p62 is overexpressed [100], and that p62 mutants with either type A (D69A/E70A) or type B mutations (K7A) deplete homodimerization *in vitro* [102]. Nevertheless, the role of p62 homo- and heterodimers and their relative abundance in physiological and pathological conditions are not yet fully clarified. Modulation of the p62-containing homo-/heterodimer ratio may be a relevant therapeutic target in several diseases in general, and in NDDs in particular.

TBS- and PB1-dependent NF-κB activation, KIR-dependent Nrf2 activation, and selective autophagy (p62 is involved in mitochondrial autophagy/mitophagy [106,107], zymogen autophagy/zymophagy [23], peroxisome autophagy/pexophagy [84], and bacterial [13] and viral [108] autophagy/xenophagy) are the main—or maybe only the best elucidated—determinants of p62 activity in physiological and pathological conditions.

The levels of p62 at any given time are influenced by its expression and by its degradation. p62 expression is increased in stress conditions, such as amino acid starvation and accumulation of ribosomal proteins [54,109], or proteasomal inhibition [110]—to compensate for reduced UPS disposal of misfolded proteins. The *p62* promoter contains a CgP island sensitive to oxidation [111]. It is oxidatively modified, and p62 expression is lowered in an age-dependent manner [112]. p62 is needed in cells to execute selective autophagy, and is degraded as a substrate during the process [53]. Thus, its lower expression impacts on the efficiency of selective autophagy. At the same time, selective autophagy impairment causes the accumulation of p62-containing, ubiquitinated inclusion bodies [112]. As p62 bodies have a short life due to their fast autophagic degradation, the presence and persistence of p62 inclusion bodies is a diagnostic indication of protein aggregation diseases [113].

Some functions of p62 are regulated by PTMs. Nucleocytoplasmic shuttling of p62 is regulated by phosphorylation of sites at or near the NLS2 domain [114]. Specifically, pS266 shows reduced nuclear import of p62, while T269 and S272 show increased import. Nucleocytoplasmic shuttling of p62 is inhibited by p62 dimerization, and contributes to the degradation of ubiquitinated nuclear proteins in the degradation-prone cytosolic environment [114]. Cyclin-dependent kinase 1 (Cdk1) is the only known T269/S272 p62 kinase, and lack of T269/S272 phosphorylation by Cdk1 leads to tumorigenesis [115]. Casein kinase 2 (CK2)-mediated phosphorylation of the S403 residue in the UBA domain of p62 has a strong aggrephagy-promoting effect [116]. pS403 p62 has a higher affinity for ubiquitinated proteins than S403-dephosphorylated p62, while dephosphorylation of dimerized/polymerized pS403-containing p62 oligomers is reduced. Thus, the pS403/S403 ratio in p62 species is shifted towards pS403 by p62 dimerization—escaping pS403 dephosphorylation—and subsequent aggrephagy—consumption of pS403 p62 [116]. TANK-binding kinase 1 (TBK1)-mediated S403 phosphorylation regulates selective xenophagy against *Mycobacterium tuberculosis* [117]. S403 phosphorylation by unknown, kinases may also happen.

In physiological conditions, an increased pS403/S403 ratio promotes p62 dimerization/p62 bodies' formation, and eventually selective autophagy. In a p62-deregulated/lower p62 expression scenario, an increased pS403/S403 ratio may promote aggrephagy and dispose of toxic protein aggregates. In a p62-deregulated/p62 accumulation/impaired autophagy scenario, an increased pS403/S403 ratio may cause further accumulation of toxic, p62- and UBQ-positive inclusion bodies. Thus, S403 phosphorylation may be a key regulatory element in physiological conditions, a neuroprotective event when p62 levels are decreased, and a potentially toxic factor in a p62 accumulation/autophagy impairment scenario.

Dysregulation of the levels and the functions of p62 is observed in several diseases [118]. Thus, p62 is a putative drug target, although p62-directed drug discovery efforts are still in their infancy. Bone and liver human diseases show accumulation of p62. Mutations in the UBA p62 domain cause *Paget's disease of bone (PDB)* [119], an osteoclastogenesis-increasing chronic disease where hyperactivated NF-κB signaling is due to the loss of p62/UBQ-TRAF6 interactions and to p62 accumulation/selective autophagy deregulation [118–120]. p62-containing inclusion bodies are found in *chronic liver diseases* [121], where their formation is due to the effect of p62 accumulation/impaired selective autophagy [109], including persistent, KIR-dependent Nrf2 hyperactivation [81]. Animal models indicate additional p62 disease connections. p62 knockout (KO) mice develop *obesity* and *insulin intolerance*, as p62-driven negative regulation of extracellular-regulated signal kinase 1 (ERK1) is lost and hyperactivation of the ERK pathway leads to adipocyte differentiation [122]. The recently discovered raptor–p62 interaction leads to mTORC1 activation, autophagy inhibition, and p62 accumulation in a positive loop leading to adipogenesis [82]. Autophagy impairment, and subsequent p62 accumulation, is frequent in *tumors* [123]. Lung [124], breast [125], pancreas [126], and brain cancer [127] are among the reported examples. Previously mentioned NF-κB [128] and mTORC1 activation [129], together with deregulation of the Wnt pathway [130], are pro-oncogenic events promoted by deregulation/accumulation of p62. Finally, autophagy impairment and p62 accumulation play a major role in the persistent proteotoxic stress experienced by cardiac tissue [131], a pathogenic factor to the development and the progression of *heart diseases*.

The link between p62 and neurodegeneration is strong. p62 KO mice show multiple pathological phenotypes [132], due to the multiple signaling roles of p62. Neurodegeneration- and selective autophagy impairment-related phenotypes can be observed [71,89]. Mutations of p62 are observed in *ALS* [132,133], and include UBA-located p62 mutations observed in PDB [119]. Other mutations are distributed throughout p62, and may affect the binding of p62 to *Tar-DNA binding protein-43 (TDP-43)* [134,135] and *SOD1* [83]. Ubiquitinated TDP-43 inclusions are found in abnormal inclusion bodies in *ALS* and *frontotemporal lobar degeneration (FTLD)* patients, and co-immunoprecipitate with p62 [134]. Deregulation of p62-mediated disposal of TDP-43 may be a causative factor for both ALS and FTLD. p62 binds to a mutated, ALS-connected SOD1 isoform through a SOD1 mutant interaction region (SMIR) [90]. The interaction is PB1-dependent, i.e., it requires oligomeric p62/p62 bodies, but is UBA-independent, and proves that p62 facilitates the disposal of aggregated proteins through several pathways [90]. p62-mediated degradation of TDP-43 and of mutant SOD1 *via* aggrephagy may represent a therapeutic option for a subpopulation of ALS, FTLD, and ALS-FTLD patients. The mTOR kinase inhibitor rapamycin activates autophagy in both ALS [136] and FTLD models [116]. Interestingly, rapamycin-mediated autophagy activation leads to therapeutic effects in the latter model [137], while ALS mice experience an acceleration of the disease [136]. p62 is present in Lewy bodies from *PD* patients [138,139], and is co-localized in neuronal and glial inclusion bodies in patients diagnosed with *Pick disease, dementia with Lewy body* and *multiple system atrophy (MSA)* [140]. p62 is involved in parkin-dependent recessive PD, most likely supporting pathologically increased mitophagy induced by K63 ubiquitination of the outer mitochondrial membrane by the E3 UBQ ligase parkin [141,142]. p62 is detected in large polyQ-containing, UBQ-free inclusions in subjects suffering from *sickle cell anemia* [143]. As to *HD*, p62 contributes to the degradation of mutant huntingtin in an UBA- and PB1-dependent manner, by creating a surrounding p62 shell around the protein and promoting its disposal by aggrephagy [74,75].

p62 has a lead role in the regulation of *AD* pathogenesis [144], although p62 mutations have not been detected yet in AD patients. Lower p62 expression levels are sometimes claimed in AD patients [112], and in an age-dependent manner in triple TG (APP-tau-presenilin 1/PS1) AD mice [112,145] due to oxidative damage. Other reports claim higher p62 levels in AD patients [146,147]. Discrepancies are most likely due to technical reasons, such as detergent-soluble and detergent-insoluble/p62 bodies-like p62 species that may be differently regulated (or even differently detected using various experimental conditions) in control and AD brains [147].

The scaffolding role of p62 influences AD neurodegeneration through at least three p62 domains. PB1-dependent p62-MEKK3 [102] or p62-atypical PKC heterodimers [103] activate NF-κB signaling. NF-κB signaling acts as a neuronal survival factor [148] and, if depleted by p62/heterodimer deficiency, could contribute to neurodegeneration. PB1-dependent activation of, *inter alia*, Akt, mitogen-associated protein kinase (MAPK) and c-Jun N-terminal kinase (JNK) in p62 KO mice may lead to neuroinflammation and beta amyloid (Aβ) pathology [149,150], while insulin resistance observed in AD mice and presumed in AD patients [151] may stem from PB1/p62 KO-dependent ERK activation [122]. NGF-dependent activation of NF-κB signaling stems from TBS domain-mediated participation of p62 to the ternary TRAF6–p62–Trka complex [80]. NGF- and TBS-dependent TRAF6–p62 interactions lead to polyubiquitination of the p75 neurotrophin receptor [152]. The AD-relevant Aβ peptide impairs NGF-mediated interactions between TRAF6–p62 and p75, preventing polyUBQ p75-mediated neuron survival [152] and possibly contributing to cholinergic dysfunctions in AD [153]. The KIR domain, finally, promotes Nrf2 activation through disruption of the Keap1–Nrf2 complex and binding sequestration of the Keap1 adaptor protein [140]. Nrf2 activation leads to AD-useful transcription of antioxidant-detoxifying genes, and Nrf2-based gene therapy shows promising cognitive improvements in a severe AD (APP-PS1) TG mouse model [154].

Tissue-selective autophagy impairment in the brain of AD patients is known [155,156]. Besides aggrephagy, another selective autophagy process—mitophagy [11]—plays a role in AD, and involves p62 in its development. Mitochondrial dysfunction is a central phenomenon in AD [157,158] and in PD [141,142]. A mitochondrial quality control based upon PINK, a kinase, and parkin, an E3 UBQ ligase, targets damaged mitochondria to mitophagy [159,160]. p62 is recruited to parkin-ubiquitinated mitochondria through its UBA domain [106] and promotes their transportation to mito-aggresomes in a PB1/p62 oligomerization-dependent manner [107]. The role of p62 in mitophagy is considered either indispensable [106] or dispensable [141]—likely due to partial functional redundancy with NBR1. p62 contributes similarly to aggrephagy and mitophagy, and it is presumable that a putative therapeutic intervention aiming at its deregulation in AD could be beneficial for both selective autophagy processes.

Several human mutations lead to p62-involving autophagic deregulation and tau pathology. Frontotemporal dementia (FTD) with the C9ORF72 hexanucleotide repeat expansion [161] shows accumulation of p62-positive neurofibrillary tangles (NFTs) and impaired protein degradation [162]. A single patient with FTD-C9ORF72 pathology crossed with a benign A239T tau mutation suffers a mixed neurodegenerative profile where tau pathology is dominant, flanked by TDP-43 and p62 pathologies [163].

A confocal/immunomicroscopy study on AD brain samples shows that inclusion bodies contain tau, hyperphosphorylated (HP)-tau, UBQ, p62, TRAF6, and the DUB UbcH7 [76]. Tau is polyubiquitinated by the K63-UBQ ligase TRAF6, recruits p62 through its UBA domain and is channeled towards degradation. TRAF6- and p62- KO mice respectively prove that TRAF6 is crucial for tau ubiquitination, and that p62 is crucial for tau elimination [97]. Namely, dysfunctional K63-polyUBQtau-p62 aggregates may be degraded by the UPS system [97], but may also become aggrephagy substrates. The switch between UPS- and autophagy-mediated tau disposal in neurons varies according to, *inter alia*, age-dependent impairment of the proteolytic quality control machinery and to the specifics of each tauopathy.

A p62-deficient mouse model shows partially overlapping neurodegeneration and obesity at various ages [71]. At 2 months of age, tau-related biochemical alterations are absent and mice do not show signs of obesity. Conversely, 2-month-old p62 KO mice show short-term memory loss, anxiety, and depression [71] with normal motor neuron functions. At 6 months of age, the obesity phenotype becomes evident [122], and the same happens for tau pathology. Chronic depletion of p62 leads to decreased brain levels of atypical PKC-ι/λ, a kinase that heterodimerizes with p62 [103]. Conversely, the PKC-regulated activity of tau kinases such as glycogen synthase kinase 3 beta (GSK-3β) and Akt is increased [71]. Tau becomes hyperphosphorylated, and HP, K63-ubiquitinated, insoluble pre-tangle tau aggregates appear in various brain areas. Loss of synapses and neuronal death are also observed at 6 months of age [71].

The wealth of material described in this section identifies p62 as a "treasure trove" target against protein misfolding diseases, neurodegeneration, and tauopathies. How to balance positive and negative effects of p62-modulating agents, and how to move from structural information regarding PB1-, UBA-, and/or LIR-centered interactions to rationally designed p62/autophagy modulators remains to be defined.

5.3.2 HDAC6

Histone deacetylase 6 (HDAC6) is a member of the HDAC family, discovered in 1999 [164], with peculiar properties with respect to other HDAC enzymes. HDAC6 is a cytosolic enzyme, and its main deacetylase function is histone-independent. The main substrate of HDAC6 is α-tubulin, influencing MT-mediated biological processes—including physiological and pathological events in the central nervous system (CNS) [165].

HDAC6 is composed of 1215 amino acids, and several functional domains (Figure 5.6).

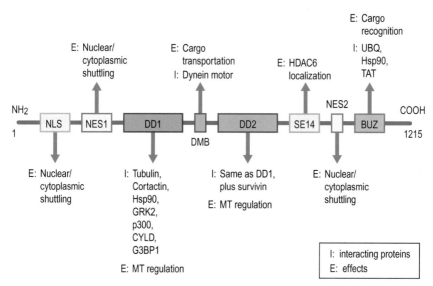

FIGURE 5.6   The aggrephagy transporter HDAC6: interactions and functions of its main domains.

α-Tubulin is the main substrate for its deacetylase activity [165,166], but other proteins can be deacetylated by HDAC6. HDAC6 binds to proteins that are not deacetylated. The domains of HDAC6 include two nuclear export systems (NES1—AA 67–76 and NES2—AA 1049–1058) and a nuclear localization system (NLS—AA 14–59), which respectively maintain most of HDAC6 in the cytoplasm [167] and, when cell proliferation is blocked, secure the presence of a fraction of HDAC6 in the nucleus [168]; eight consecutive Ser-Glu-containing tetradecapeptide repeats (SE14—AA 884–1022), which also ensure a stable localization of HDAC6 in the cytoplasm [169]; and a dynein binding motor domain (DBM—AA 439–503), which binds to dynein motors [40].

HDAC6 is the only HDAC family member endowed with two functional catalytic domains (DD1—AA 87–447 and DD2—AA 482–800). DD2 is responsible for most of HDAC6 deacetylase activity on histone and α-tubulin substrates [170,171], although DD1 is also catalytically active [165]. The spacer between DD domains [172] influences the efficiency of deacetylation for histone and α-tubulin substrates. Other HDAC6 deacetylation substrates include the F-actin binding protein cortactin [173], the chaperone Hsp90 [174], the redox regulatory proteins peroxiredoxins [175], the nuclear factor Ku70 [176], the anti-apoptotic protein survivin [177], the intracellular signal transducer β-catenin [178] and the regulatory *trans*-activator of transcription (TAT) HIV protein [179].

Another HDAC6 peculiarity is a zinc finger, UBQ-binding domain (BUZ—AA 1131–1192). The BUZ domain, whose X-ray structure either alone or complexed with UBQ is available [180], contains Cys- and His-rich regions and has a strong affinity for mono- and polyUBQ proteins [181].

HDAC6 plays a major role in selective autophagy, in particular in aggrephagy [25,182]. Initiation of aggrephagy, i.e., the early aggregation of misfolded polyUBQ proteins in pre-aggresomes/microaggregates, is p62-dependent [109]. The subsequent transport of pre-aggresomes to MTOCs in non-neuronal cells, and to local nucleation sites in neurons (see Section 5.3.1) is HDAC6-dependent [36]. In fact, HDAC6 KO cells are characterized by dispersed, UBQ-containing pre-aggresomes [40].

The BUZ domain of HDAC6 strongly binds to ubiquitinated pre-aggresomes (1:1 stoichiometry, K_d = 60 nM compared to K_d values between 5 and 500 μM for other UBQ-binding proteins) [183]. The binding strength secures the HDAC6–pre-aggresome connection, and ensures a high UBQ content to HDAC6-containing microaggregates [184]. The AAA ATPase VCP/p97 [45] regulates HDAC6–UBQ interactions, as it is able to displace HDAC6 from pre-aggresomes and to bind to it, while the 19S proteasome UBQ-binding protein Rnp10 binds to pre-aggresomes and leads them to UPS degradation [183]. Aggrephagy is favored *vs*. UPS when the

HDAC6-VCP/p97 equilibrium is shifted towards HDAC6, or when UPS degradation is impaired.

The BUZ domain of HDAC6 specifically binds to C-terminal diGly UBQ motifs, similar to other BUZ-containing, UBQ-binding proteins [185]. Crystal structures of the BUZ domain alone and of its complex with either FL UBQ or C-terminal UBQ justify the molecular interaction through a hydrogen bond network that does not involve any UBQ residue beyond the C-terminal diGly [186]. Pre-aggresomes are formed through aggregation of polyUBQ proteins, where C-terminal diGly UBQ motifs are covalently bound to the misfolded proteins. Surprisingly, unconjugated UBQ is found in pre-aggresomes [186].

The DUB ataxin-3 co-localizes with HDAC6 and dynein in pre-aggresomes, and is found after UPS impairment in aggresomes containing the mutant cystic fibrosis transmembrane regulator protein CFTRΔ508. Silencing of ataxin-3 significantly decreases aggresome formation [187]. Ataxin-3 is recruited to polyUBQ pre-aggresomes and trims UBQ chains, so as to expose unconjugated C-terminal diGly UBQ motifs on the surface. Eventually, exposed C-terminal diGly UBQ motifs strongly bind to HDAC6 *via* the BUZ domain and enter the aggrephagy pathway [186].

The transport of HDAC6/polyUBQ-containing pre-aggresomes to aggresome assembly points is HDAC6-dependent. HDAC6 is a microtubule-associated deacetylase that binds to the dynein retrograde motor complex [168]. The HDAC6–dynein interaction is weak in ordinary conditions, but is enhanced by UPS impairment [40]. The DBM domain of HDAC6 is sufficient to bind to the dynein motor complex, even when the BUZ domain of HDAC6 connects the protein to ubiquitinated pre-aggresomes. The HDAC6–dynein interaction is promoted by CK2, which connects HDAC6 and dynein in a phosphorylation-independent manner [187]. Additionally, CK2-mediated phosphorylation of the DMB domain of HDAC6 at Ser458 decreases HDAC6 deacetylase activity and prevents aggresome formation. CK2 suppression by RNA interference and CK2 phosphorylation-defective mutation of HDAC6 (S458A) have the same effect [188].

Once complexed with HDAC6, the MT-bound dynein motor transports microaggregates to the growing aggresomes. The process is HDA1 deacetylase activity-dependent, as MT deacetylation facilitates pre-aggresome transport [40]. The build-up of APs around aggresomes is initiated by the interaction of p62 with LC3 [40] (see also Section 5.3.1). Mature APs then fuse with LSs, providing needed proteolytic enzymes and acidic pH for misfolded protein degradation [96]. The AP–LS fusion in aggrephagy is HDAC6-dependent, and relies upon deacetylation of cortactin, a regulator of cell motility [189].

Cortactin binds to HDAC6 through interactions between the double repeat/F-actin-binding domain of cortactin and the DD1-DBM-DD2 region of HDAC6 [173]. HDAC6, and maybe also the deacetylase silent information regulator 2 (Sirt2), deacetylate cortactin on nine Lys residues in the double repeat cortactin domain. Deacetylated cortactin interacts with F-actin and promotes cell migration. Lys/Gln deacetylation-insensitive mutations of four or more double repeat Lys residues inhibit cortactin translocation, prevent the cortactin–F-actin interaction and reduce cell motility [173]. HDAC6 co-localizes with key factors in signaling-dependent actin remodeling at actin-associated membrane ruffles, and regulates their activities [190].

Fibroblasts from HDAC6 KO mice show defects in AP–LS fusion, and an accumulation of APs [191]. Impaired fusion is observed also using a cell-free assay. F-actin structures surround HDAC6-induced APs from WT mice, but are largely missing in HDAC6 KO-derived samples. F-actin polymerization stimulates the fusion of APs with LSs, and cortactin—a substrate of HDAC6—is recruited by HDAC6 and deacetylated to promote F-actin polymerization [191]. Cortactin-dependent F-actin remodeling, and eventually HDAC6 regulation, is needed only for aggresome-containing, aggrephagy-targeted APs (selective degradation of misfolded proteins), while starvation-induced, non-selective MA is HDAC6-independent [191].

HDAC6 promotes the degradation of misfolded proteins through aggrephagy (namely, pre-aggresome transport and AP–LS fusion). Examples include mutated CFTR [40,183,186,187,192,193], α-synuclein [40,194,195], huntingtin [193], polyQ-ataxin-1 [186], polyQ-ataxin-3 [196], and DJ-1 [197]. HDAC6 shows effects due to either deacetylation or interaction with other proteins. A detailed list of HDAC6-interacting partners may be found in a recent review [182]. Some of them are mentioned here, although unrelated to aggrephagy, because they impact on neuronal functions in general, and on neurodegenerative processes in particular.

Augmented MT acetylation on Lys40, i.e., inhibition of HDAC6, causes the increase of both dynein and kinesin-1 recruitment (≈three-fold) to MTs [198]. Consequently, MT-dependent anterograde and retrograde cargo transport is stimulated. MT stability is not affected by the Lys40 acetylation status [198]. Increased kinesin-1-mediated, anterograde transport of the cargo protein JNK-interacting protein 1 (JIP1) towards neuritic tips is dependent on MT acetylation [199]. Another report [200] shows the increased anterograde/retrograde, kinesin-1/dynein-dependent transport of the UBQ E3 ligase parkin–HDAC6 complex along MT structures. MT acetylation on Lys40/HDAC6 inhibition prevents the transport of parkin [200]. Surprisingly, chemical inhibition of HDAC6 with small molecules decreases the rates of MT growth, while HDAC6 ablation does not influence MT growth [201]. Single DD domain-mutated HDAC6 decreases MT growth, while HDAC6 bearing mutations in both DDs mimicks HDAC6 KO. A capping function of binding-capable, catalytically inactive

HDAC6 at MT ends is suggested to explain the slower MT growth [201]. The relationship between MT acetylation and transport is probably cellular context- and cargo-dependent, requiring additional investigations to be further clarified.

The tubulin polymerization-promoting protein 1/p25 (TPPP-1/p25) binds to HDAC6 and to MTs, showing a number of HDAC6-dependent, MT- and cell motility-related effects [202]. TPPP-1 induces MT bundling [203], promotes α-tubulin acetylation through HDAC6 binding (K_d = 112 nM) and inhibition of its deacetylase activity. MT growth velocity is reduced, as with chemical inhibition of HDAC6 [201], and cell motility is decreased. MT stabilization is independent from HDAC6 inhibition and stems from MT bundling [202]. The Rho-associated coiled-coil kinase (ROCK) phosphorylates TPPP-1 at Ser32, Ser107, and Ser159 residues [204]. ROCK-mediated phosphorylation of TPPP-1 prevents its binding to/inhibition of HDAC6, and prevents MT acetylation and the reduction of cell motility and invasiveness. MT bundling-dependent effects of TPPP-1 on MT stability and bundling are unaffected by ROCK phosphorylation [204]. Conversely, cyclin B/cyclin-dependent kinase 1 (Cdk1) phosphorylates TPPP-1 at Thr14, Ser18, Ser45, and Ser160 residues, and regulates MT stability and bundling, without altering its affinity for HDAC6 [205]. TPPP-1 is dynamically phosphorylated by ROCK and Cdk1, alone or in combination, to stimulate cell proliferation through regulation of HDAC6 deacetylation, MT polymerization and growth, and cell motility [205].

Hsp90 is an HDAC6 substrate [206]. Chemical and/or small interfering RNA (siRNA) inhibition of HDAC6 causes Hsp90 acetylation, leading to the reduction of Hsp90–ATP binding, of the Hsp90 chaperone function, and of the levels of some of its client proteins, possibly through their polyubiquitination and UPS-driven disposal. The effect of pan-HDAC and HDAC6-selective inhibitors on Hsp90 is different [206]. UPS inhibition induces heat shock factor-1 (HSF-1) activation, and subsequent heat shock gene response to fight the accumulation of polyUBQ proteins [207]. HDAC6 promotes HSF-1 activation in a BUZ-dependent (but deacetylation-independent) manner [208]. VCP/p97 is also essential for HSF-1 activation. HDAC6 detects an increase of UBQ proteins through its BUZ domain, and in conjunction with VCP/p97 it binds to Hsp90, dissociating it from the Hsp90–HSF-1 complex. The freed HSF-1 induces the accumulation, *inter alia*, of Hsp70 and Hsp25 [208]. HDAC6- and p97-mediated HSF-1 stimulation takes place only when UPS is impaired, counteracting the impaired degradation of misfolded proteins. Conversely, misfolded protein-unrelated, heat shock-induced HSF-1 activation is unaffected by HDAC6 and p97 [208].

HDAC6-mediated Hsp90 deacetylation increases its affinity for the co-chaperone p23, promotes Hsp90–glucocorticoid receptor (GR) association, and eventually regulates GR ligand binding, nuclear translocation, and transcriptional activation [174,209]. Mice with selective HDAC6 in-activation in serotonin neurons show GR response reduction associated with increased resilience and antidepressant effects [210]. The effect is due to the blockage of acute stress-induced increase in α-amino-3-hydroxy-5-methyl-4-isoxazolepropionic acid receptor (AMPAR) surface expression, and to serum and glucocorticoid-inducible kinase 1/3 (SGK1/3) up-regulation [211].

Salt-inducible kinase 2 (SIK2) is an HDAC6 substrate activated by Lys53 deacylation [212]. Acetylated, catalytically inactive SIK2 is translocated to APs, where it is activated by HDAC6 and contributes to AP maturation. SIK2 knockdown causes accumulation of APs and impairment of AP–LS fusion, and impairs the clearance of TDP-43 inclusion bodies by aggre-phagy [212].

Mitochondrial movement in neurons is influenced by HDAC6 [213]. Chemical inhibition of HDAC6 increases mitochondrial motility and kinesin-1 association with MTs through MT acetylation [200].

Peroxiredoxins I and II (Prx I, II) are antioxidant enzymes and HDAC6 substrates involved in redox regulation [213]. HDAC6 deacetylation of Prx I on Lys196 and of Prx II on Lys197 decreases their H_2O_2-reducing activity, and HDAC inhibition leads to increased antioxidant activity. The former effect may be beneficial to treat Prx-overexpressing cancer, while the latter may be useful in NDDs [175]—see below.

The E3 UBQ ligase tripartite motif-containing 50 (TRIM50) associates with the aggresome when cells are treated with a UPS inhibitor [214]. TRIM50 is transported to aggresomes by HDAC6, and HDAC6 KO cells show scattered TRIM50 bodies. TRIM50 promotes the sequestration of polyUBQ proteins into aggresomes and their clearance. TRIM50 binds to p62 and stimulates the accumulation of p62 and HDAC6 into aggresomes, while being itself an aggrephagy substrate [214].

Transportation of protein cargos and deacetylation by HDAC6 are es-sential for the formation of stress granules (SGs), a reversible assembly of translationally stalled mRNAs formed under cellular stress conditions [215]. HDAC6 binds to rat sarcoma (Ras)-GTPase-activating protein SH3 domain-binding protein 1 (G3BP1), and phosphorylation of G3BP1 nega-tively regulates the HDAC6–G3BP1 interaction. HDAC6 binds to G3BP1 through its DD domains, but catalytically inactive, single point-mutated HDAC6 also binds to G3BP1. HDAC6 is recruited to SGs with G3BP1, and both its deacetylase activity and its BUZ domain are essential for SG for-mation [215]. The dynein motor- and MT-dependent SG assembly process, thus, may be similar to the aggresome assembly.

HDAC6-mediated deacetylation of the cytosolic DNA repair protein Ku70 stimulates its binding to the pro-apoptotic protein Bax [216]. Chemical and siRNA inhibition of HDAC6 leads to an increase of acetylated Ku70, to the release of Bax and to Bax-dependent apoptotic death in neuroblastoma cancer cells [216].

HDAC6 is a molecular target pursued in various therapeutic indications [182]. HDAC6 is a key regulator of signaling pathways linked to *cancer* [217], and its direct inhibition counteracts the aberrant expression levels of HDAC6 [218]. HDAC6 inhibition can be indirectly exploited through its regulatory function on the activation of CYLD in tumor growth [219], on the oncogenic Ras signaling pathway in tumorigenesis [220], and on breast cancer metastasis suppressor 1 (BRMS1) in metastasis [221]. HDAC6 inhibition shows Hsp90/HSF-1-mediated beneficial effects against *autoimmunity* and allograft rejection [222]. HSF-1 activation leads to Hsp70 up-regulation, which promotes Foxp3$^+$ T-regulatory cell (Treg) survival and suppressive functions when cells are under stress. Hsp90 inhibitors are effective *in vivo* in models of cholitis and transplant rejection, and combination treatments using Hsp90 and HDAC6 inhibitors reduce the needed drug dosage [222]. Hsp90-mediated effects on glucocorticoid receptors (GRs) [174,209] mirror a key role for HDAC6 in hepatic-glucocorticoid-induced gluconeogenesis by controlling GR nuclear translocation [223]. Thus, HDAC6 may represent a novel therapeutic target for glucocorticoid-induced *diabetes*.

HDAC6 is a popular target in *CNS* [184,224], showing conflicting effects that may suggest disease-specific HDAC6-targeted therapeutic interventions [225]. Neuroprotective effects may be due to HDAC6-mediated recruitment and transportation of polyUBQ microaggregates to aggresomes [40], and to AP–LS fusion [173] in aggrephagy and mitophagy [200]; to chaperone upregulation and stimulation of protein refolding *via* Hsp90-mediated HSF-1 activation [208]; to HDAC6 deacetylation-dependent increased recruitment of molecular motors onto MTs [200]; and to Ku70-mediated anti-apoptotic [216] and anti-inflammatory effects in CNS [226]. Conversely, excessive MT and cortactin deacetylation may lead to the impairment of MT trafficking and stability [225], to reduced recruitment of molecular motors onto MTs [198], and to impairment of protein cargo [199] and mitochondrial [213] transport; Hsp90 deacetylation may lead to GR-mediated disturbance of synaptic functions [211]; Prx deacetylation may reduce their anti-oxidant properties [175]; and HDAC6-mediated neuroinflammation may take place [227]. HDAC6 effects, such as decreased α-tubulin and cortactin acetylation, may switch from an early neuroprotective—motor complex recruitment, autophagy stimulation—to a later neurotoxic phenotype—impairment of MT stability and trafficking, and autophagy stalling [225]. Aggrephagy itself may be either neuroprotective or neurotoxic, depending on the progression and the nature of each NDD [25]. HDAC6 KO mice show hyperacetylated α-tubulin, but remain viable and fertile [228]; exhibit α-tubulin-dependent antistress and antidepressant behaviors [210,211,229]; and do not show impairments on brain morphology, motor coordination, or hippocampus-dependent cognitive function [230]. Conversely, other studies on HDAC6 KO mice report age-dependent neurodegeneration and neuronal apoptotic cell death [192].

HDAC6 is a component of Lewy bodies, aggresome-like structures characteristic of PD and *dementia with Lewy bodies*, extracted from brain sections of patients [40,231]. It is also found in glial cytoplasmic inclusions from patients suffering from *MSA* [231], another α-synucleinopathy. An *in vitro* model of PD-like α-synucleinopathy shows HDAC6 accumulation and relocation to MTOC structures together with LSs, indicating the activation and progression of aggrephagy [195]. Conversely, HDAC6 deficiency caused by siRNA or by chemical inhibition increases the levels of α-synuclein, shows accumulation of autophagosomal vacuoles, impairs the progression of aggrephagy, and makes cells more sensitive to 1-methyl-4-phenylpyridinium (MPP$^+$) toxic stress [195]. An *in vivo Drosophila* model shows that HDAC6 deficiency exacerbates α-synuclein toxicity and promotes α-synuclein-mediated locomotor dysfunction [194]. HDAC6 interacts with neurotoxic, soluble α-synuclein oligomers and promotes their aggregation into insoluble aggregates to limit their neurotoxicity. Conversely, HDAC6 deficiency causes a reduction in insoluble α-synuclein aggregates and an impairment of the protein quality control (PQC) system [194].

Parkin, a UBQ E3 ligase, ubiquitinates a mutated DJ-1 protein isoform in an autosomal recessive form of PD, recruits HDAC6, and promotes the transportation of polyUBQ-mutated DJ-1 microaggregates to aggresomes [197]. HDAC6 strongly binds parkin through its DD1 and DD2 domains in a BUZ domain-independent manner [200]. HDAC6 acts as a UPS impairment-sensing protein, and delivers parkin to aggresomes when UPS is impaired. Meanwhile, parkin ubiquitinates misfolded proteins on their K63 residues. Ubiquitinated proteins are then recruited by HDAC6 through its BUZ domain in a synergistic, aggregate-forming and -growing process. The process is reversible, MT- and motor protein-dependent. The HDAC6–parkin complexes are transported towards aggresomes by the dynein motor complex when UPS is impaired. Restoring UPS activity *in vitro* causes kinesin-1-dependent transport of HDAC6–parkin complexes to the cell periphery [200]. As to PD, either parkin mutations [232] or HDAC6 deficiency, caused by siRNA inactivation or by chemical inhibition [200], triggers a reduction in Lewy bodies through aggrephagy impairment. Parkin and HDAC6 play similar roles in mitophagy, and in mito-aggresome formation and clearance [233,234].

p97/VCP is mutated in the multifactorial *inclusion body myopathy, Paget disease of the bone and frontotemporal dementia (IBMPFD, Paget disease)* [235]. IBMPFD tissues contain UBQ inclusion bodies, show an impaired microaggregate-to-aggresome process and an increased sensitivity to UPS inhibition. Exogenous protein aggregates added to mutated p97-expressing cells are trapped with mutated p97 and are not transported to aggresome-forming sites. Addition of exogenous HDAC6 restores MT-dependent transport of microaggregates, and stimulates aggrephagy and misfolded protein clearance by counteracting the effect of mutated p97 [236]. Thus, aggrephagy/mitophagy stimulation through HDAC6 activation appears to be a suitable therapeutic strategy against PD and IBMPFD.

HDAC6 has an essential role in the aggregation, transport, and clearance of polyQ-huntingtin in an *HD* cellular model. Dynein motor-mediated retrograde transportation and AP–LS fusion are HDAC6-dependent steps in the UPS impairment-induced clearance of polyQ-huntingtin by aggrephagy [193]. Conversely, MT-dependent transportation of organelles in HD is impaired [237] and MT acetylation is decreased in *postmortem* brain samples from advanced HD patients [198]. Chemical inhibition of HDAC6 causes an increased MT acetylation on Lys40, the recruitment of kinesin-1 and dynein motors to MTs, and the consequent, amplified MT transportation in a huntingtin/aggrephagy-independent manner. Namely, HDAC6 inhibition leads to the restoration of axonal transport of brain-derived neurotrophic factor (BDNF) in polyQ-huntingtin containing cortical neurons, preventing the reduced trophic support and neuronal cell death observed in HD [198]. A therapeutic balance between aggrephagy/ HDAC6 stimulation and cargo transport/HDAC6 inhibition seems to privilege the latter, at least in advanced HD. Conversely, when UPS is genetically impaired in an *in vivo* model (*Drosophila*) of *spinobulbar muscular atrophy* (*SBMA*), another polyQ disease, compensatory autophagy is induced by expression of HDAC6, and insoluble protein aggregates are cleared through aggrephagy [238]. Surprisingly, KO of HDAC6 in R6/2 TG HD mice does promote tubulin acetylation on Lys40, but does not show any biochemical (BDNF transportation, polyQ-huntingtin aggregation) or behavioral (forelimb grip strength, rotarod test) improvement in terms of the onset and progression of the disease [239].

Mutations in the 27 kDa small HSP gene (HSPB1) cause a subtype of either *Charcot–Marie tooth disease* (*CMT*), a peripheral nervous system disease characterized by dysfunctional axonal transport, or distal *hereditary motor neuron disease* (*HMN*) [240]. TG mice bearing neuronal CMT-causing HSPB1 mutations show a CMT-like phenotype, including severe axonal transport deficits, and decreased MT acetylation [241]. Chemical inhibition of HDAC6 in HSPB1-mutated mice restores axonal transport of mitochondria, increases the level of MT acetylation, and partially restores a normal neuronal phenotype [241]. Selective HDAC6 inhibitors may offer a therapeutic opportunity against CMT and HMN.

TDP-43 is a nuclear RNA-binding protein with a role in transcriptional repression, exon splicing, and RNA stabilization [242]. Ubiquitinated TDP-43 is abundant in cytosolic inclusions from patients suffering from *FTLD* and *ALS*. TDP-43 mutations produce polyUBQ TDP-43-containing inclusions in a subset of FTLD and ALS [243]. TDP-43 knockdown downregulates HDAC6 through reduced translation of HDAC6 mRNA in cells and in a *Drosophila* model [196]. Transfection with an aggregate-prone polyQ ataxin-3 construct in TDP-43 KO cells shows decreased aggresome formation and aggrephagy clearance. Transfection of the same cells with exogenous TDP-43 restores aggrephagy clearance to basal levels [196]. Fused in sarcoma/translated in liposarcoma (FUS/TLS), a structurally related RNA-binding protein, is also mutated in ALS and causes aggrephagy-related phenotypes [244]. FUS/TLS co-localizes with TDP-43S (the smaller of two TDP-43 complexes with distinct cellular functions) [245]. FUS/TLS and TDP-43 bind HDAC6 mRNA on overlapping sites, and regulate HDAC6 expression [245]. TDP-43 knockdown in neurons reduces HDAC6 levels, and causes HDAC6-mediated reduction of neurite outgrowth upon BDNF stimulation [246]. HDAC6 KO shows a similar phenotype. Accordingly, the addition of either exogenous TDP-43 or HDAC6 increases neurite outgrowth [246]. TDP-43 regulates parkin expression through binding to parkin mRNA [247]. Parkin polyubiquitinates TDP-43, and a complex between HDAC6, parkin, and TDP-43 may lead to the cytosolic translocation of polyUBQ–TDP-43, which is then resistant to autophagy and to UPS clearance [248].

Axonal transport deficiencies in ALS are recapitulated in a mutant, aggregation-prone SOD1 mouse model [249]. Deletion of HDAC6 in these KO mice slows the progression of motor neuron degeneration. A prevalence of positive (restoration of axonal transport) *vs.* negative (aggrephagy impairment) effects is observed for HDAC6 inhibition in ALS. Another α-tubulin deacetylase, Sirt2, does not affect the disease phenotype of mutant SOD1 mice [249]. A pan-HDAC inhibitor shows increased MT acetylation, restored axonal transport, reduced motor neuron loss in the spinal cord, reduced denervation at neuromuscular junctions (NMJs), and reduced skeletal muscle atrophy. HDAC6 inhibition is at least partially responsible for the observed phenotype [250]. HDAC6 binds to mutant, aggregation-prone SOD1 isoforms through oligomerization-prone SMIR-like motifs [251], as shown for p62–SOD1 interactions [90]. Namely, a SMIR sequence (Leu816-Leu835, C-SMIR, immediately before the SE14 domain of HDAC6) is a BUZ-independent, selective binder for non-ubiquitinated, mutant SOD1 proteins in an ALS model [251]. C-SMIR binds mutant SOD1 after homodimerization, as it happens for the structurally similar SMIR sequence in p62 [90]. A second SMIR motif in HDAC6 (N-SMIR, His275-Gly301) can bind mutant SOD1, but its inability to induce homodimerization requires preexistent oligomerization of HDAC6 to induce the N-SMIR–mutant SOD1 interaction. The HDAC6–mutant SOD1 protein-protein interaction (PPI) leads to the sequestration of HDAC6 in SOD1 inclusion bodies, and to increased α-tubulin acetylation [251]. HDAC6 ablation increases UBQ-lacking SOD1 aggregation in a BUZ domain-independent manner. HDAC6–

SOD1 binding impairs HDAC6 deacetylation and leads to a therapeutically useful increase in axonal transport. Prolonged inhibition of HDAC6 deacetylation and generation of SOD1 inclusion bodies may, though, cause significant side effects related to aggrephagy and UPS impairment [251]. Thus, it is reasonable to think that inhibitors of the HDAC6–mutant SOD1 PPI (possibly also inhibiting the SMIR-based p62–mutant SOD1 PPI [252]) could be therapeutically active against ALS.

HDAC6 is overexpressed (51% increase in the cerebral cortex and 91% increase in the hippocampus) in *AD* brains, compared to age-matched control brains [253]. Another study [254] does not report increased HDAC6 levels in the temporal neocortex of AD patients, possibly due to different experimental protocols.

Tau and HDAC6 interact in cells and in human brain tissues. Tau–HDAC6 immunoprecipitation is more abundant in AD brain samples. The tau–HDAC6 interaction is BUZ domain- and UBQ-independent, and involves respectively the SE14 domain of HDAC6 (AA 840–1090) and the MTBRs of tau [253]. Catalytically inactive DD1-DD2 double HDAC6 mutants bind tau, and chemical inhibition of HDAC6 deacetylation by tubacin does not prevent the tau–HDAC6 interaction. Chemical UPS inhibition causes an increased co-localization of tau–HDAC6 complexes and their perinuclear localization, pointing to an aggregation-transport role for HDAC6 that promotes tau clearance—a putative therapeutic role for HDAC6 in UPS-impaired AD tissues [253]. Chemical inhibition of HDAC6 deacetylase activity does not prevent the perinuclear localization of tau–HDAC6 complexes in UPS-impaired cells, but it indirectly influences tau phosphorylation. It reduces the abundance of the pathological pT231 epitope—possibly a neurotoxic, HDAC6-mediated effect in *tauopathies* [253]. In fact, tau–HDAC6 PPIs stimulate pathological hyperphosphorylation of tau, and the abundance of HDAC6 in clinical isolates from AD patients underlines its multiple neurotoxic effects [253].

Increased MT acetylation is observed in AD neurons and NFTs [255]. MT acetylation is reduced in tau KO mice and is increased in tau-overexpressing cells, when compared with control samples. Tau acts as a selective HDAC6 deacetylase inhibitor, either in a soluble/BUZ domain-independent manner or in a UBQ-containing, insoluble PHF/BUZ domain-dependent manner [255]. UPS impairment-determined aggrephagy seems also to be affected by tau through its interaction with HDAC6, which inhibits aggresome formation and aggrephagy progression [256].

The role of aggrephagy and HDAC6 in tau clearance in UPS-impaired conditions is studied in tau-overexpressing cells [256]. Chemical UPS inhibition stimulates the formation of spherical, tau-containing, aggresome-like structures, which include vimentin and are surrounded by dynein motors and mitochondria. Aggresome-like structures are reversibly dismantled by washout of UPS inhibitors, and preferentially include pathologically hyperphosphorylated tau [256]. HDAC6 is present in these structures. siRNA-mediated HDAC6 silencing in tau-overexpressing cells with functioning UPS leads to the disappearance of insoluble tau, while UPS inhibition causes an increase of insoluble tau [256] with respect to non-HDAC6-silenced cells. The former phenotype may reflect UPS-driven elimination of soluble tau, while the latter may indicate an overall impairment of tau degradation and an increase of dispersed tau-containing microaggregates, which do not merge into aggresomes and cannot be cleared *via* autophagy. The existence of a robust assay format [256] suitable for high throughput screening (HTS) campaigns aiming to identify small molecule modulators of the HDAC6–tau PPI, and of the following neurotoxic cascade of events, should be exploited for drug discovery purposes. Small molecule modulators should then allow to better define the interacting regions in the SE14/HDAC6 and MTBR/tau domains, eventually leading to the rational design/structural optimization of drug-like inhibitors of the HDAC6–tau PPI.

Ubiquitinated HDAC6 binds to C-terminus of Hsc70 interacting protein (CHIP) an Hsp90 co-chaperone previously described in detail (see Section 3.3), and is degraded *via* the UPS. CHIP silencing leads to increased HDAC6 levels [257]. HDAC6 promotes tau accumulation indirectly, *via* deacetylation of Hsp90 and increase of Hsp90/p23 complex-mediated protection of tau from UPS degradation. Conversely, HDAC6 inhibition leads to acetylated Hsp90-mediated tau clearance by UPS [257]. In primary cortical neurons, chemical HDAC6 inhibition causes increased MT acetylation, decreased p23 levels, increased HSF1-mediated gene transcription, and decreased levels of pathological tau isoforms [257]. CHIP may play a role in HDAC6 regulation *via* the UPS, while CHIP/UPS-impaired environments may switch HDAC6 towards neurotoxic scenarios. A synergistic combination of HDAC6–Hsp90 inhibitors against tauopathies is thus conceivable [257].

A *Drosophila* model, expressing either WT or mutated human tau in muscle cells, shows defects in MT density and fragmentation [258]. A genetic screen for mutations influencing the phenotype highlights HDAC6 KO flies as fully rescued in terms of MT defects in muscle cells. Moreover, HDAC6 KO flies completely restore NMJ defects in a deacetylase-dependent manner. A chemical HDAC6 inhibitor shows similar effects [258]. Chemical or genetic HDAC6 silencing is equally active against WT and mutated tau, and does not influence tau degradation, providing a suitable *in vivo* screen for targets and compounds influencing tau-induced MT defects [258].

HDAC6 deacetylase activity, and its effects on MT transport of cargo proteins and organelles, is regulated by phosphorylation [213]. Namely, GSK-3β is a positive regulator of HDAC6 catalytic activity through phosphorylation of the Ser22 residue of rodent HDAC6. Conversely, Akt activation and consequent GSK-3β inhibition leads to HDAC6 inhibition [213].

GSK-3β is a well-known tau kinase with therapeutic potential against tauopathies and its activity is increased in early AD [259]. Kinesin-1-mediated mitochondrial transport in hippocampal neurons is enhanced by HDAC6 inhibition, either directly with HDAC6 inhibitors or through kinase regulation. Abnormal mitochondrial transport is observed in AD and in other NDDs, and its HDAC6 inhibition-promoted rescue should have therapeutic effects [213].

HDAC6 inhibition shows positive effects on β-amyloid pathology in AD. The double TG APP-PS1 mice show severe, β-amyloid-driven neuropathology. The triple TG APP-PS1-HDAC KO mice [230] show restoration of associative function and of spatial memory, likely due to increased Lys40 acetylation of MTs and to restoration of AD-impaired mitochondrial trafficking. The triple TG mice are also resistant to exogenous Aβ-induced mitochondrial trafficking impairment [230]. Another TG APP-PS1 model shows rescue of impaired mitochondrial trafficking by chemical inhibition of HDAC6 [260]. The effect is MT deacetylation-dependent. HDAC6 inhibition restores both velocity and motility of mitochondrial transport, and increases mitochondrial elongation [260]. Finally, chemical HDAC6 inhibition provides learning and memory improvements in triple TG APP-tau-PS1 mice [261]. Biochemical effects of chemical HDAC6 inhibition include reduction of Aβ levels, transcriptional regulation of Aβ synthesis and degradation, and decrease of tau phosphorylation at the Thr181 epitope [261].

Positive effects on impaired MT transport and on tau clearance aggregation seem to prevail on negative effects on aggrephagy for HDAC6 inhibitors against tauopathies, and in particular AD [262]. Additionally, regulation of abnormal synaptic functions [211] and reduction of neuroinflammatory responses [227] may also derive from HDAC6 inhibition in AD. Conversely, the observed excess of Aβ-containing APs in AD [263] requires a more detailed evaluation of the role of autophagy in AD and tauopathies.

5.4 DISEASE-MODIFYING COMPOUNDS

This chapter deals with mid-late steps leading to neuropathological alterations related to protein misfolding and aggregation in general, and to tau and/or tau-connected events in particular. Two molecular targets are chosen and discussed in detail. Several other targets are only mentioned here, mostly for reasons of space. Forty-two compounds/scaffolds acting on p62 and HDAC6 are diffusely covered in a chemistry-oriented companion book [264] devoted to disease-modifying compounds, and are briefly characterized in Table 5.1. Each compound class is numbered as in the chemistry-oriented companion book, and its chemical core is structurally

TABLE 5.1   Compounds 5.8–5.47: Chemical Class, Target, Developing Organization, Development Status

Number	Chemical cpd./class	Target	Organization	Dev. status
5.8	Stearyl-NH-DPETGEL peptide amide	p62	University College, London	DD
5.9	Indolinones, SU6668	TBK-1, other kinase inhibitor	Sugen/Pfizer	Ph II
5.10	Aminopyrimidines, BX795	TBK-1, other kinase inhibitor	Novartis	Ph I
5.11	Aminopyrimidines, MRT67307	TBK-1-IKKε inhibitors	Dundee University	LO
5.12	Aminopyrimidines	TBK-1-IKKε inhibitors	Dundee University	PE
5.13	6-aminopyrazolopyrimidines	TBK-1-IKKε inhibitors	SouthWestern Medical Center, Dallas, TX	DD
5.14	Azabenzimidazoles	TBK-1-IKKε inhibitors	AstraZeneca	LO
5.15	Pyrimidines	TBK-1 inhibitors	University of North Carolina	DD
5.16	Pyridopyrimidines	TBK-1 inhibitors	University of North Carolina	DD
5.17	Tricyclic pyridothiophenes	TBK-1 inhibitors	University of North Carolina	DD
5.18	Amlexanox	TBK-1-IKKε inhibitor	University of Michigan	MKTD
5.19	Hydroxamic acids, SAHA	Broad HDAC inhibitors	Merck Sharp & Dohme	MKTD
5.20	Cyclic depsipeptides, romidepsin, FK-228	Broad HDAC inhibitor	Celgene	MKTD
5.21	Panobinostat, LBH-589	Broad HDAC inhibitor	Novartis	Ph III
5.22	Abexinostat, PCI-24781	Broad HDAC inhibitor	Pharmacyclics	Ph II
5.23	Givinostat, ITF2357	Broad HDAC inhibitor	Italfarmaco	Ph II
5.24	Entinostat, MS-275	Class I-specific HDAC inhibitor	Syndax Pharmac.	Ph II

TABLE 5.1   Compounds 5.8–5.47: Chemical Class, Target, Developing Organization, Development Status (*cont.*)

Number	Chemical cpd./class	Target	Organization	Dev. status
5.25	Mocetinostat, MGCD0103	Class I-specific HDAC inhibitor	MethylGene	Ph I
5.26	Tubacin	HDAC6 inhibitor	Harvard University	PE
5.27	Rocilinostat, ACY-1215	HDAC6 inhibitor	Acetylon Pharmac.	Ph I/II
5.28	M344	Broad HDAC inhibitor	Munster University, Germany	PE
5.29a,b	ST80 (a)	HDAC6 inhibitor	Munster University, Germany	LO
5.30, 5.31	Isoxazole hydoxamates	Class I-specific HDAC inhibitors	University of Illinois	LO
5.32	Triazolylphenyl hydroxamates	Broad HDAC inhibitors	University of Illinois	LO
5.33	C1A	Broad HDAC inhibitor	Imperial College, UK	LO
5.34	Furylamine hydroxamates	Broad HDAC inhibitors	China Pharm. University	DD
5.35	Arylalkene hydroxamates	HDAC6–HDAC8 inhibitor	Broad Institute, Boston, USA	LO
5.36	Dihydroqui-noxalinone hydroxamates	HDAC6 inhibitor	MethylGene	LO
5.37	Pyrroline hydroxamates	HDAC6 inhibitor	Broad Institute, Boston, USA	DD
5.38a,b	Tubastatin (a)	HDAC6 inhibitor	University of Illinois	PE
5.39	Tubathian A	HDAC6 inhibitor	University of Ghent, Belgium	LO
5.40	Nexturastat	HDAC6 inhibitor	University of Illinois	DD
5.41	ACY-738	HDAC6 inhibitor	University of Pennsylvania	LO
5.42	Quinazolin-4-one hydroxamates	HDAC6 inhibitor	Taiwan University	PE

(*Continued*)

TABLE 5.1   Compounds 5.8–5.47: Chemical Class, Target, Developing Organization, Development Status *(cont.)*

Number	Chemical cpd./class	Target	Organization	Dev. status
5.43	HPOB	HDAC6 inhibitor	Sloan Kettering Cancer Center, USA	LO
5.44	Thiols	Broad HDAC inhibitors	Nagoya University	DD
5.45, 5.46	Mercaptoamides, DMA-CB	Broad HDAC inhibitors	University of Illinois	PE
5.47	Trifluoroacetylthio-phenes	Broad HDAC inhibitors	Angeletti, Rome, Italy	DD

Not progressed, NP; early discovery, DD; lead optimization, LO; preclinical evaluation, PE; clinical Phase I–II–III, Ph I–Ph III; marketed, MKTD; food supplement, FS.

defined; its mechanism of action and molecular target are mentioned; the public or private laboratory that develops the compound is listed; and the development status—according to publicly available information—is provided.

References

1. He, C.; Klionsky, D. J. Regulation mechanisms and signaling pathways of autophagy. *Annu. Rev. Genet.* **2009**, *43*, 67–93.
2. Choi, A. M. K.; Ryter, S. W.; Levine, B. Autophagy in human health and disease. *N. Engl. J. Med.* **2013**, *368*, 651–662.
3. Li, W.; Yang, Q.; Mao, Z. Chaperone-mediated autophagy: machinery, regulation and biological consequences. *Cell. Mol. Life Sci.* **2011**, *68*, 749–763.
4. Chiang, H. L.; Terlecky, S. R.; Plant, C. P.; Dice, J. F. A role for a 70-KDa heat shock protein in lysosomal degradation of intracellular proteins. *Science* **1989**, *246*, 382–385.
5. Cuervo, A. M. Chaperone-mediated autophagy: selectivity pays off. *Trends Endocrinol. Metab.* **2010**, *21*, 142–150.
6. Orenstein, S. J.; Cuervo, A. M. Chaperone-mediated autophagy: molecular mechanisms and physiological relevance. *Semin. Cell. Dev. Biol.* **2010**, *21*, 719–726.
7. Cuervo, A. M.; Dice, J. F. A receptor for the selective uptake and degradation of proteins by lysosomes. *Science* **1996**, *273*, 501–503.
8. Salvador, N.; Aguado, C.; Horst, M.; Knecht, E. Import of a cytosolic protein into lysosomes by chaperone-mediated autophagy depends on its folding state. *J. Biol. Chem.* **2000**, *275*, 27447–27456.
9. Bandyopadhyay, U.; Kaushik, S.; Varticovski, L.; Cuervo, A. M. The chaperone mediated autophagy receptor organizes in dynamic protein complexes at the lysosomal membrane. *Mol. Cell. Biol.* **2008**, *28*, 5747–5763.
10. Kaushik, S.; Cuervo, A. M. Chaperone-mediated autophagy: a unique way to enter the lysosome world. *Trends Cell Biol.* **2012**, *22*, 407–417.
11. Tolkovsky, A. M. Mitophagy. *Biochim. Biophys. Acta* 1793, **2009**, 1508–1515.
12. Ding, W.-X.; Yin, X.-M. Mitophagy: mechanisms, pathophysiological roles, and analysis. *Biol. Chem.* **2012**, *393*, 547–564.

13. Zheng, Y. T.; Shahnazari, S.; Brech, A.; Lamark, T.; Johansen, T.; Brumell, J. H. The adaptor protein p62/SQSTM1 targets invading bacteria to the autophagy pathway. *J. Immunol.* **2009**, *183*, 5909–5916.
14. Wileman, T. Autophagy as a defence against intracellular pathogens. *Essays Biochem.* **2013**, *55*, 153–163.
15. Ogata, M.; Hino, S.-i.; Saito, A.; Morikawa, K.; Kondo, S.; Kanemoto, S., et al. Autophagy is activated for cell survival after endoplasmic reticulum stress. *Mol. Cell. Biol.* **2006**, *26*, 9220–9231.
16. Bernales, S.; McDonald, K. L.; Walter, P. Autophagy counterbalances endoplasmic reticulum expansion during the unfolded protein response. *PLoS Biol.* **2006**, *4*, e423.
17. Kraft, C.; Deplazes, A.; Sohrmann, M.; Peter, M. Mature ribosomes are selectively degraded upon starvation by an autophagy pathway requiring the Ubp3p/Bre5p ubiquitin protease. *Nat. Cell Biol.* **2008**, *10*, 602–610.
18. Cebollero, E.; Reggiori, F.; Kraft, C. Reticulophagy and ribophagy: regulated degradation of protein production factories. *Int. J. Cell Biol.* **2012**, *2012*, 182834.
19. Singh, R.; Kaushik, S.; Wang, Y.; Xiang, Y.; Novak, I.; Komatsu, M., et al. Autophagy regulates lipid metabolism. *Nature* **2009**, *458*, 1131–1135.
20. Singh, R.; Cuervo, M. A. Lipophagy: connecting autophagy and lipid metabolism. *Int. J. Cell Biol.* **2012**, *2012*, 282041.
21. Iwata, J.; Ezaki, J.; Komatsu, M.; Yokota, S.; Ueno, T.; Tanida, I., et al. Excess peroxisomes are degraded by autophagic machinery in mammals. *J. Biol. Chem.* **2006**, *281*, 4035–4041.
22. Till, A.; Lakhani, R.; Burnett, S. F.; Subramani, S. Pexophagy: the selective degradation of peroxisomes. *Int. J. Cell Biol.* **2012**, 512721.
23. Grasso, D.; Ropolo, A.; Lo Re, A.; Boggio, V.; Molejon, M. I.; Iovanna, J. L., et al. Zymophagy, a novel selective autophagy pathway mediated by VMP1-USP9x-p62, prevents pancreatic cell death. *J. Biol. Chem.* **2011**, *286*, 8308–8324.
24. Vaccaro, M. I. Zymophagy: selective autophagy of secretory granules. *Int. J. Cell Biol.* **2012**, 396705.
25. Lamark, T.; Johansen, T. Aggrephagy: selective disposal of protein aggregates by macroautophagy. *Int. J. Cell Biol.* **2012**, 736905.
26. Tyedmers, J.; Moegk, A.; Bukau, B. Cellular strategies for controlling protein aggregation. *Nat. Rev. Mol. Cell Biol.* **2010**, *11*, 777–788.
27. Kopito, R. R. Aggresomes, inclusion bodies and protein aggregation. *Trends Cell Biol.* **2000**, *10*, 524–530.
28. de Calignon, A.; Fox, L. M.; Pitstick, R.; Carlson, G. A.; Bacskai, B. J.; Spires-Jones, T. L.; Hyman, B. T. Caspase activation precedes and leads to tangles. *Nature* **2010**, *464*, 1201–1204.
29. Reggiori, F.; Komatsu, M.; Finley, K.; Simonsen, A. Selective types of autophagy. *Int. J. Cell Biol.* **2012**, 219625.
30. Boland, B.; Kumar, A.; Lee, S.; Platt, F. M.; Wegiel, J.; Huang Yu, W.; Nixon, R. A. Autophagy induction and autophagosome clearance in neurons: relationship to autophagic pathology in Alzheimer's disease. *J. Neurosci.* **2008**, *28*, 6926–6937.
31. Wong, E.; Cuervo, A. M. Autophagy gone awry in neurodegenerative diseases. *Nat. Neurosci. Rev.* **2010**, *13*, 805–811.
32. Cuervo, A. M.; Bergamini, E.; Brunk, U. T.; Droge, W.; French, M.; Terman, A. Autophagy and aging: the importance of maintaining "clean" cells. *Autophagy* **2005**, *1*, 131–140.
33. Johnston, J. A.; Ward, C. L.; Kopito, R. R. Aggresomes: a cellular response to misfolded proteins. *J. Cell Biol.* **1998**, *143*, 1883–1898.
34. Kaganovich, D.; Kopito, R. R.; Frydman, J. Misfolded proteins partition between two distinct quality control compartments. *Nature* **2008**, *454*, 1088–1095.
35. Kirkin, V.; McEwan, D. G.; Novak, I.; Dikic, I. A role for ubiquitin in selective autophagy. *Mol Cell.* **2009**, *34*, 259–269.

36. Yao, T.-P. The role of ubiquitin in autophagy-dependent protein aggregate processing. *Genes & Cancer* **2010**, *1*, 779–786.
37. Stiess, M.; Maghelli, N.; Kapitein, L. C.; Gomis-Rüth, S.; Wilsch-Bräuninger, M.; Hoogenraad, C. C., et al. Axon extension occurs independently of centrosomal microtubule nucleation. *Science* **2010**, *327*, 704–707.
38. McNaught, K. S.; Shashidharan, P.; Perl, D. P.; Jenner, P.; Olanow, C. W. Aggresome-related biogenesis of Lewy bodies. *Eur. J. Neurosci.* **2002**, *16*, 2136–2148.
39. Perrot, R.; Eyer, J. Neuronal intermediate filaments and neurodegenerative disorders. *Brain Res. Bull.* **2009**, *80*, 282–295.
40. Kawaguchi, Y.; Kovacs, J. J.; McLaurin, A.; Vance, J. M.; Ito, A.; Yao, T. P. The deacetylase HDAC6 regulates aggresome formation and cell viability in response to misfolded protein stress. *Cell* **2003**, *115*, 727–738.
41. McCollum, A. K.; Casagrande, G.; Kohn, E. C. Caught in the middle: the role of Bag3 in disease. *Biochem. J.* **2010**, *425*, e1–e3.
42. Gamerdinger, M.; Carra, S.; Behl, C. Emerging roles of molecular chaperones and co-chaperones in selective autophagy: focus on BAG proteins. *J. Mol. Med.* **2011**, *89*, 1175–1182.
43. Gamerdinger, M.; Kaya, A. M.; Wolfrum, U.; Clement, A. M.; Behl, C. BAG3 mediates chaperone-based aggresome targeting and selective autophagy of misfolded proteins. *EMBO Reports* **2011**, *12*, 149–156.
44. Kettern, N.; Dreiseidler, M.; Tawo, R.; Hohfeld, J. Chaperone-assisted degradation: multiple paths to destruction. *Biol. Chem.* **2010**, *391*, 481–489.
45. Yamanaka, K.; Sasagawa, Y.; Ogura, T. Recent advances in p97/VCP/Cdc48 cellular functions. *Biochim. Biophys. Acta* **1823**, *2012*, 130–137.
46. Tresse, E.; Salomons, F. A.; Vesa, J.; Bott, L. C.; Kimonis, V.; Yao, T. P., et al. VCP/p97 is essential for maturation of ubiquitin-containing autophagosomes and this function is impaired by mutations that cause IBMPFD. *Autophagy* **2010**, *6*, 217–227.
47. Ju, J. S.; Weihl, C. C. Inclusion body myopathy, Paget's disease of the bone and frontotemporal dementia: a disorder of autophagy. *Hum. Mol. Gen.* **2010**, *19*, R38–R45.
48. Buchberger, A. From UBA to UBX: new words in the ubiquitin vocabulary. *Trends Cell Biol.* **2002**, *12*, 216–221.
49. Rothenberg, C.; Srinivasan, D.; Mah, L.; Kaushik, S.; Peterhoff, C. M.; Ugolino, J., et al. Ubiquilin functions in autophagy and is degraded by chaperone mediated autophagy. *Hum. Mol. Gen.* **2010**, *19*, 3219–3232.
50. Stieren, E. S.; El Ayadi, A.; Xiao, Y.; Siller, E.; Landsverk, M. L.; Oberhauser, A. F., et al. Ubiquilin-1 is a molecular chaperone for the amyloid precursor protein. *J. Biol. Chem.* **2011**, *286*, 35689–35698.
51. Shin, J. p62 and the sequestosome, a novel mechanism for protein metabolism. *Arch. Pharm. Res.* **1998**, *21*, 629–633.
52. Gamerdinger, M.; Hajieva, P.; Kaya, A. M.; Wolfrum, U.; Hartl, F. U.; Behl, C. Protein quality control during aging involves recruitment of the macroautophagy pathway by BAG3. *EMBO J.* **2009**, *28*, 889–901.
53. Johansen, T.; Lamark, T. Selective autophagy mediated by autophagic adapter proteins. *Autophagy* **2011**, *7*, 279–296.
54. Kraft, C.; Peter, M.; Hofmann, K. Selective autophagy: ubiquitin-mediated recognition and beyond. *Nat. Cell Biol.* **2010**, *12*, 836–841.
55. Kachaner, D.; Genin, P.; Laplantine, E.; Weil, R. Toward an integrative view of Optineurin functions. *Cell Cycle* **2012**, *11*, 2808–2818.
56. Korac, J.; Schaeffer, V.; Kovacevic, I.; Clement, A. M.; Jungblut, B.; Behl, C., et al. Ubiquitin-independent function of optineurin in autophagic clearance of protein aggregates. *J. Cell Sci.* **2013**, *126*, 580–592.
57. Maruyama, H.; Morino, H.; Ito, H.; Izumi, Y.; Kato, H.; Watanabe, Y., et al. Mutations of optineurin in amyotrophic lateral sclerosis. *Nature* **2010**, *465*, 223–226.

58. Schwab, C.; Yu, S.; McGeer, E. G.; McGeer, P. L. Optineurin in Huntington's disease intranuclear inclusions. *Neurosci. Lett.* **2012**, *506*, 149–154.

59. Osawa, T.; Mizuno, Y.; Fujita, Y.; Takatama, M.; Nakazato, Y.; Okamoto, K. Optineurin in neurodegenerative diseases. *Neuropathology* **2011**, *31*, 569–574.

60. Gibbings, D.; Mostowy, S.; Jay, F.; Schwab, Y.; Cossart, P.; Voinnet, O. Selective autophagy degrades DICER and AGO2 and regulates miRNA activity. *Nat. Cell Biol.* **2012**, *14*, 1314–1321.

61. von Muhlinen, N.; Akutsu, M.; Ravenhill, B. J.; Foeglein, A.; Bloor, S.; Rutherford, T. J., et al. LC3C, bound selectively by a noncanonical LIR motif in NDP52, is required for antibacterial autophagy. *Mol. Cell* **2012**, *48*, 329–342.

62. Thurston, T. L.; Wandel, M. P.; von Muhlinen, N.; Foeglein, A.; Randow, F. Galectin 8 targets damaged vesicles for autophagy to defend cells against bacterial invasion. *Nature* **2012**, *482*, 414–418.

63. Filimonenko, M.; Isakson, P.; Finley, K. D.; Anderson, M.; Jeong, H.; Melia, T. J., et al. The selective macroautophagic degradation of aggregated proteins requires the PI3P-binding protein Alfy. *Mol. Cell* **2010**, *38*, 265–279.

64. Isakson, P.; Holland, P.; Simonsen, A. The role of ALFY in selective autophagy. *Cell Death Differ.* **2013**, *20*, 12–20.

65. Clausen, T. H.; Lamark, T.; Isakson, P.; Finley, K.; Larsen, K. B.; Brech, A., et al. p62/SQSTM1 and ALFY interact to facilitate the formation of p62 bodies/ALIS and their degradation by autophagy. *Autophagy* **2010**, *6*, 330–344.

66. Simonsen, A.; Birkeland, H. C.; Gillooly, D. J.; Mizushima, N.; Kuma, A.; Yoshimori, T., et al. Alfy, a novel FYVE domain-containing associated with protein granules and autophagic membranes. *J. Cell. Sci.* **2004**, *117*, 4239–4251.

67. Cuervo, A. M.; Stefanis, L.; Fredenburg, R.; Lansbury, P. T.; Sulzer, D. Impaired degradation of mutant α-synuclein by chaperone-mediated autophagy. *Science* **2004**, *305*, 1292–1295.

68. Sarkar, S.; Ravikumar, B.; Floto, R. A.; Rubinsztein, D. C. Rapamycin and mTOR-independent autophagy inducers ameliorate toxicity of polyglutamine-expanded huntingtin and related proteinopathies. *Cell Death Differ.* **2009**, *16*, 46–56.

69. Wang, Y.; Martinez-Vicente, M.; Kruger, U.; Kaushik, S.; Wong, E.; Mandelkow, E. M., et al. Tau fragmentation, aggregation and clearance: the dual role of lysosomal processing. *Hum. Mol. Genet.* **2009**, *18*, 4153–4170.

70. Dickey, C. A.; Kamal, A.; Lundgren, K.; Klosak, N.; Bailey, R. M.; Dunmore, J., et al. The high-affinity HSP90-CHIP complex recognizes and selectively degrades phosphorylated tau client proteins. *J. Clin. Invest.* **2007**, *117*, 648–658.

71. Ramesh Babu, J.; Lamar Seibenhener, M.; Peng, J.; Strom, A. L.; Kemppainen, R.; Cox, N., et al. Genetic inactivation of p62 leads to accumulation of hyperphosphorylated tau and neurodegeneration. *J. Neurochem.* **2008**, *106*, 107–120.

72. Joung, I.; Strominger, J. L.; Shin, J. Molecular cloning of a phosphotyrosine-independent ligand of the p56lck SH2 domain. *Proc. Natl. Acad. Sci. U.S.A.* **1996**, *93*, 5991–5995.

73. Shin, J. p62 and the sequestosome, a novel mechanism for protein metabolism. *Arch. Pharm. Res.* **1998**, *21*, 629–633.

74. Bjørkøy, G.; Lamark, T.; Brech, A.; Outzen, H.; Perander, M.; Oeervatn, A., et al. p62/SQSTM1 forms protein aggregates degraded by autophagy and has a protective effect on huntingtin-induced cell death. *J. Cell Biol.* **2005**, *171*, 603–614.

75. Pankiv, S.; Clausen, T. H.; Lamark, T.; Brech, A.; Bruun, J. A.; Outzen, H., et al. p62/SQSTM1 binds directly to Atg8/LC3 to facilitate degradation of ubiquitinated protein aggregates by autophagy. *J. Biol. Chem.* **2007**, *282*, 24131–24145.

76. Rechsteiner, M.; Rogers, S. W. PEST sequences and regulation of proteolysis. *Trends Biochem. Sci.* **1996**, *21*, 267–271.

77. Yu, H. B.; Kielczewska, A.; Rozek, A.; Takenaka, S.; Li, Y.; Thorson, L., et al. Sequestosome-1/p62 is the key intracellular target of innate defence regulator peptide. *J. Biol. Chem.* **2009**, *284*, 36007–36011.

78. Jiang, J.; Parameshwaran, K.; Seibenhener, M. L.; Kang, M. G.; Suppiramaniam, V.; Huganir, R. L., et al. AMPA receptor trafficking and synaptic plasticity require SQSTM1/p62. *Hippocampus* **2009**, *19*, 392–406.
79. Geetha, T.; Seibenhener, M. L.; Chen, L.; Madura, K.; Wooten, M. W. p62 serves as a shuttling factor for TrkA interaction with the proteasome. *Biochem. Biophys. Res. Commun.* **2008**, *374*, 33–37.
80. Wooten, M. W.; Geetha, T.; Seibenhener, M. L.; Babu, J. R.; Diaz-Meco, M. T.; Moscat, J. The p62 scaffold regulates nerve growth-induced NF-kB activation by influencing TRAF6 polyubiquitination. *J. Biol. Chem.* **2005**, *280*, 35625–35629.
81. Komatsu, M.; Kurokawa, H.; Waguri, S.; Taguchi, K.; Kobayashi, A.; Ichimura, Y., et al. The selective autophagy substrate p62 activates the stress responsive transcription factor Nrf2 through inactivation of Keap1. *Nat. Cell Biol.* **2010**, *12*, 213–223.
82. Duran, A.; Amanchy, R.; Linares, J. F.; Joshi, J.; Abu-Baker, S.; Porollo, A., et al. p62 is a key regulator of nutrient sensing in the mTORC1 pathway. *Mol. Cell* **2011**, *44*, 134–146.
83. Gal, J.; Ström, A. L.; Kilty, R.; Zhang, F.; Zhu, H. p62 accumulates and enhances aggregate formation in model systems of familial amyotrophic lateral sclerosis. *J. Biol. Chem.* **2007**, *282*, 11068–11077.
84. Kim, P. K.; Hailey, D. W.; Mullen, R. T.; Lippincott-Schwartz, J. Ubiquitin signals autophagic degradation of cytosolic proteins and peroxisomes. *Proc. Natl. Acad. Sci. U.S.A.* **2008**, *105*, 20567–20574.
85. Vadlamudi, R. K.; Joung, I.; Strominger, J. L.; Shin, J. p62, a phosphotyrosine-independent ligand of the SH2 domain of p56lck, belongs to a new class of ubiquitin binding proteins. *J. Biol. Chem.* **1996**, *271*, 20235–20237.
86. Tan, J. M.; Wong, E. S.; Dawson, V. L.; Dawson, T. M.; Lim, K. L. Lysine 63-linked polyubiquitin potentially partners with p62 to promote the clearance of protein inclusions by autophagy. *Autophagy* **2008**, *4*, 251–253.
87. Tan, J. M.; Wong, E. S.; Kirkpatrick, D. S.; Pletnikova, O.; Ko, H. S.; Tay, S. P., et al. Lysine 63-linked ubiquitination promotes the formation and autophagic clearance of protein inclusions associated with neurodegenerative diseases. *Hum. Mol. Genet.* **2008**, *17*, 431–439.
88. Moscat, J.; Diaz-Meco, M. T.; Wooten, M. W. Signal integration and diversification through the p62 scaffold protein. *Trends Biochem. Sci.* **2007**, *32*, 95–100.
89. Wooten, M. W.; Geetha, T.; Babu, J. R.; Seibenhener, M. L.; Peng, J.; Cox, N., et al. Essential role of sequestosome 1/p62 in regulating accumulation of Lys63-ubiquitinated proteins. *J. Biol. Chem.* **2008**, *283*, 6783–6789.
90. Gal, J.; Ström, A. L.; Kwinter, D. M.; Kilty, R.; Zhang, J.; Shi, P., et al. Sequestosome 1/p62 links familial ALS mutant SOD1 to LC3 via an ubiquitin independent mechanism. *J. Neurochem.* **2009**, *111*, 1062–1073.
91. Watanabe, Y.; Tanaka, M. p62/SQSTM1 in autophagic clearance of a non-ubiquitinated substrate. *J. Cell Sci.* **2011**, *124*, 2692–2701.
92. Kirkin, V.; Lamark, T.; Sou, Y.-S.; Bjørkøy, G.; Nunn, J. L.; Bruun, J.-A., et al. A role for NBR1 in autophagosomal degradation of ubiquitinated substrates. *Mol. Cell* **2009**, *33*, 505–516.
93. Itakura, E.; Mizushima, N. p62 targeting to the autophagosome formation site requires self-oligomerization but not LC3 binding. *J. Cell Biol.* **2011**, *192*, 17–27.
94. Behrends, C.; Sowa, M. E.; Gygi, S. P.; Harper, J. W. Network organization of the human autophagy system. *Nature* **2010**, *466*, 68–76.
95. Ichimura, Y.; Kumanomidou, T.; Sou, Y. S.; Mizushima, T.; Ezaki, J.; Ueno, T., et al. Structural basis for sorting mechanism of p62 in selective autophagy. *J. Biol. Chem.* **2008**, *283*, 22847–22857.
96. Ravikumar, B.; Sarkar, S.; Davies, J. E.; Futter, M.; Garcia-Arencibia, M.; Green-Thompson, Z. W., et al. Regulation of mammalian autophagy in physiology and pathophysiology. *Physiol. Rev.* **2010**, *90*, 1383–1435.

97. Babu, J. R.; Geetha, T.; Wooten, M. W. Sequestosome 1/p62 shuttles polyubiquitinated tau for proteasomal degradation. *J. Neurochem.* **2005**, *94*, 192–203.
98. Seibenhener, M. L.; Babu, J. R.; Geetha, T.; Wong, H. C.; Krishna, N. R.; Wooten, M. W. Sequestosome 1/p62 is a polyubiquitin chain binding protein involved in ubiquitin proteasome degradation. *Mol. Cell. Biol.* **2004**, *24*, 8055–8068.
99. Long, J.; Garner, T. P.; Pandya, M. J.; Craven, C. J.; Chen, P.; Shaw, B., et al. Dimerisation of the UBA domain of p62 inhibits ubiquitin binding and regulates NFκB signalling. *J. Mol. Biol.* **2010**, *396*, 178–194.
100. Lamark, T.; Perander, M.; Outzen, H.; Kristiansen, K.; Overvatn, A.; Michaelsen, E., et al. Interaction codes within the family of mammalian Phox and Bem1p domain-containing proteins. *J. Biol. Chem.* **2003**, *278*, 34568–34581.
101. Sumimoto, H.; Kamakura, S.; Ito, T. Structure and function of the PB1 domain, a protein interaction module conserved in animals, fungi, amoebas, and plants. *Science STKE* **2007**, *401*, re6.
102. Nakamura, K.; Kimple, A. J.; Siderovski, D. P.; Johnson, G. L. PB1 domain interaction of p62/sequestosome 1 and MEKK3 regulates NF-κB activation. *J. Biol. Chem.* **2010**, *285*, 2077–2089.
103. Sanz, L.; Diaz-Meco, M. T.; Nakano, H.; Moscat, J. The atypical PKC-interacting protein p62 channels NF-κB activation by the IL-1-TRAF6 pathway. *EMBO J.* **2000**, *19*, 1576–1586.
104. Joung, I.; Kim, H. J.; Kwon, Y. K. p62 modulates Akt activity via association with PKCζ in neuronal survival and differentiation. *Biochem. Biophys. Res. Commun.* **2005**, *334*, 654–660.
105. Kim, Y.; Park, M. K.; Uhm, D. Y.; Shin, J.; Chung, S. Modulation of delayed rectifier potassium channels by α1-adrenergic activation via protein kinase Cζ and p62 in PC12 cells. *Neurosci. Lett.* **2005**, *387*, 43–48.
106. Geisler, S.; Holmström, K. M.; Skujat, D.; Fiesel, F. C.; Rothfuss, O. C.; Kahle, P. J.; Springer, W. PINK1/Parkin mediated mitophagy is dependent on VDAC1 and p62/SQSTM1. *Nat. Cell Biol.* **2010**, *12*, 119–131.
107. Okatsu, K.; Saisho, K.; Shimanuki, M.; Nakada, K.; Shitara, H.; Sou, Y., et al. p62/SQSTM1 cooperates with Parkin for perinuclear clustering of depolarized mitochondria. *Genes Cell* **2010**, *15*, 887–900.
108. Orvedahl, A.; MacPherson, S.; Sumpter, R.; Talloczy, Z.; Zou, Z.; Levine, B. Autophagy protects against sindbis virus infection of the central nervous system. *Cell Host Microbe* **2010**, *7*, 115–127.
109. Komatsu, M.; Waguri, S.; Koike, M.; Sou, Y. S.; Ueno, T.; Hara, T., et al. Homeostatic levels of p62 control cytoplasmic inclusion body formation in autophagy-deficient mice. *Cell* **2007**, *131*, 1149–1163.
110. Nagaoka, U.; Kim, K.; Jana, N. R.; Doi, H.; Maruyama, M.; Mitsui, K., et al. Increased expression of p62 in expanded polyglutamine-expressing cells and its association with polyglutamine inclusions. *J. Neurochem.* **2004**, *91*, 57–68.
111. Vadlamudi, R. K.; Shin, J. Genomic structure and promoter analysis of the p62 gene encoding a nonproteasomal multiubiquitin chain binding protein. *FEBS Lett.* **1998**, *435*, 138–142.
112. Du, Y.; Wooten, M. C.; Gearing, M.; Wooten, M. W. Age associated oxidative damage to the p62 promoter: implications for Alzheimer disease. *Free Radic. Biol. Med.* **2009**, *46*, 492–501.
113. Kuusisto, E.; Kauppinen, T.; Alafuzoff, I. Use of p62/SQSTM1 antibodies for neuropathological diagnosis. *Neuropathol. Appl. Neurobiol.* **2008**, *34*, 169–180.
114. Pankiv, S.; Lamark, T.; Bruun, J. A.; Overvatn, A.; Bjørkøy, G.; Johansen, T. Nucleocytoplasmic shuttling of p62/SQSTM1 and its role in recruitment of nuclear polyubiquitinated proteins to promyelocytic leukemia bodies. *J. Biol. Chem.* **2010**, *285*, 5941–5953.

115. Linares, J. F.; Amanchy, R.; Diaz-Meco, M. T.; Moscat, J. Phosphorylation of p62 by cdk1 controls the timely transit of cells through mitosis and tumor cell proliferation. *J. Biol. Chem.* 2010, *285*, 5941–5953.
116. Matsumoto, G.; Wada, K.; Okuno, M.; Kurosawa, M.; Nukina, N. Serine 403 phosphorylation of p62/SQSTM1 regulates selective autophagic clearance of ubiquitinated proteins. *Mol. Cell* 2011, *44*, 279–289.
117. Pilli, M.; Arko-Mensah, J.; Ponpuak, M.; Roberts, E.; Master, S.; Mandell, M. A., et al. TBK-1 promotes autophagy mediated antimicrobial defense by controlling autophagosome maturation. *Immunity* 2012, *37*, 223–234.
118. Liu, X.; Gal, J.; Zhu, H. Sequestosome 1/p62: a multi-domain protein with multi-faceted functions. *Front. Biol.* 2012, *7*, 189–201.
119. Helfrich, M. H.; Hocking, L. J. Genetics and aetiology of Pagetic disorders of bone. *Arch. Biochem. Biophys.* 2008, *473*, 172–182.
120. Goode, A.; Layfield, R. Recent advances in understanding the molecular basis of Paget disease of bone. *J. Clin. Pathol.* 2010, *63*, 199–203.
121. Strnad, P.; Zatloukal, K.; Stumptner, C.; Kulaksiz, H.; Denk, H. Mallory-Denk-bodies: lessons from keratin containing hepatic inclusion bodies. *Biochim. Biophys. Acta* **1782**, 2008, 764–774.
122. Rodriguez, A.; Durán, A.; Selloum, M.; Champy, M. F.; Diez-Guerra, F. J.; Flores, J. M., et al. Mature-onset obesity and insulin resistance in mice deficient in the signaling adapter p62. *Cell Metab.* 2006, *3*, 211–222.
123. Puissant, A.; Fenouille, N.; Auberger, P. When autophagy meets cancer through p62/SQSTM1. *Am. J. Cancer Res.* 2012, *2*, 397–413.
124. Inoue, D.; Suzuki, T.; Mitsuishi, Y.; Miki, Y.; Suzuki, S.; Sugawara, S., et al. Accumulation of p62/SQSTM1 is associated with poor prognosis in patients with lung adenocarcinoma. *Cancer Sci.* 2012, *103*, 760–766.
125. Rolland, P.; Madjd, Z.; Durrant, L.; Ellis, I. O.; Layfield, R.; Spendlove, I. The ubiquitin-binding protein p62 is expressed in breast cancers showing features of aggressive disease. *Endocr. Relat. Cancer* 2007, *14*, 73–80.
126. Ling, J.; Kang, Y.; Zhao, R.; Xia, Q.; Lee, D. F.; Chang, Z., et al. KrasG12D-induced IKK2/beta/NF-kappaB activation by IL-1alpha and p62 feedforward loops is required for development of pancreatic ductal adenocarcinoma. *Cancer Cell* 2012, *21*, 105–120.
127. Galavotti, S.; Bartesaghi, S.; Faccenda, D.; Shaked-Rabi, M.; Sanzone, S.; McEvoy, A., et al. The autophagy associated factors DRAM1 and p62 regulate cell migration and invasion in glioblastoma stem cells. *Oncogene* 2013, *36*, 699–712.
128. Duran, A.; Linares, J. F.; Galvez, A. S.; Wikenheiser, K.; Flores, J. M.; Diaz-Meco, M. T.; Moscat, J. The signaling adaptor p62 is an important NFkappaB mediator in tumorigenesis. *Cancer Cell* 2008, *13*, 343–354.
129. Parkhitko, A.; Myachina, F.; Morrison, T. A.; Hindi, K. M.; Auricchio, N.; Karbowniczek, M., et al. Tumorigenesis in tuberous sclerosis complex is autophagy and p62/sequestosome 1 (SQSTM1)-dependent. *Proc. Natl. Acad. Sci. U.S.A.* 2011, *108*, 12455–12460.
130. Gao, C.; Cao, W.; Bao, L.; Zuo, W.; Xie, G.; Cai, T., et al. Autophagy negatively regulates Wnt signalling by promoting Dishevelled degradation. *Nat. Cell. Biol.* 2010, *12*, 781–790.
131. Su, H.; Wang, X. p62 stages an interplay between the ubiquitin-proteasome system and autophagy in the heart of defense against proteotoxic stress. *Trends Cardiovasc Med.* 2011, *21*, 224–228.
132. Fecto, F.; Yan, J.; Vemula, S. P.; Liu, E.; Yang, Y.; Chen, W., et al. SQSTM1 mutations in familial and sporadic amyotrophic lateral sclerosis. *Arch. Neurol.* 2011, *68*, 1440–1446.
133. Teyssou, E.; Takeda, T.; Lebon, V.; Boillee, S.; Doukoure, B.; Bataillon, G., et al. Mutations in SQSTM1 encoding p62 in amyotrophic lateral sclerosis: genetics and neuropathology. *Acta Neuropathol.* 2013, *125*, 511–522.

134. Tanji, K.; Zhang, H. X.; Mori, F.; Kakita, A.; Takahashi, H.; Wakabayashi, K. p62/ sequestosome 1 binds to TDP-43 in brains with frontotemporal lobar degeneration with TDP-43 inclusions. *J. Neurosci. Res.* **2012**, *90*, 2034–2042.

135. Pikkarainen, M.; Hartikainen, P.; Alafuzoff, I. Neuropathologic features of fronto-temporal lobar degeneration with ubiquitin-positive inclusions visualized with ubiquitin-binding protein p62 immunohistochemistry. *J. Neuropathol. Exp. Neurol.* **2008**, *67*, 280–298.

136. Zhang, X.; Li, L.; Chen, S.; Yang, D.; Wang, Y.; Zhang, X., et al. Rapamycin treatment augments motor neuron degeneration in SOD1G93A mouse model of amyotrophic lateral sclerosis. *Autophagy* **2011**, *7*, 412–425.

137. Wang, I.-F.; Guo, B.-S.; Liu, Y.-C.; Wu, C.-C.; Yang, C.-H.; Tsai, K.-J.; Chen, C.-K. J. Autophagy activators rescue and alleviate pathogenesis of a mouse model with pro-teinopathies of the TAR DNA-binding protein 43. *Proc. Natl. Acad. Sci. U.S.A.* **2012**, *109*, 15024–15029.

138. Kuusisto, E.; Parkkinen, L.; Alafuzoff, I. Morphogenesis of Lewy bodies: Dissimilar incorporation of α-synuclein, ubiquitin, and p62. *J. Neuropathol. Exp. Neurol.* **2003**, *62*, 1241–1253.

139. Nakaso, K.; Yoshimoto, Y.; Nakano, T.; Takeshima, T.; Fukuhara, Y.; Yasui, K., et al. Transcriptional activation of p62/A170/ZIP during the formation of the aggregates: possible mechanisms and the role in Lewy body formation in Parkinson's disease. *Brain Res.* **2004**, *1012*, 42–51.

140. Kuusisto, E.; Salminen, A.; Alafuzoff, I. Ubiquitin-binding protein p62 is present in neuronal and glial inclusions in human tauopathies and synucleinopathies. *NeuroReport* **2001**, *12*, 2085–2090.

141. Narendra, D. P.; Kane, L. A.; Hauser, D. N.; Fearnley, I. M.; Youle, R. J. p62/SQSTM1 is required for Parkin-induced mitochondrial clustering but not mitophagy; VDAC1 is dispensable for both. *Autophagy* **2010**, *6*, 1090–1106.

142. Rakovic, A.; Shurkewitsch, K.; Seibler, P.; Gruenewald, A.; Zanon, A.; Hagenah, J., et al. Phosphatase and Tensin Homolog (PTEN)-induced putative kinase 1 (PINK1)-dependent ubiquitination of endogenous Parkin attenuates mitophagy. *J. Biol. Chem.* **2013**, *288*, 2223–2237.

143. Pikkarainen, M.; Hartikainen, P.; Soininen, H.; Alafuzoff, I. Distribution and pattern of pathology in subjects with familial or sporadic late-onset cerebellar ataxia as assessed by p62/sequestosome immunohistochemistry. *Cerebellum* **2011**, *10*, 720–731.

144. Salminen, A.; Kaarniranta, K.; Haapasalo, A.; Hiltunen, M.; Soininen, H.; Alafuzoff, I. Emerging role of p62/sequestosome-1 in the pathogenesis of Alzheimer's disease. *Progr. Neurobiol.* **2012**, *96*, 87–95.

145. Du, Y.; Wooten, M. C.; Wooten, M. W. Oxidative damage to the promoter region of SQSTM1/p62 is common to neurodegenerative disease. *Neurobiol. Dis.* **2009**, *35*, 302–310.

146. Bartlett, B. J.; Isakson, P.; Lewerenz, J.; Sanchez, H.; Kotzebue, R. W.; Cumming, R., et al. p62, Ref(2)P and ubiquitinated proteins are conserved markers of neuronal aging, aggregate formation and progressive autophagic defects. *Autophagy* **2011**, *7*, 572–583.

147. Tanji, K.; Maruyama, A.; Odagiri, S.; Mori, F.; Itoh, K.; Kakita, A., et al. Keap1 is localized in neuronal and glial cytoplasmic inclusions in various neurodegenerative diseases. *J. Neuropathol Exp. Neurol.* **2012**, *72*, 18–28.

148. Kaltschmidt, B.; Kaltschmidt, C. NF-κB in the nervous system. *Cold Spring Harb. Perspect. Biol.* **2009**, *1*, a001271.

149. Zhu, X.; Lee, H. G.; Raina, A. K.; Perry, G.; Smith, M. A. The role of mitogen activated protein kinase pathways in Alzheimer's disease. *Neurosignals* **2002**, *11*, 270–281.

150. Mehan, S.; Meena, H.; Sharma, D.; Sankhla, R. JNK: a stress-activated protein kinase therapeutic strategies and involvement in Alzheimer's and various neurodegenerative abnormalities. *J. Mol. Neurosci.* **2011**, *43*, 376–390.

151. de la Monte, S. M. Insulin resistance and Alzheimer's disease. *BMB Rep.* **2009**, *42*, 475–481.
152. Geetha, T.; Zheng, C.; McGregor, W. C.; White, B. D.; Diaz-Meco, M. T.; Moscat, J.; Ramesh Babu, J. TRAF6 and p62 inhibit amyloid β-induced neuronal death through p75 neurotrophin receptor. *Neurochem. Int.* **2012**, *61*, 1289–1293.
153. Coulson, E. J.; May, L. M.; Sykes, A. M.; Hamlin, A. S. The role of the p75 neurotrophin receptor in cholinergic dysfunction in Alzheimer's disease. *Neuroscientist* **2009**, *15*, 317–323.
154. Kanninen, K.; Heikkinen, R.; Malm, T.; Rolova, T.; Kuhmonen, S.; Leinonen, H., et al. Intrahippocampal injection of lentiviral vector expressing Nrf2 improves spatial learning in mouse model of Alzheimer's disease. *Proc. Natl. Acad. Sci. U.S.A.* **2009**, *106*, 16505–16510.
155. Nixon, R. A.; Wegiel, J.; Kumar, A.; Yu, W. H.; Peterhoff, C.; Cataldo, A.; Cuervo, A. M. Extensive involvement of autophagy in Alzheimer disease: an immuno-electron microscopy study. *J. Neuropathol. Exp. Neurol.* **2005**, *64*, 113–122.
156. Nixon, R. A. Autophagy, amyloidogenesis and Alzheimer disease. *J. Cell. Sci.* **2007**, *120*, 4081–4091.
157. Moreira, P. I.; Carvalho, C.; Zhu, X.; Smith, M. A.; Perry, G. Mitochondrial dysfunction is a trigger of Alzheimer's disease pathophysiology. *Biochim. Biophys. Acta* **1802**, *2010*, 2–10.
158. Swerdlow, R. H.; Burns, J. M.; Khan, S. M. The Alzheimer's disease mitochondrial cascade hypothesis. *J. Alzheimer's Dis.* **2010**, *20* (Suppl. 2), S265–S279.
159. Whitworth, A. J.; Pallanck, L. J. The PINK1/Parkin pathway: a mitochondrial quality control system? *J. Bioenerg. Biomembr.* **2009**, *41*, 499–503.
160. Narendra, D. P.; Jin, S. M.; Tanaka, A.; Suen, D. F.; Gautier, C. A.; Shen, J., et al. PINK1 is selectively stabilized on impaired mitochondria to activate Parkin. *PLoS Biol.* **2010**, *8*, 1000298.
161. Mahoney, C. J.; Beck, J.; Rohrer, J. D.; Lashley, T.; Mok, K.; Shakespeare, T., et al. Frontotemporal dementia with the C9ORF72 hexanucleotide repeat expansion: clinical, neuroanatomical and neuropathological features. *Brain* **2012**, *135*, 736–750.
162. Bieniek, K. F.; Murray, M. E.; Rutherford, N. J.; Castanedes-Casey, M.; DeJesus-Hernandez, M.; Liesinger, A. M., et al. Tau pathology in frontotemporal lobar degeneration with C9ORF72 hexanucleotide repeat expansion. *Acta Neuropathol.* **2013**, *125*, 289–302.
163. King, A.; Al-Sarraj, S.; Troakes, C.; Smith, B. N.; Maekawa, S.; Iovino, M., et al. Mixed tau, TDP-43 and p62 pathology in FTLD associated with a C9ORF72 repeat expansion and p. Ala239Thr MAPT (tau) variant. *Acta Neuropathol.* **2013**, *125*, 303–310.
164. Grozinger, C. M.; Hassig, C. A.; Schreiber, S. L. Three proteins define a class of human histone deacetylases related to yeast Hda1p. *Proc. Natl. Acad. Sci. U.S.A.* **1999**, *96*, 4868–4873.
165. Hubbert, C.; Guardiola, A.; Shao, R.; Kawaguchi, Y.; Ito, A.; Nixon, A., et al. HDAC6 is a microtubule-associated deacetylase. *Nature* **2002**, *417*, 455–458.
166. Zhang, Y.; Li, N.; Caron, C.; Matthias, G.; Hess, D.; Khochbin, S.; Matthias, P. HDAC-6 interacts with and deacetylates tubulin and microtubules in vivo. *EMBO J.* **2003**, *22*, 1168–1179.
167. de Ruijter, A. J.; van Gennip, A. H.; Caron, H. N.; Kemp, S.; van Kuilenburg, A. B. Histone deacetylases (HDACs): characterization of the classical HDAC family. *Biochem. J.* **2003**, *370*, 737–749.
168. Liu, Y.; Peng, L.; Seto, E.; Huang, S.; Qiu, Y. Modulation of histone deacetylase 6 (HDAC6) nuclear import and tubulin deacetylase activity through acetylation. *J. Biol. Chem.* **2012**, *287*, 29168–29174.
169. Bertos, N. R.; Gilquin, B.; Chan, G. K.; Yen, T. J.; Khochbin, S.; Yang, X. J. Role of the tetradecapeptide repeat domain of human histone deacetylase 6 in cytoplasmic retention. *J. Biol. Chem.* **2004**, *279*, 48246–48254.

170. Haggarty, S. J.; Koeller, K. M.; Wong, J. C.; Grozinger, C. M.; Schreiber, S. L. Domain-selective small-molecule inhibitor of histone deacetylase 6 (HDAC6)-mediated tubulin deacetylation. *Proc. Natl. Acad. Sci. U.S.A.* **2003**, *100*, 4389–4394.

171. Zou, H.; Wu, Y.; Navre, M.; Sang, B. C. Characterization of the two catalytic domains in histone deacetylase 6. *Biochem. Biophys. Res. Commun.* **2006**, *341*, 45–50.

172. Zhang, Y.; Gilquin, B.; Khochbin, S.; Matthias, P. Two catalytic domains are required for protein deacetylation. *J. Biol. Chem.* **2006**, *281*, 2401–2404.

173. Zhang, X.; Yuan, Z.; Zhang, Y.; Yong, S.; Salas-Burgos, A.; Koomen, J., et al. HDAC6 modulates cell motility by altering the acetylation level of cortactin. *Mol. Cell* **2007**, *27*, 197–213.

174. Kovacs, J. J.; Murphy, P. J.; Gaillard, S.; Zhao, X.; Wu, J. T.; Nicchitta, C. V., et al. HDAC6 regulates Hsp90 acetylation and chaperone-dependent activation of glucocorticoid receptor. *Mol. Cell* **2005**, *18*, 601–607.

175. Parmigiani, R. B.; Xu, W. S.; Venta-Perez, G.; Erdjument-Bromage, H.; Yaneva, M.; Tempst, P.; Marks, P. A. HDAC6 is a specific deacetylase of peroxiredoxins and is involved in redox regulation. *Proc. Natl. Acad. Sci. U.S.A.* **2008**, *105*, 9633–9638.

176. Subramanian, C.; Jarzembowski, J. A.; Opipari, A. W., Jr.; Castle, V. P.; Kwok, R. P. HDAC6 deacetylates Ku70 and regulates Ku70-Bax binding in neuroblastoma. *Neoplasia* **2011**, *13*, 726–734.

177. Riolo, M. T.; Cooper, Z. A.; Holloway, M. P.; Cheng, Y.; Bianchi, C.; Yakirevich, E., et al. Histone deacetylase 6 (HDAC6) deacetylates survivin for its nuclear export in breast cancer. *J. Biol. Chem.* **2012**, *287*, 10885–10893.

178. Mak, A. B.; Nixon, A. M. L.; Kittanakom, S.; Stewart, J. M.; Chen, G. I.; Curak, J., et al. Regulation of CD133 by HDAC6 promotes β-catenin signaling to suppress cancer cell differentiation. *Cell Reports* **2012**, *2*, 951–963.

179. Huo, L.; Li, D.; Sun, X.; Shi, X.; Karna, P.; Yang, W., et al. Regulation of Tat acetylation and transactivation activity by the microtubule-associated deacetylase HDAC6. *J. Biol. Chem.* **2011**, *286*, 9280–9286.

180. Ouyang, H.; Ali, Y. O.; Ravichandran, M.; Dong, A.; Qiu, W.; MacKenzie, F., et al. Protein aggregates are recruited to aggresome by histone deacetylase 6 via unanchored ubiquitin C termini. *J. Biol. Chem.* **2012**, *287*, 2317–2327.

181. Hook, S. S.; Orian, A.; Cowley, S. M.; Eisenman, R. N. Histone deacetylase 6 binds polyubiquitin through its zinc finger (PAZ domain) and copurifies with deubiquitinating enzymes. *Proc. Natl. Acad. Sci. U.S.A.* **2002**, *99*, 13425–13430.

182. Li, Y.; Shin, D.; Kwon, S. H. Histone deacetylase 6 plays a role as a distinct regulator of diverse cellular processes. *FEBS Lett.* **2013**, *280*, 775–793.

183. Boyault, C.; Gilquin, B.; Zhang, Y.; Rybin, V.; Garman, E.; Meyer-Klaucke, W., et al. HDAC6-p97/VCP controlled polyubiquitin chain turnover. *EMBO J.* **2006**, *25*, 3357–3366.

184. Li, G.; Jiang, H.; Chang, M.; Xie, H.; Hu, L. HDAC6 α-tubulin deacetylase: a potential therapeutic target in neurodegenerative diseases. *J. Neurosci.* **2011**, *304*, 1–8.

185. Bonnet, J.; Romier, C.; Tora, L.; Devys, D. Zinc-finger UBPs: regulators of deubiquitination. *Trends Biochem. Sci.* **2008**, *33*, 369–375.

186. Ouyang, H.; Ali, Y. O.; Ravichandran, M.; Dong, A.; Qiu, W.; MacKenzie, F., et al. Protein aggregates are recruited to aggresome by histone deacetylase 6 via unanchored ubiquitin C termini. *J. Biol. Chem.* **2012**, *287*, 2317–2327.

187. Burnett, B. G.; Pittman, R. N. The polyglutamine neurodegenerative protein ataxin 3 regulates aggresome formation. *Proc. Natl. Acad. Sci. U.S.A.* **2005**, *102*, 4330–4335.

188. Watabe, M.; Nakaki, T. Protein kinase CK2 regulates the formation and clearance of aggresomes in response to stress. *J. Cell. Sci.* **2011**, *124*, 1519–1532.

189. Wu, H.; Parsons, J. T. Cortactin, an 80/85-kilodalton pp60src substrate, is a filamentous actin-binding protein enriched in the cell cortex. *J. Cell Biol.* **1993**, *120*, 1417–1426.

190. Gao, Y.-S.; Hubbert, C. C.; Lu, J.; Lee, Y.-S.; Lee, J.-Y.; Yao, T.-P. Histone deacetylase 6 regulates growth factor-induced actin remodeling and endocytosis. *Mol. Cell. Biol.* **2007**, *27*, 8637–8647.

191. Lee, J. Y.; Koga, H.; Kawaguchi, Y.; Tang, W.; Wong, E.; Gao, Y. S., et al. HDAC6 controls autophagosome maturation essential for ubiquitin-selective quality-control autophagy. *EMBO J.* **2010**, *29*, 969–980.

192. Cebotaru, L.; Vij, N.; Ciobanu, I.; Wright, J.; Flotte, T.; Guggino, W. B. Cystic fibrosis transmembrane regulator missing the first four transmembrane segments increases wild type and ΔF508 processing. *J. Biol. Chem.* **2008**, *283*, 21926–21933.

193. Iwata, A.; Riley, B. E.; Johnston, J. A.; Kopito, R. R. HDAC6 and microtubules are required for autophagic degradation of aggregated huntingtin. *J. Biol. Chem.* **2005**, *280*, 40282–40292.

194. Du, G.; Liu, X.; Chen, X.; Song, M.; Yan, Y.; Jiao, R.; Wang, C.-C. *Drosophila* histone deacetylase 6 protects dopaminergic eurons against α-synuclein toxicity by promoting inclusion formation. *Mol. Biol. Cell* **2010**, *21*, 2128–2137.

195. Su, M.; Shi, J.-J.; Yang, Y.-P.; Li, J.; Zhang, Y.-L.; Chen, J., et al. HDAC6 regulates aggresome-autophagy degradation pathway of α-synuclein in response to MPP⁺-induced stress. *J. Neurochem.* **2011**, *117*, 112–120.

196. Fiesel, F. C.; Voigt, A.; Weber, S. S.; Van den Haute, C.; Waldenmaier, A.; Görner, K., et al. Knockdown of transactive response DNA-binding protein (TDP-43) downregulates histone deacetylase 6. *EMBO J.* **2010**, *29*, 209–221.

197. Olzmann, J. A.; Li, L.; Chudaev, M. V.; Chen, J.; Perez, F. A.; Palmiter, R. D.; Chin, L.-S. Parkin mediated K63-linked polyubiquitination targets misfolded DJ-1 to aggresomes via binding to HDAC6. *J. Cell. Biol.* **2007**, *178*, 1025–1038.

198. Dompierre, J. P.; Godin, J. D.; Charrin, B. C.; Cordelieres, F. P.; King, S. J.; Humbert, S.; Saudou, F. Histone deacetylase 6 inhibition compensates for the transport deficit in Huntington's disease by increasing tubulin acetylation. *J. Neurosci.* **2007**, *27*, 3571–3583.

199. Reed, N. A.; Cai, D.; Blasius, T. L.; Jih, G. T.; Meyhofer, E.; Gaertig, J.; Verhey, K. J. Microtubule acetylation promotes kinesin-1 binding and transport. *Curr. Biol.* **2006**, *16*, 2166–2172.

200. Jiang, Q.; Ren, Y.; Feng, J. Direct binding with histone deacetylase 6 mediates the reversible recruitment of parkin to the centrosome. *J. Neurosci.* **2008**, *28*, 12993–13002.

201. Zilberman, Y.; Ballestrem, C.; Carramusa, L.; Mazitschek, R.; Khochbin, S.; Bershadsky, A. Regulation of microtubule dynamics by inhibition of the tubulin deacetylase HDAC6. *J. Cell Sci.* **2009**, *122*, 3531–3541.

202. Tökési, N.; Lehotzky, A.; Horváth, I.; Szabó, B.; Oláh, J.; Lau, P.; Ovádi, J. TPPP/p25 promotes tubulin acetylation by inhibiting histone deacetylase 6. *J. Biol. Chem.* **2010**, *285*, 17896–17906.

203. Lehotzky, A.; Tirián, L.; Tökési, N.; Lénárt, P.; Szabó, B.; Kovács, J.; Ovádi, J. Dynamic targeting of microtubules by TPPP/p25 affects cell survival. *J. Cell Sci.* **2004**, *117*, 6249–6259.

204. Schofield, A. V.; Steel, R.; Bernard, O. Rho-associated coiled-coil kinase (ROCK) protein controls microtubule dynamics in a novel signaling pathway that regulates cell migration. *J. Biol. Chem.* **2012**, *287*, 43620–43629.

205. Schofield, A. V.; Gamell, C.; Suryadinata, R.; Sarcevic, B.; Bernard, O. Tubulin polymerization promoting protein 1 (Tppp1) phosphorylation by rho-associated coiled-coil kinase (Rock) and cyclin-dependent kinase 1 (Cdk1) inhibits microtubule dynamics to increase cell proliferation. *J. Biol. Chem.* **2013**, *288*, 7907–7917.

206. Bali, P.; Pranpat, M.; Bradner, J.; Balasis, M.; Fiskus, W.; Guo, F., et al. Inhibition of histone deacetylase 6 acetylates and disrupts the chaperone function of heat shock protein 90. *J. Biol. Chem.* **2005**, *280*, 26729–26734.

207. Goldberg, A. L. Protein degradation and protection against misfolded or damaged proteins. *Nature* **2003**, *426*, 895–899.

208. Boyault, C.; Zhang, Y.; Fritah, S.; Caron, C.; Gilquin, B.; Kwon, S. H., et al. HDAC6 controls major cell response pathways to cytotoxic accumulation of protein aggregates. *Genes Dev.* **2007**, *21*, 2172–2181.

209. Murphy, P. J.; Morishima, Y.; Kovacs, J. J.; Yao, T. P.; Pratt, W. B. Regulation of the dynamics of hsp90 action on the glucocorticoid receptor by acetylation/deacetylation of the chaperone. *J. Biol. Chem.* **2005**, *280*, 33792–33799.

210. Espallergues, J.; Teegarden, S. L.; Veerakumar, A.; Boulden, J.; Challis, C.; Jochems, J., et al. HDAC6 regulates glucocorticoid receptor signaling in serotonin pathways with critical impact on stress resilience. *J. Neurosci.* **2012**, *32*, 4400–4416.

211. Lee, J. B.; Wei, J.; Liu, W.; Cheng, J.; Feng, J.; Yan, Z. Histone deacetylase 6 gates the synaptic action of acute stress in prefrontal cortex. *J Physiol.* **2012**, *590*, 1536–1545.

212. Yang, F.-C.; Tan, B. C.-M.; Chen, W.-H.; Lin, Y.-H.; Huang, J.-Y.; Chang, H.-Y., et al. Reversible acetylation regulates salt-inducible kinase (SIK2) and its function in autophagy. *J. Biol. Chem.* **2013**, *288*, 6227–6237.

213. Chen, S.; Owens, G. C.; Makarenkova, H.; Edelman, D. B. HDAC6 regulates mitochondrial transport in hippocampal neurons. *PLoS One* **2010**, *5*, e10848.

214. Fusco, C.; Micale, L.; Egorov, M.; Monti, M.; D'Addetta, E. V.; Augello, B., et al. The E3-ubiquitin ligase TRIM50 interacts with HDAC6 and p62, and promotes the sequestration and clearance of ubiquitinated proteins into the aggresome. *PLoS One* **2012**, *7*, e40440.

215. Kwon, S.; Zhang, Y.; Matthias, P. The deacetylase HDAC6 is a novel critical component of stress granules involved in the stress response. *Genes Dev.* **2007**, *21*, 3381–3394.

216. Subramanian, C.; Jarzembowski, J. A.; Opipari, A. W., Jr.; Castle, V. P.; Kwok, R. P. HDAC6 deacetylates Ku70 and regulates Ku70-Bax binding in neuroblastoma. *Neoplasia* **2011**, *13*, 726–734.

217. Aldana-Masangkay, G. I.; Sakamoto, K. M. The role of HDAC6 in cancer. *J. Biomed. Biotechnol.* **2011**, *2011*, 875824.

218. Sakuma, T.; Uzawa, K.; Onda, T.; Shiiba, M.; Yokoe, H.; Shibahara, T.; Tanzawa, H. Aberrant expression of histone deacetylase 6 in oral squamous cell carcinoma. *Int. J. Oncol.* **2006**, *29*, 117–124.

219. Wickstrom, S. A.; Masoumi, K. C.; Khochbin, S.; Fassler, R.; Massoumi, R. CYLD negatively regulates cell-cycle progression by inactivating HDAC6 and increasing the levels of acetylated tubulin. *EMBO J.* **2010**, *29*, 131–144.

220. Lee, Y. S.; Lim, K. H.; Guo, X.; Kawaguchi, Y.; Gao, Y.; Barrientos, T., et al. The cytoplasmic deacetylase HDAC6 is required for efficient oncogenic tumorigenesis. *Cancer Res.* **2008**, *68*, 7561–7569.

221. Hurst, D. R.; Mehta, A.; Moore, B. P.; Phadke, P. A.; Meehan, W. J.; Accavitti, M. A., et al. Breast cancer metastasis suppressor 1 (BRMS1) is stabilized by the Hsp90 chaperone. *Biochem. Biophys. Res. Commun.* **2006**, *348*, 1429–1435.

222. de Zoeten, E. F.; Wang, L.; Butler, K.; Beier, U. H.; Akimova, T.; Sai, H., et al. Histone deacetylase 6 and heat shock protein 90 control the functions of Foxp3$^+$ T-regulatory cells. *Mol. Cell. Biol.* **2011**, *31*, 2066–2078.

223. Winkler, R.; Benz, V.; Clemenz, M.; Bloch, M.; Foryst-Ludwig, A.; Wardat, S., et al. Histone deacetylase 6 (HDAC6) is an essential modifier of glucocorticoid-induced hepatic gluconeogenesis. *Diabetes* **2012**, *61*, 513–523.

224. Simões-Pires, C.; Zwick, V.; Nurisso, A.; Schenker, E.; Carrupt, P. A.; Cuendet, M. HDAC6 as a target for neurodegenerative diseases: what makes it different from the other HDACs? *Mol. Neurodeg.* **2013**, *8*, 7.

225. d'Ydewalle, C.; Bogaert, E.; Van Den Bosch, L. HDAC6 at the intersection of neuroprotection and neurodegeneration. *Traffic* **2012**, *13*, 771–779.

226. Beurel, E. HDAC6 regulates LPS-tolerance in astrocytes. *PLoS One* **2011**, *6*, e25804.

227. Shakespear, M. R.; Halili, M. A.; Irvine, K. M.; Fairlie, D. P.; Sweet, M. J. Histone deacetylases as regulators of inflammation and immunity. *Trends Immunol.* **2011**, *32*, 335–343.

228. Zhang, Y.; Kwon, S.; Yamaguchi, T.; Cubizolles, F.; Rousseaux, S.; Kneissel, M., et al. Mice lacking histone deacetylase 6 have hyperacetylated tubulin but are viable and develop normally. *Mol. Cell. Biol.* **2008**, *28*, 1688–1701.

229. Fukada, M.; Hanai, A.; Nakayama, A.; Suzuki, T.; Miyata, N.; Rodriguiz, R. M., et al. Loss of deacetylation activity of Hdac6 affects emotional behavior in mice. *PLoS One* **2012**, *7*, e30924.

230. Govindarajan, N.; Rao, P.; Burkhardt, S.; Sananbenesi, F.; Schlüter, O. M.; Bradke, F., et al. Reducing HDAC6 ameliorates cognitive deficits in a mouse model for Alzheimer's disease. *EMBO Mol. Med.* **2012**, *5*, 52–63.

231. Miki, Y.; Mori, F.; Tanji, K.; Kakita, A.; Takahashi, H.; Wakabayashi, K. Accumulation of histone deacetylase 6, an aggresome-related protein, is specific to Lewy bodies and glial cytoplasmic inclusions. *Neuropathology* **2011**, *31*, 561–568.

232. Savitt, J. M.; Dawson, V. L.; Dawson, T. M. Diagnosis and treatment of Parkinson disease: molecules to medicine. *J. Clin. Invest.* **2006**, *116*, 1744–1754.

233. Vives-Bauza, C.; Zhou, C.; Huang, Y.; Cui, M.; de Vries, R. L.; Kim, J., et al. PINK1-dependent recruitment of Parkin to mitochondria in mitophagy. *Proc. Natl. Acad. Sci. U.S.A.* **2010**, *107*, 378–383.

234. Lee, J. Y.; Nagano, Y.; Taylor, J. P.; Lim, K. L.; Yao, T. P. Disease-causing mutations in parkin impair mitochondrial ubiquitination, aggregation, and HDAC6-dependent mitophagy. *J. Cell. Biol.* **2010**, *189*, 671–679.

235. Hubbers, C. U.; Clemen, C. S.; Kesper, K.; Boddrich, A.; Hofmann, A.; Kamarainen, O., et al. Pathological consequences of VCP mutations on human striated muscle. *Brain* **2006**, *130*, 381–393.

236. Ju, J.-S.; Miller, S. E.; Hanson, P. I.; Weihl, C. C. Impaired protein aggregate handling and clearance underlie the pathogenesis of p97/VCP-associated disease. *J. Biol. Chem.* **2008**, *283*, 30289–30299.

237. Trushina, E.; Dyer, R. B.; Badger, J. D., II.; Ure, D.; Eide, L.; Tran, D. D., et al. Mutant huntingtin impairs axonal trafficking in mammalian neurons in vivo and in vitro. *Mol. Cell. Biol.* **2004**, *24*, 8195–8209.

238. Pandey, U. B.; Nie, Z.; Batlevi, Y.; McCray, B. A.; Ritson, G. P.; Nedelsky, N. B., et al. HDAC6 rescues neurodegeneration and provides an essential link between autophagy and the UPS. *Nature* **2007**, *447*, 859–863.

239. Bobrowska, A.; Paganetti, P.; Matthias, P.; Bates, G. P. Hdac6 knock-out increases tubulin acetylation but does not modify disease progression in the R6/2 mouse model of Huntington's disease. *PLoS One* **2011**, *6*, e20696.

240. Evgrafov, O. V.; Mersiyanova, I.; Irobi, J.; Van Den Bosch, L.; Dierick, I.; Leung, C. L., et al. Mutant small heat-shock protein 27 causes axonal Charcot-Marie-Tooth disease and distal hereditary motor neuropathy. *Nat. Genet.* **2004**, *36*, 602–606.

241. d'Ydewalle, C.; Krishnan, J.; Chiheb, D. M.; Van Damme, P.; Irobi, J.; Kozikowski, A. P., et al. HDAC6 inhibitors reverse axonal loss in a mouse model of mutant HSPB1-induced Charcot-Marie-Tooth disease. *Nat. Med.* **2011**, *17*, 968–974.

242. Neumann, M.; Sampathu, D. M.; Kwong, L. K.; Truax, A. C.; Micsenyi, M. C.; Chou, T. T., et al. Ubiquitinated TDP-43 in frontotemporal lobar degeneration and amyotrophic lateral sclerosis. *Science* **2006**, *314*, 130–133.

243. Sreedharan, J.; Blair, I. P.; Tripathi, V. B.; Hu, X.; Vance, C.; Rogelj, B., et al. TDP-43 mutations in familial and sporadic amyotrophic lateral sclerosis. *Science* **2008**, *319*, 1668–1672.

244. Kwiatkowski, T. J., Jr.; Bosco, D. A.; Leclerc, A. L.; Tamrazian, E.; Vanderburg, C. R.; Russ, C., et al. Mutations in the FUS/TLS gene on chromosome 16 cause familial amyotrophic lateral sclerosis. *Science* **2009**, *323*, 1205–1208.

245. Kim, S. H.; Shanware, N. P.; Bowler, M. J.; Tibbetts, R. S. Amyotrophic lateral sclerosis-associated proteins TDP-43 and FUS/TLS function in a common biochemical complex to co-regulate HDAC6 mRNA. *J. Biol. Chem.* **2010**, *285*, 34097–34105.

246. Fiesel, F. C.; Schurr, C.; Weber, S. S.; Kahle, P. J. TDP-43 knockdown impairs neurite outgrowth dependent on its target histone deacetylase 6. *Mol. Neurodegener.* **2011**, *6*, 64.

247. Polymenidou, M.; Lagier-Tourenne, C.; Hutt, K. R.; Huelga, S. C.; Moran, J.; Liang, T. Y., et al. Long pre-mRNA depletion and RNA missplicing contribute to neuronal vulnerability from loss of TDP-43. *Nat. Neurosci.* **2011**, *14*, 459–468.
248. Hebron, M. L.; Lonskaya, I.; Sharpe, K.; Weerasinghe, P. P. K.; Algarzae, N. K.; Shekoyan, A. R.; Moussa, C. E.-H. Parkin ubiquitinates Tar-DNA binding protein-43 (TDP-43) and promotes its cytosolic accumulation via interaction with histone deacetylase 6 (HDAC6). *J. Biol. Chem.* **2013**, *288*, 4103–4115.
249. Taes, I.; Timmers, M.; Hersmus, N.; Bento-Abreu, A.; Van Den Bosch, L.; Van Damme, P., et al. Hdac6 deletion delays disease progression in the SOD1G93A mouse model of ALS. *Hum. Mol. Genet.* **2013**, *22*, 1783–1790.
250. Yoo, Y.-E.; Ko, C.-P. Treatment with trichostatin A initiated after disease onset delays disease progression and increases survival in a mouse model of amyotrophic lateral sclerosis. *Exp. Neurol.* **2011**, *231*, 147–159.
251. Gal, J.; Chen, J.; Barnett, K. R.; Yang, L.; Brumley, E.; Zhu, H. HDAC6 regulates mutant SOD1 aggregation through two SMIR motifs and tubulin acetylation. *J. Biol. Chem.* **2013**, *288*, 15035–15045.
252. Fan, W.; Tang, Z.; Chen, D.; Moughon, D.; Ding, X.; Chen, S., et al. Keap1 facilitates p62-mediated ubiquitin aggregate clearance via autophagy. *Autophagy* **2010**, *6*, 614–621.
253. Ding, H.; Dolan, P. J.; Johnson, G. V. Histone deacetylase 6 interacts with the microtubule-associated protein tau. *J. Neurochem.* **2008**, *106*, 2119–2130.
254. Odagiri, S.; Tanji, S.; Mori, F.; Miki, Y.; Kakita, A.; Takahashi, H.; Wakabayashi, K. Brain expression level and activity of HDAC6 protein in neurodegenerative dementia. *Biochem. Biophys. Res. Commun.* **2013**, *430*, 394–399.
255. Perez, M.; Santa-Maria, I.; Gomez de Barreda, E.; Zhu, X.; Cuadros, R.; Cabrero, J. R., et al. Tau-an inhibitor of deacetylase HDAC6 function. *J. Neurochem.* **2009**, *109*, 1756–1766.
256. Guthrie, C. R.; Kraemer, B. C. Proteasome inhibition drives HDAC6-dependent recruitment of tau to aggresomes. *J. Mol. Neurosci.* **2011**, *45*, 32–41.
257. Cook, C.; Gendron, T. F.; Scheffel, K.; Carlomagno, Y.; Dunmore, J.; Deture, M.; Petrucelli, L. Loss of HDAC6, a novel CHIP substrate, alleviates abnormal tau accumulation. *Hum. Mol. Genet.* **2012**, *21*, 2936–2945.
258. Xiong, Y.; Zhao, K.; Wu, J.; Xu, Z.; Jin, S.; Zhang, Y. Q. HDAC6 mutations rescue human tau-induced microtubule defects in Drosophila. *Proc. Natl. Acad. Sci. U.S.A.* **2013**, *110*, 4604–4609.
259. Leroy, K.; Yilmaz, Z.; Brion, J. P. Increased level of active GSK-3β in Alzheimer's disease and accumulation in argyrophilic grains and in neurones at different stages of neurofibrillary degeneration. *Neuropathol. Appl. Neurobiol.* **2007**, *33*, 43–55.
260. Kim, C.; Choi, H.; Jung, E. S.; Lee, W.; Oh, S.; Jeon, N. L.; Mook-Jung, I. HDAC6 inhibitor blocks amyloid beta-induced impairment of mitochondrial transport in hippocampal neurons. *PLoS One* **2012**, *7*, e42983.
261. Sung, Y. M.; Lee, T.; Yoon, H.; DiBattista, A. M.; Song, J. M.; Sohn, Y., et al. Mercaptoacetamide-based class II HDAC inhibitor lowers Aβ levels and improves learning and memory in a mouse model of Alzheimer's disease. *Exp. Neurol.* **2013**, *239*, 192–201.
262. Zhang, L.; Sheng, S.; Qin, C. The role of HDAC6 in Alzheimer's disease. *J. Alzheimer's Dis.* **2013**, *33*, 283–295.
263. Barnett, A.; Brewer, G. J. Autophagy in aging and Alzheimer's disease: pathologic or protective? *J. Alzheimer's Dis.* **2011**, *23*, 385–394.
264. Seneci, P. Chemical modulators of protein misfolding and neurodegenerative disease. Elsevier, accepted for publication, **2015**.

CHAPTER

6

Assembly and Disassembly
of Protein Aggregates
Unraveling the Maze

6.1 INTRODUCTION

Protein aggregation was repeatedly mentioned throughout the earlier chapters, as an overall (neuro)toxic process that is physiologically countered by protein quality control (PQC) pathways acting either at the early/soluble/monomeric-small oligomeric stage (Hsp70, Hsp90/chaperones, Chapter 2; ubiquitin–proteasome system (UPS), Chapter 3), or at the late/insoluble/aggregate stage (macroautophagy (MA), Chapter 4; selective autophagy/aggrephagy, Chapter 5). The earlier mechanisms aim to rescue and refold the misfolded protein copies, while the latter drive toxic, unrecoverable, large protein oligomers, and/or aggregates towards degradation. This chapter will cover the *aggregation* and *disaggregation* of soluble protein oligomers and insoluble aggregates, examining their therapeutic and pathological consequences. It will describe the influence of protein solubility, and of post-translational modifications (PTMs), on the assembly of insoluble protein aggregates in basal and pathological environments.

6.2 DISORDERED PROTEIN AGGREGATES AND ORDERED AMYLOID FIBRILS

Protein aggregation [1,2] is influenced by a number of physicochemical and biological factors in humans. Its detailed description is useful to select the steps which may, if properly regulated, lead to the prevention of toxic aggregate formation. It is useful also to appreciate how different species of protein aggregates observed in human cells and tissues are formed. Their sensitivity to any disaggregation-driven approach determines the pathological impact

Molecular Targets in Protein Misfolding and Neurodegenerative Disease. DOI: 10.1016/B978-0-12-800186-8.00006-7

and the chance of disaggregation-targeted intervention in proteinopathies in general, and in neurodegenerative diseases (NDDs) in particular.

A tendency to aggregate is common to all proteins, and depends on the arrangement of the polypeptide regions of each protein [1]. *Ordered aggregates* are (un)branched amyloid fibrils composed of a few thousand copies of a protein, arranged in a repeating cross-β-sheet structure that spans a few nanometers (nm) [3]. Most proteins form ordered amyloid structures *in vitro* [4]. Ordered aggregation *in vivo*, leading to chronic amyloidosis, occurs only for a small number of proteins [5]. At least 30 human amyloid proteins are found in the extracellular space (e.g., β-amyloid fibrils), while at least one amyloid formation, tau neurofibrillary tangles (NFTs), is found intracellularly [6]. Among them, some have functional roles in basal conditions (e.g., actin [7]). Most of them cause pathological consequences, and constitute the structural core of proteinopathies [8]. *Disordered aggregates* composed of misfolded, non-organized cytosolic proteins [9] are formed due to acute cellular stimuli (i.e., stress-caused denaturation, lack of assembly partners). The relocation of disordered aggregates and amyloid structures in different cellular organelles [10] drives their rescuing/disposal through different PQC pathways. Namely, disordered/misfolded species localize in a nucleus indentation named the *juxtanuclear quality control (JUNQ)* compartment. JUNQs are populated by soluble, ubiquitinated species that are either rescued through chaperone-driven PQC, or disposed of through the UPS [10] (see Chapters 2 and 3). Conversely, amyloid aggregates are directed to peripheral perivacuolar compartments named *inclusion protein deposits (IPODs)*. IPODs are populated by insoluble, non-ubiquitinated aggregates that are disposed of through autophagic processes [10] (see Chapters 4 and 5).

CNS-related proteinopathies/NDDs, caused either by disordered aggregates (i.e., superoxide dismutase 1 (SOD1) in amyotrophic lateral sclerosis (ALS) [11], or huntingtin in Huntington disease (HD) [12]), or by ordered amyloid aggregates (e.g., β-amyloid in Alzheimer's disease (AD) [13], or prion protein in prion disease [14]) result from the saturation and impairment of the PQC machinery. Unprocessable protein species in JUNQs may evolve into insoluble aggregates and be directed to IPODs [15], or may *per se* constitute the structural core of NDDs [10]. The most therapeutically relevant proteinopathies/NDDs are briefly described in Chapter 1, focusing on tauopathies.

The transition from a soluble, functional protein to misfolded, aggregation-prone species, and eventually to ordered or disordered aggregates, is promoted by protein-dependent (structural features, genetic or PTMs) and protein-independent factors (cellular, environmental stimuli) [16,17]. As to *structural features*, intrinsically disordered proteins (IDPs [18]) sample a larger conformational space, including several metastable, aggregation-prone conformations. Proteinopathy-prone IDPs include Aβ [19], α-synuclein [20], and tau [21]. The sequence of amino acids in a protein may promote aggregation [22] through β-sheet motifs (i.e., NFGAIL in human islet amyloid polypeptide (IAPP) [23], KLVFFA in β-amyloid [24], and VQIVY in tau [25]). As to *genetic modifications*, point mutations may significantly increase the aggregation tendency of a protein (i.e., the I31E/E22G mutation in the $A\beta_{42}$ peptide [26], the A30P and A53T mutations in α-synuclein [27], and the P301L mutation in tau [28]). Trinucleotide repeat disorders [29] are characterized by an expansion of trinucleotide repeats that leads to aggregation-prone, neurotoxic mutant proteins (i.e., CAG/polyQ repeats in huntingtin [30] and polyA repeats in polyA disorders [31]). As to *PTMs*, they cause aggregation and proteinopathies by influencing, *inter alia*, the solubility and the conformational stability of proteins (i.e., the hyperphosphorylation of tau [32], the proteolytic cleavage of $A\beta_{42}$ peptide precursors [33], and the oxidative/nitrative stress of the prion protein [34]).

Protein-dependent aggregation factors increase the probability for a functional or mutated protein to switch to an aggregation-promoting conformation. The formation of insoluble aggregates is further accelerated by acute or chronic *environmental changes*. Aggregate formation is artificially induced by increased temperature [35], controlled stirring [36], sample concentration [37], pH lowering [38], presence [39] or absence [40] of specific ions, and of organic solvents [41]. These conditions are reproduced in cellular environments (e.g., disturbed metal homeostasis in AD [42], fast increase in local insulin concentration after its i.v. injection [43]). Chronic environmental effects play a major role in proteinopathy-dependent NDDs. Chronic exposure to high protein concentrations (e.g., β2-microglobulin in long-term hemodialysis patients [44]), or to pathologic protein modifiers (e.g., pesticides and herbicides in Parkinson's disease (PD) [45], reactive oxygen species (ROS) in PD and AD [46]) leads to protein aggregation.

The *aging process* is the most common cause of protein aggregation [47], to the extent that the latter is a diagnostic marker for the former process in living organisms [48]. Aging-dependent protein aggregation is a slow, continuous process determined by subtle chronic changes in cellular components that impact on pathology-unrelated and -related proteins. A "vicious circle" scenario gradually increases the amount of protein aggregates, stressing the capacity of the PQC machinery and causing further misfolding and aggregation. PQC is further impaired through a loss of efficiency in UPS [49], chaperone-mediated autophagy (CMA [50]) and unfolded protein response (UPR [51]). An increased level of transcriptional and translational errors [52] and of oxidative stress [53] contributes to aging in general, and to protein misfolding and aggregation in particular.

Protein solubility and concentration depend respectively on structural and genetic factors, and influence the aggregation propensity of any protein. A recent report [54] defines a *supersaturation score* for more than

16,000 human proteins in their folded/functional state (σ_f), and for more than 6000 unfolded human proteins (σ_u). The score indicates the tendency of a protein to aggregate/become insoluble, and takes into account its intrinsic propensity to aggregate and its concentration in cellular environments. Aggregation-prone proteins involved in NDDs show the highest σ_u scores, and do aggregate once unfolded protein copies escape the PQC system due to aging-related factors. Accordingly, an unbiased biochemical screen identifies AD, PD, and HD as the disease pathways containing the larger number of proteins with high σ_f and σ_u scores [54].

Let us now move to the single molecule level, to see how the molecular events take place to lead to an ordered aggregate. Native proteins tend to aggregate in a concentration-dependent manner [55]. Their aggregation process is reversible in its early phases. It becomes irreversible when covalent modifications stabilize the aggregate [56], or when a large number of molecular interactions take place in an ordered amyloid structure [57].

Single protein copies initiate several NDDs, as they start the pathology-determining protein aggregation process [58]. Their aggregation through multiple pathways to yield key aggregation nuclei is summarized in Figure 6.1 [59].

A small percentage of misfolded proteins is present at any time. The PQC machinery takes care of them in basal conditions, and native functional proteins carry out their roles. If endogenous (i.e., mutations) or exogenous conditions (i.e., stress) alter the physiological environment, an increased concentration of misfolded, aggregation-prone protein copies is observed (steps 1a–c, Figure 6.1). In addition to *random errors in transcription and folding* (step 1a), an *abnormal PTM pattern* may modify the protein features (different conformational preferences, and/or reduced binding with endogenous ligands, and/or reduced solubility). Thus, non-native, PTM-modified, aggregation-prone protein copies accumulate in the cell (e.g., phosphorylated α-synuclein in synucleinopathies [60]) (step 1b). *Mutations* in a protein-coding gene may lead to mutated, aggregation-prone protein species (e.g., an ALS-associated mutant TAR DNA-binding protein 43 (TDP-43) [61]) (step 1c). The disordered aggregation of misfolded, monomeric (MM) species yields a *disordered aggregate* (DA, step 2). Ordered aggregation takes place with a non-structured, largely unclarified process when overexpression of the aggregation-prone protein and impairment of the PQC machinery happen simultaneously.

The gradual concentration increase of soluble, misfolded protein copies (steps 1a,b), or the translation of aggregation-prone mutant proteins (step 1c) initiates a chronic process, which becomes apparent (first symptoms of a proteinopathy/NDD in the central (CNS) and peripheral (PNS) nervous system) years after the first biochemical abnormalities take place. The process may start with a *conformational activation*, which increases the aggregation tendency of a monomer MM by exposing one or more of its β-sheet ordered

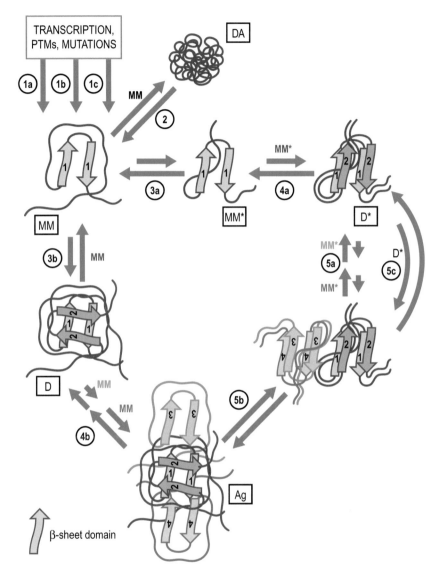

FIGURE 6.1   Protein aggregation: nucleation of amyloidogenic proteins.

sequences (step 3a, activated MM*). Then, a thermodynamically unfavorable interaction between two MM*s leads to a proto-amyloidogenic template D* (step 4a). Further unfavorable interactions between D*s and additional MM*s build up a small, β-sheet-rich nucleus of aggregation (step 5a, AN*). Different aggregation-prone proteins form varying AN*s in terms of size (number of monomers), shape (interaction mode between monomers), or composition (conformational rearrangement of monomers). The four-monomer nucleus depicted in Figure 6.1, thus, is only an average representation.

The nucleation process is extremely slow, as dissociation of small D*s is favored *vs.* the addition of new MM*s to complete the AN* assembly [62]. Thus, nucleation corresponds to a long, phenotype-lacking lag phase in developing proteinopathies/NDDs [16]. AN*s composed by different amyloidogenic proteins share their nature as the last thermodynamically unfavorable intermediates on the path to aggregation, due to their ordered β-sheet structure that facilitates the addition of new MM*s—see below.

Alternatively to the conformational switch depicted in step 3a, a misfolded, disordered MM protein may slowly aggregate in a stepwise process to yield non-structured dimers D (step 3b) and disordered aggregates Ag (step 4b). Ags then undergo a conformational switch to the β-sheet-enriched aggregation nucleus AN* (step 5b) [57]. The same AN*, finally, may result from the condensation of two or more proto-amyloidogenic D*s (step 5c, Figure 6.1).

The growth of AN*s to mature amyloid fibrils is depicted in Figure 6.2 [59]. The fast addition of monomers takes place at the ends of the pre-organized structure of the nuclei [63] in a growth-exponential phase. Its

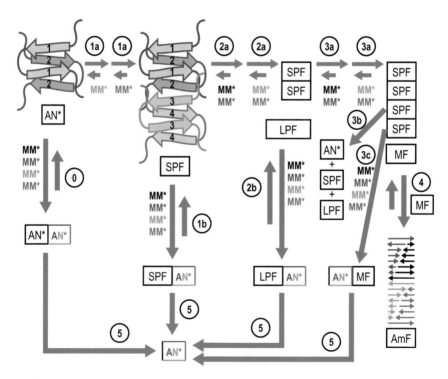

FIGURE 6.2   Protein aggregation: from aggregation nuclei/AN*s to mature fibrils/MFs.

favorable kinetics are modulated by the intrinsic β-sheet propensity of a protein (higher propensity/faster generation of amyloid fibrils). Aggregation usually happens through rod-like/short (SPFs, step 1a) [64] and worm-like/long protofibrils (LPFs, step 2a) [65], which stabilize the intermediate aggregates, evolve into mature fibrils (MFs, step 3a), and eventually, by lateral association, into amyloid fibers (AmFs, step 4) [64].

Aggregation nuclei may be formed through mechanisms other than primary nucleation (Figure 6.1). Established nuclei, early, or late pre-aggregates act as templates for secondary nucleation events (respectively steps 0, 1b, and 2b, Figure 6.2) [66]. Once formed, the new AN*s disengage themselves from the templating aggregates (step 5) and re-enter the aggregation cycle. Mature fibrils undergo fragmentation during the growth of amyloid fibers [67], resulting in different fragments (e.g., AN*s, SPFs, and LPFs, step 3b) that act as seeds for secondary nucleation. Seeding-dependent nucleation is influenced by the susceptibility of amyloid fibrils to fragmentation events, which may even lead to an "infection-like" transmission of amyloid fragments to neighboring tissues (e.g., prion proteins [68]). Cell-to-cell propagation of NDD-related Aβ, tau, and α-synuclein is postulated [69], most likely through the neuronal connections among affected areas [70,71] (see Section 6.2.1). Secondary nucleation events may increase due to fibril branching-creation of new reactive ends that grow into novel AN*s (step 3c), followed by release of the forked AN*s (step 5, Figure 6.2) [72] that re-enter the aggregation cycle.

Finally, nucleation may also accelerate due to the presence of aggregation-promoting biological surfaces that interact with native proteins [73]. The interaction promotes a conformational switch to a non-native, aggregation-prone protein monomer (e.g., the interaction of β2-microglobulin with collagen in dialysis-related amyloidosis [74]), followed by its facilitated nucleation on the biological surface.

Proteins aggregate through a wealth of molecular mechanisms. They are influenced by changes in the cellular environments and produce different meta-stable early intermediates and/or mature fibrils [65] depending on their thermodynamic stability. *Polymorphism* is observed for a large number of aggregation-prone proteins, creating aggregate strains with significantly diverse properties [75] that influence, *inter alia*, their interaction network in cellular environments [76].

The kinetics of multiple primary and secondary nucleation events are the rate-limiting steps of the overall aggregation process. The complexity of such a network, briefly summarized in Figures 6.1 and 6.2, requires accurate and specific *analytical and separative methods* to determine the involved protein and ligand species, their molecular interactions, and the molecular events that determine the aggregation kinetics and the nature of the resulting, insoluble protein aggregates [77]. Soluble, misfolded early intermediates and downstream, insoluble fibrils and fibers need different experimental techniques to be characterized and isolated. The formation of transient oligomeric intermediates, their conformational changes to ordered β-sheet oligomers, and their evolution into protofibrillar structures are difficult events to be detected in real time [77]. Popular separative techniques include native [78] or denaturing electrophoresis [79], size exclusion chromatography (SEC) [80], and analytical ultracentrifugation [81]. Spectroscopic and spectrometric methods used to characterize soluble and insoluble amyloid structures include solution [82] and solid state [83] nuclear magnetic resonance (NMR) spectroscopy, turbidimetry [84], dynamic light scattering (DLS) [85], electron spray ionization (ESI) [86], and matrix-assisted laser desorption ionization (MALDI) mass spectrometry (MS) [87]. Ultra-imaging methods include electron microscopy (EM) [88] and atomic force microscopy (AFM) [89]. Finally, fluorescent chemical probes such as Congo Red [90] and thioflavin T [91], and antibodies [92] discriminate among soluble and insoluble aggregate species in NDDs. A recent, extensive review thoroughly covers these and other analytical and separative methods [77].

Detailed understanding of molecular mechanisms leading to protein aggregation is the gateway to anti-aggregation therapeutics. Such therapeutical agents may be immunotherapeutics/biologicals [93] or small molecules [94]. The former entities are not dealt with in this Chapter, while the latter are briefly introduced in Section 6.4, and thoroughly presented in a chemistry-oriented companion book [95]. A key question to be answered concerns the toxicity of each intermediate on the path to protein aggregation, and the consequences of blocking a specific molecular reaction/process with a small molecule—i.e., preventing the formation of a potentially toxic intermediate, or causing the accumulation of an even more toxic precursor [96].

The end results of aggregation, i.e., *insoluble aggregates*, were initially believed to be the toxic species in proteinopathies [97,98]. This assumption is now challenged by preclinical (e.g., rescuing of neurons by precipitation of tau NFTs in AD models [99], a lower huntingtin inclusion body content in severely affected neurons in HD models [100]) and clinical observations (e.g., absence of AD-like cognitive deficits in Aβ plaque-bearing individuals [101], poor correlation between disease advancement and Aβ plaque content in isolates from AD patients [102], a higher Lewy body content in viable neurons in isolates from PD patients [103], a higher huntingtin inclusion body content in viable neurons in isolates from HD patients [104]).

The consensus is now for *non-fibrillary, soluble oligomers* to be the most neurotoxic species [57,105]. Their stability is limited, and often their characterization depends on the conversion into stable derivatives that can be structurally described [106]. Polymorphic Aβ oligomers with varying size and structure (e.g., annular [107], globular species [108]) show varying neurotoxicity. Neurotoxic Aβ nonamers and dodecamers are observed in the brain of AD mice [109], while Aβ tetramers are highly toxic *in vitro* [110].

Prion [111] and α-synuclein oligomers [112] are also classified in terms of shape, size, and neurotoxicity.

Interestingly, a single monoclonal antibody (mAb) is capable of recognizing several Aβ-, α-synuclein-, and polyQ-containing neurotoxic oligomers [113]. Preincubation of the same mAb with any of these oligomers prevents their neurotoxicity when administered to neuronal cell lines, supporting the theory of an NDD-shared, pathology-inducing mechanism [113]. Possibly, the toxicity of soluble oligomers arises from the exposure of structurally similar, hydrophobic, "sticky" sequences in their non-native conformational states. Their stickiness stimulates the aberrant interaction with a large set of cellular components [114]. Among them, are elements of the PQC machinery (e.g., chaperones [115], MA proteins [116], and UPS components [117]), whose aggregation with soluble oligomers impairs proteostasis, and transcription factors [118]. A recent study [119] defines a metastable "β protein interactome," composed of aggregation-prone IDPs and multi-domain, slow-folding proteins. A "vicious circle" scenario caused by soluble oligomers, aging, and PQC impairment leads to multiple toxic events and to the collapse of key cellular functions [119]. The aggregation of soluble amyloid oligomers into insoluble fibers, thus, may be seen as an endogenous rescue mechanism to dispose of highly toxic species, due to the lower toxicity of insoluble aggregates [120].

The interaction between aggregation-prone protein species and *biological membranes* has a major impact on the cytotoxicity of the former and the operational efficiency of the latter species [121,122]. A simplified depiction, shown in Figure 6.3, illustrates how protein aggregation and protein–membrane binding are connected processes, causing multiple toxic events [123].

Soluble proteins may either fold properly (step 1a, Figure 6.3) and perform their physiological functions (step 2a), or misfold due to aggregation-inducing factors (step 1b) and further aggregate into soluble small oligomers (step 2b) that progress to amyloid aggregates through a process thoroughly described in Figure 6.1.

Soluble proteins, though, may also bind to ubiquitous lipid membranes by assuming an α-helix conformation (step 1c) [124]. The cationic nature of membrane-binding sequences in NDD-causing proteins (e.g., Aβ, α-synuclein) favors their binding to anionic lipids, such as phosphatidylserine [125]. Their electrostatic interactions cause the exposure of hydrophobic, aggregation-prone epitopes [126]. Once the local concentration of aggregation-prone proteins bound to the lipid surface increases, their aggregation happens through an α-helix/β-sheet conformational switch that leads to small β-sheet-rich membrane-bound oligomers (step 2c). Their further aggregation to amyloid fibrils (step 3, Figure 6.3) is actively promoted by the structure of specific membrane regions [127]. Lipid membranes include *lipid rafts*, ordered domains with a higher content of sphingolipids, gangliosides, and cholesterol (respectively blue heads, fucsia heads, and

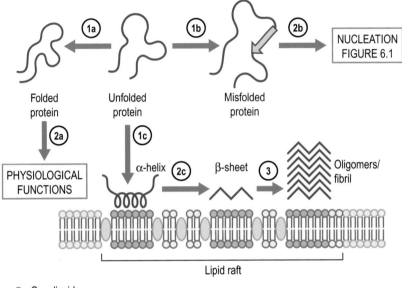

FIGURE 6.3   Interactions between amyloid protein and lipid rafts: membrane-assisted oligomerization.

orange ovals, Figure 6.3) [128]. Lipid rafts act as dynamic/floating organization centers for membrane signaling and trafficking [129]. They are privileged binding domains for amyloid proteins in the brain, where they play an important role in neuroprotection (preserved raft structure) [130] and neurodegeneration (raft disorganization) [131]. Lipid rafts contain the highest concentration of gangliosides and cholesterol in neuronal lipid membranes [132,133]. Physiological levels of gangliosides are neuroprotective, as their ablation in knockout (KO) mice causes PD-like pathologies [134], and their decrease is observed in aging, and in AD-affected, brains [135].

The sialyl ganglioside GM1 (**GM1**, Figure 6.4) is the most negatively charged ganglioside in neuronal lipid rafts [121]. GM1 clusters position sialic acid moieties in a regularly spaced arrangement that promotes the binding of β-sheet-rich amyloid peptide conformers [136], such as prions [137] and Aβ peptides [138]. GM1-rich membrane domains are amyloid binding sites in model systems [139], and GM1–Aβ conjugates are observed in isolates from AD patients [140]. An excess of GM1 is observed in cortical lipid rafts from AD patients, and is found in senile Aβ plaques

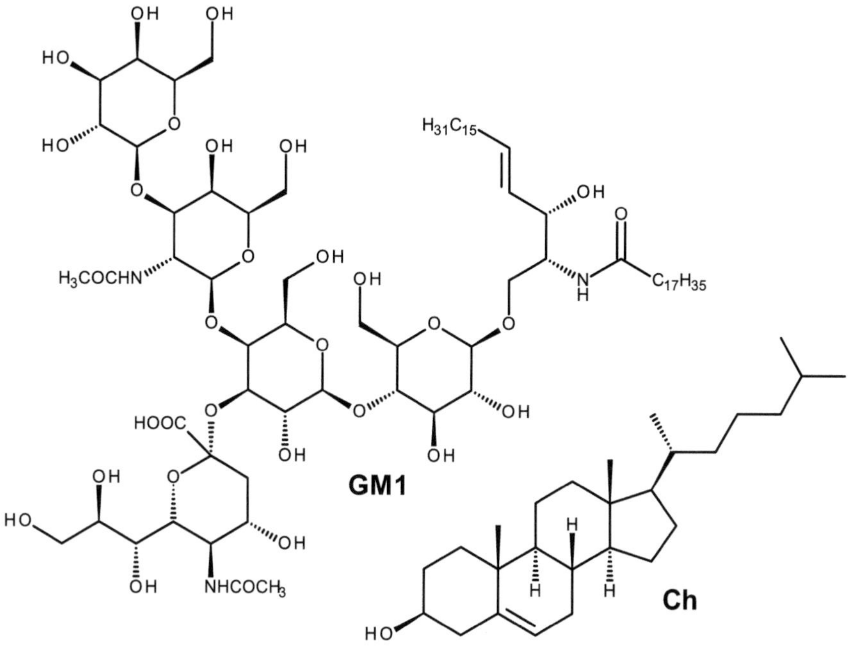

FIGURE 6.4   Sialyl ganglioside GM1 and cholesterol Ch: chemical structures.

[141,142]. A balance between neurotoxic and neuroprotective roles of gangliosides, and of GM1 in particular, is likely (when altered by external factors) to determine pathological conditions.

Cholesterol (**Ch**, Figure 6.4) interacts with Aβ through a cholesterol-binding domain [143], and modulates GM1–amyloid peptide interactions [144]. Cholesterol induces a conformational switch in GM1 and similar gangliosides, that accelerates the interaction between GM1 and Aβ peptides [144]. The effect of cholesterol is mainly kinetic, as the GM1–Aβ interaction energy is only marginally affected. The regulatory role of cholesterol on ganglioside–amyloid protein interactions is complex, as its role in organizing/rigidifying the lipid raft structure may also prevent ganglioside–amyloid interactions in cholesterol-rich lipid rafts [145]. *In vitro* models show a concentration-dependent effect on Aβ aggregation (10% cholesterol levels in a lipid bilayer cause membrane-bound Aβ aggregation that is not observed when cholesterol levels are raised to 30% [122]). Thus, cholesterol is needed to promote GM1-mediated Aβ aggregation in dynamic lipid rafts, but it causes lipid raft thickening and prevents amyloid binding when its concentration exceeds a threshold. The combination of neuroprotective/GM1-switching effects and neurotoxic/lipid raft stiffening effects by cholesterol is mirrored by contrasting observations in AD patients. Reduced [146] and increased [147] cholesterol levels are reported in AD patients from different studies. Interestingly, the observed alteration of cholesterol metabolism in AD leads to simultaneous sterol-lowering and -increasing effects [148]. Decryption of the multiple roles of cholesterol in normal brain functioning, in neuroprotection, and neurotoxicity is needed to identify novel therapeutic avenues against NDDs.

Membrane–amyloid interactions eventually cause membrane destabilization and leakage. Repeated α-helix refolding/membrane binding events (step 1a, Figure 6.5) and β-sheet conformational switches (step 2a) lead to a local high concentration of small, β-sheet-rich, membrane-bound oligomers. Their proximity facilitates the formation of oligomeric nuclei (step 3a) and medium-sized protofibrils (step 4a) [123], similar to what happens in solution (MM to MM* to D* to AN* to protofibrils, upper Figure 6.5).

When the concentration of membrane-bound, medium-sized protofibrils is sufficiently high, they assemble into a supramolecular *annular protofibril structure* and insert themselves through the membrane (step 5a) [149]. Their pore-like nature causes bacterial toxin-like permeabilization of the membranes [150], and in particular the disturbance of Ca^{2+} homeostasis [151]. Annular, pore-like amyloid structures are formed also in solution from protofibrils (step 5b) [152], but their insertion into lipid membranes (step 6a, Figure 6.5) is much less efficient than the membrane-assisted process [153].

Annular protofibrils can be characterized in model bilayer membrane systems by high-resolution optical imaging [154], AFM [155], and EM [156].

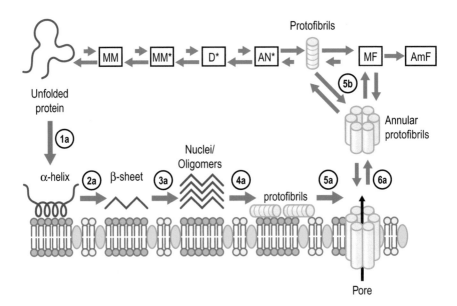

FIGURE 6.5   Annular protofibrils as lipid membrane-spanning pores.

Small–medium-size tetra- to 13-mers induce pore formation, while smaller oligomers and larger fibrils do not [157]. Their inner diameter varies between 1 and 2.5 nm, while external diameters range between 7 and 10 nm [154,158]. Small molecules and ions rapidly permeate pore-rich membranes, while larger molecules show slower kinetics [159]. A pore-like induction of ion channel currents is observed with wild-type (WT) [160] and mutant [161] Aβ, prion [162], WT [163] and mutant [164] α-synuclein, and polyQ oligomers [165]. Currents are inhibited by aggregation inhibitors [166], by ion chelators [167], and by Zn^{2+} [168].

Annular structures are observed *in vivo*. Pore-like Aβ structures are detected by transmission electron microscopy (TEM) in brain tissue membranes from AD patients [169]. Annular α-synuclein is isolated from *postmortem* samples of patients suffering from multiple system atrophy (MSA) [170]. Their structure is similar to artificial pores in model bilayer membranes, although they are larger (>10 nm inner pores, up to 50 nm external diameters).

Pores formed exclusively by amyloid peptides (*barrel-stave pores*, hydrophobic contacts between lipid rafts and amyloids, hydrophilic regions in the inner pore, top left, Figure 6.6), or by membrane–amyloid complexes (toroidal pores, hydrophilic contacts between the lipid head groups and the hydrophilic regions of amyloids, top right, Figure 6.6) are observed respectively with the amyloidogenic bacterial toxins alamethicin [171] and magainin [172].

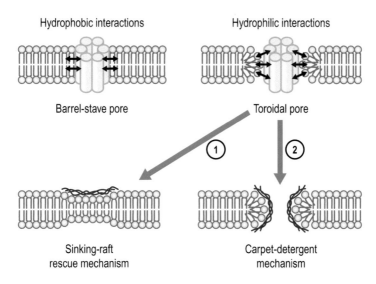

FIGURE 6.6   Evolution of annular pores: transient and permanent-disruptive consequences.

Toroidal pores made by neuronal amyloid proteins are known [173]. They are metastable structures, where the peptide–membrane interaction forces a curvature in the surrounding membrane region. Toroidal pores may collapse and restore a lipid bilayer with membrane-bound peptide oligomers (sinking-raft model, rescue mechanism, step 1) [174]. They may also lead to the detachment of a lipid–amyloid peptide aggregate, and eventually to the collapse of the membrane (carpet model, destabilization mechanism, step 2, Figure 6.6) [175]. The latter model accounts for the detergent-like effects observed with a number of pore-forming amyloid proteins [123].

Another lipid extraction and membrane collapse process may coexist with the toroidal pore-driven mechanism of membrane-targeted toxicity (Figure 6.7).

Once the critical concentration of membrane-bound, medium-sized protofibrillar oligomers is reached, the resulting annular, pore-like structures span the membrane and act as channels, as described (step 1, Figure 6.7). They may also rearrange and bind to the outer membrane (step 2), so that their interaction with the lipids causes their extraction into an amyloid–lipid aggregate (step 3) [176]. Lipid aggregates detach themselves from the membrane as amyloid fibrils (step 4), cause its progressive thinning and increase its permeability to ions and small molecules (step 5, Figure 6.7) [177]. The end result of a "toroidal pore-carpet" model (Figure 6.6) and of an "outer membrane binding–membrane thinning and leakage" model (Figure 6.7) is the same, i.e., membrane collapse and cellular toxicity. The former process is characterized by easily detectable pores, and by previously mentioned single ion channel-like currents [160–165]. Visualization of an "outer membrane binding–membrane thinning and leakage" process is more complex. Its occurrence is indirectly confirmed by a

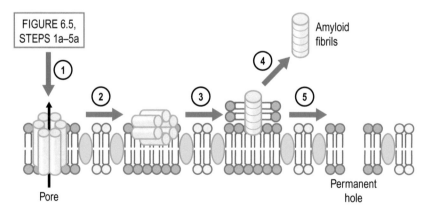

FIGURE 6.7   Alternative pore evolution: outer membrane binding–membrane thinning and leakage.

concentration-dependent increase of conductivity sometimes observed for the whole membrane [178,179]. The pore-like process appears to be more relevant for amyloid aggregates, but the simultaneous occurrence of both mechanisms is observed for amyloid islet protein in type 2 diabetes (hIAPP) [180]. Namely, the N-terminal amino acid (AA) 1–19 sequence of hIAPP is responsible for pore formation and initial membrane leakage [181], while the aggregation-prone AA 20–29 sequence induces membrane disassembly by outer layer binding [182]. Both mechanisms should play a role for most amyloidogenic proteins.

Targeting small molecules against the aggregation process of NDD-related proteins is reported in literature [183,184]. Further elucidation of several molecular mechanisms and "clean" targeting of toxic events (i.e., lowering the concentration of toxic species, increasing the concentration of non-toxic species) are goals for the future. As they may significantly differ among each aggregation-prone protein, the next section deals with a specific example—tau—and examines its neurotoxic aggregation process in detail.

6.2.1 The Target: Interfering with (Neuro)toxic Tau Species in the Aggregation Process

Tau and tauopathies are thoroughly described in Sections 1.3 and 1.4 in Chapter 1. Here we focus on the aggregation process of tau, on its intermediate and final stages, and on the identification of prospective points of intervention for anti-aggregation drugs targeted against tau [185,186].

Tau is a highly hydrophilic protein, with around 200 polar/charged residues out of 441 AAs (longer tau isoform, 4R2N: 80 Ser or Thr, 56 Asp or Glu, 58 Lys or Arg) [187]. Tau has an overall basic character, but negatively (the N-terminus) and positively charged domains (the microtubule/MT binding regions (MTBRs)) influence its interactions with MTs and other binding partners. Tau is an abundant neuronal IDP, with a 27-fold larger volume than a globular protein with a similar weight [188]. Its high solubility and its engagement in multiple cellular interactions make tau aggregation-resistant in basal conditions [189]. Tau aggregation is induced by *conformational switches* caused by PTMs [190], described in Section 1.3 (see Figure 1.5). Among them, O-glycosylation, dephosphorylation and prolyl isomerization are anti-aggregating PTMs (stimulation required to inhibit/prevent aggregation), while tau hyperphosphorylation and acetylation are aggregation-inducing PTMs (inhibition required to inhibit/prevent aggregation) [190]. Tau aggregation may be promoted by *mutations* in its structure [191]. Section 1.3 describes how tau mutations (illustrated in Figure 1.6) are the causative factors of several tauopathies, and how each of them causes changes either in MT binding, or in aggregation propensity for mutated tau. Several mutations are used in transgenic (TG) models that recapitulate neurodegenerative tauopathies [192]. Finally, tau aggregation is promoted by its interactions with *lipid membranes* [189], in a process portrayed later in this section.

Clinically isolated NFTs are composed by paired helical filaments (PHFs) made primarily of tau proteins [193]. PHFs are structurally characterized by AFM [194]. They are double helical fibrils made of two twisted ribbons of varying length/periodicity (65 to 75 nm) and height (8 to 20 nm) [195]. Each 2N4R tau fibril is composed of densely packed MTBRs (AAs 244–369) arranged in a β-sheet, core amyloid structure [196]. Two "hot spots" ([306]VQIVYK/PHF6 in R3, and 4R tau-specific [275]VQIINK/PHF6* in R2) are identified in MTBRs as independent aggregation inducers [197]. In basal conditions, tau assumes a "paper clip" conformation where the C- and N-domains of tau shield the hot spots and prevent aggregation [198]. Phosphorylation and proteolytic cleavage, *inter alia*, cause the switch to hot spot-exposing, aggregation-prone tau conformers [199].

Largely unstructured N-terminal (AAs 1–120), middle basic (AAs 121–243), and C-terminal domains (AAs 370–441) form a highly mobile, disordered *fuzzy coat* [200] around the rigid MTBR core of tau fibrils. Figure 6.8 represents the section of a 2N4R fibril, containing four tau monomers.

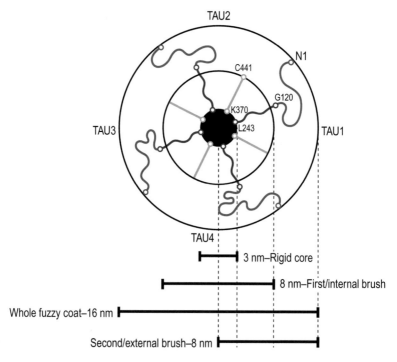

FIGURE 6.8   Tau structure: rigid core and fuzzy coat.

The rigid fibril core (black circle, diameter ≈3 nm) is surrounded by an ≈8 nm-wide, denser brush layer made by the C-terminus (green line) and by part of the positively charged, middle basic domain (violet line). An external, softer, negatively charged brush layer (width ≈8 nm) is mostly made by the N-terminal domain of tau (fucsia line, Figure 6.8). The fuzzy coat behaves as a polyelectrolyte polymer brush. It promotes pathological, pH-dependent cellular interactions [201], and protects the amyloid core from disassembly [195].

In vitro tau aggregation at physiological concentrations requires months to take place [202]. *Polyanionic inducers* attract the positively charged MT-BRs and promote their packing. They are used to accelerate tau aggregation, so to make it suitable for *in vitro* studies [203]. Among them, the sulfated glycosaminoglycan heparin shows multiple, low-affinity tau binding sites identified by NMR [204] and calorimetry [205]. Heparin binds to positively charged regions in MTBRs, in the middle basic domain and in the N-terminus of tau [204]. A single, high-affinity binding site (K_D ≈ 20 nM) [21] is detected by single molecule Förster resonance energy transfer (FRET). Heparin binding causes the loss of interactions between MTBRs and C- and N-tau termini, the exposure of the MTBRs and their packing in a putative aggregation-prone conformation [21].

MTBR-containing tau fragments are used to study the tau–heparin interaction [206,207]. In particular, heparin-induced aggregation of a His-tagged tau repeat domain (tauRD) construct (AAs 244–378, MTBRs) is studied by AFM and TEM [208]. The proposed nucleation-like mechanism is depicted in Figure 6.9.

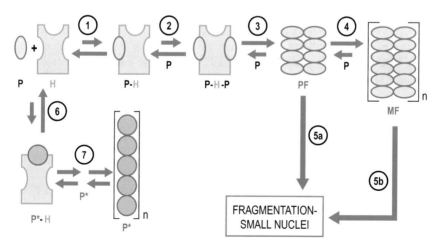

FIGURE 6.9 Nucleation-like aggregation of tau: on-pathway and off-pathway mechanisms.

The rate-limiting step of an on-pathway to AD-like fibrils is the formation of an *aggregation heteronucleus* (P–H–P) made by sequential addition of a tauRD monomer (P) onto heparin (H) (step 1—P–H—and step 2—P–H–P). The two Ps in P–H–P assume the aggregation-prone conformation, and bind to H with moderate affinity. Once a P–H–P heteronucleus is formed, thermodynamically favored monomer addition (elongation to proto-fibrils, PFs, step 3, and to mature fibrils, MFs, step 4) proceeds faster. The low content of H (<1:20 with respect to tau) in such fibrils [207] may in-dicate that H kinetically promotes the rate-limiting heteronucleation step, while elongation is H-independent [208]. Alternatively, elongation causes further conformational rearrangements in tauRDs that stimulate the re-lease of H from growing fibrils [208]. The highest rate for fibril formation observed at a 2:1 P:H ratio confirms the heterotrimeric P–H–P structure as the rate-limiting intermediate. The elongation phase is further accelerated by secondary nucleation events, mostly due to protofibril and/or fibril fragmentation (steps 5a and 5b, Figure 6.9) [209]. The increase in nuclei/ short protofibrils as secondary nucleation seeds for fibril formation expo-nentially accelerates the fibrillization process. Similar kinetic effects are obtained by external seeding with sonicated fibrils [209].

An excess of H slows the fibrillization process by increasing the con-centration of heterodimeric P–H, and depleting the P pool to form the P–H–P heteronucleus [208]. An excess of P also reduces fibrillization, and a proposed mechanism [208] involves the formation of a tightly bound off-pathway P*–H heterodimer (step 6, Figure 6.9). The P*H heterodimer can-not re-enter the on-pathway, and rather grows into an alternative proto-fibril structure P_n^* (step 7, Figure 6.9). The rod-like, shorter and thinner P_n^* protofibrils are extremely unstable, and do not evolve into stable fibrils. They rapidly assemble from, and disassemble into, P*–H heterodimers. The off-pathway is based upon a tightly bound P*–H intermediate, but the continuous dissociation of off-pathway protofibrils and the slow formation of on-pathway, fibril-productive P–H–Ps drives the equilibrium towards AD-like fibrils and causes the disappearance of short-lived protofibrils [210].

The generality of such a mechanism is questionable, as small struc-tural differences—absence of the His tag, shorter (AAs 244–372) tauRD sequence—prevent the observation of an off-pathway to protofibrils [209]. Different, putative off-pathway species are observed in other heparin-induced aggregation experiments [211,212]. On-pathway full-length (FL) tau fibrils induced *in vitro* by heparin are structurally different from fibrils that constitute PHFs from the frontal cortex of AD patients [213]. Namely, the former are thinner (avg. 16.9 *vs.* 19.5 nm width) and longer ribbons (avg. 129 *vs.* 84.5 nm periodicity). Heparin-induced fibrils are more sta-ble to proteolysis, and are more heterogeneous [213]. Nevertheless, *in vitro* assays recapitulating tau aggregation are suitable tools to identify anti-aggregating, or disassembling agents [185,214]. It is crucial to define which on-pathway and/or off-pathway aggregation steps lead to neurotoxic species, to target them, and to reduce their impact. Conversely, the steps leading to non-toxic or even neuroprotective species should not be disturbed. Much remains to be done, but our limited knowledge about authentic *in vivo* aggregation of tau isoforms in NDDs provides some guidance.

The earliest pathological state observed in humans entails MT-unbound, soluble, hyperphosphorylated (HP) tau, recognized by phosphorylation-specific tau antibodies but unreactive with β-sheet conformation-specific dyes [215]. HP-tau species range from monomers to small oligomers that have not yet switched to an amyloidosis-promoting conformation [216]. Amyloidogenic small–medium soluble oligomers, protofibrils, and fibril fragments (either on-pathway or off-pathway to mature fibrils) that react with β-sheet conformation-specific dyes are the next observed tauopathy stage in humans [215,216]. Mature, large NFTs are the easily detectable (imaging, silver staining) end result of the aggregation process [217].

The obvious suspects—NFTs, insoluble tau aggregates—were long believed to be the major determinants of neurotoxic effects in tauopathies [218]. Recent observations challenge this belief, and revisit the toxic effects of tau species on neurons [219–221].

The toxicity of tau may depend on *loss-of-function* (LOF) and *gain-of-function* (GOF) factors [222]. The slow aggregation of tau eventually depletes the pool of soluble, functional tau. Consequently, its functional interactions are impaired in an LOF scenario. The structure and integrity of MTs rely on tau to dynamically adapt to neuronal environments. LOF/MT-dependent neurotoxicity could be expected [223], but some evidence argues against it. Four independent tau KO mice are perfectly viable [219], and three of them do not show behavioral deficits even at older age [224,225]—a compensatory role for other MT-associated proteins in MT stabilization is postulated. Tau KO mice are more resistant to drug-induced seizures [226]. Hippocampal slices from tau KO mice are more resistant to gamma-aminobutyric acid A ($GABA_A$) receptor antagonist-induced epileptiform bursting [227], and tau KO neurons show normal and viable MTs [228]. Moreover, an excess of tau, rather than its reduction, leads to the impairment of axonal transport [229]. Excessive stabilization of MTs by tau and its competition with kinesin and dynein motors in terms of MT binding are possible explanations. In general, tau LOF does not appear to significantly contribute to neurotoxicity.

Tau GOF consists of the interaction(s) between tau species depicted in Figure 6.9 and any cellular component. Matching the reactivity of cell cultures, tissues, or whole organisms with tau species-specific antibodies and dyes, and the insurgence or the progression of neurotoxicity provides indications about toxic aggregate species [230]. Neuronal viability is an obvious toxicity marker, as is the operational efficiency of neurons (i.e., their functional integration in neuronal networks/circuits) and, on a whole organism scale, the observation of behavioral deficits.

The presence of NFTs, and their load in brain samples from controls and AD patients, correlates more with AD state and progression than β-amyloid plaques [231]. The correlation is qualitative, as neuronal death largely exceeds the NFT load in general [232], and significant neurotoxicity is observed in brain areas with limited NFT load [233]. Several tauopathy-recapitulating TG mice are available [192], providing evidence on NFT toxicity—or lack thereof. An inducible mouse model overexpressing the aggregation-prone P301L mutated tau ($\approx$15-fold increase) causes up to 70% neuronal loss in the *cornus ammonis* 1 (CA1) hippocampal region, gross brain atrophy and behavioral impairment, combined with progressive NFT formation [234]. Tau switch-off (down to $\approx$2.5-fold decrease) does not prevent further accumulation of NFTs, but rescues the memory functions and stops neuronal loss in TG mice [234]. Another TG model can be forced to overexpress either human ΔK280 tau/pro-aggregation mutated tauRD, or human I277P-I308P tau/anti-aggregation tauRD [235]. The former leads to synaptotoxicity/impaired neuronal functionality, memory deficits, and NFT accumulation. The latter does not show neurotoxicity [235]. Switching off the pro-aggregation tau transgene for 4 months rescues electrophysiological and behavioral abnormalities, while NFTs—now composed mostly of endogenous murine tau—are still abundantly present [236]. *In vivo* two-photon imaging shows the rapid formation of NFTs in viable neurons of TG mice overexpressing human P301L tau [237]. NFT-bearing neurons in the visual cortex area are integrated in a functional neuronal network, are capable of integrating dendritic inputs, and respond to visual stimuli [238]. Thus, NFTs are presumably involved in tau toxicity, but should not constitute the main target for disease-modifying agents—they probably represent an endogenous rescuing mechanism to dispose of strongly toxic oligomeric species in the neuronal environment [239].

Conversely, tau oligomers show neurotoxicity in preclinical and clinical environments. Dying neurons in aged old mice transfected with human FL tau show multiple abnormalities by EM, but no NFTs [240]. Purified tau trimers added to neuronal cultures show neurotoxicity at low nM concentrations (FL 2N4R trimers are more potent than their 1N4R counterparts) [241]. Monomeric and dimeric HP species (MW = 64, 140, and 170 kDa) are found in brain tissues from mice overexpressing human P301L tau. Their levels correlate with memory loss at various ages/disease stages [242,243]. Entorhinal cortex (EC)-targeted overexpression of human P301L tau (16 month-old, no NFTs) leads to electrophysiological deficits, possibly due to the accumulation of oligomeric soluble tau in soma, dendrites, and axons [244]. Dorsal root ganglia (DRG) neurons from TG mice overexpressing human P301S tau show age-dependent tau oligomer-specific antibody reactivity that parallels the progression of tau pathology (i.e., impaired axonal transport and mitochondrial trafficking) [245]. Synaptotoxicity, neuronal death, and absence of NFTs are observed in P301L-overexpressing mice [246]. Conversely, FL human tau-overexpressing TG mice exhibit tangles and limited neuronal death, but do not show eletrophysiological deficits (marginal toxicity caused by endogenous rescuing) [246]. Neurodegeneration is observed *in vivo* before NFT formation in *Drosophila* models [247,248].

Neurotoxic tau oligomers are observed in clinical isolates from AD patients. On-pathway granular oligomers, composed by ≈40 tau monomers, are detected in heparin-driven tau polymerization [211]. They are observed at high concentrations in the frontal cortex of early AD patients (Braak stage 1) before NFT formation, suggesting their usefulness as pre-symptomatic diagnostics [249]. Their levels are inversely proportional to the levels of heat shock proteins (Hsps), suggesting an impairment of the PQC [250]. Small oligomeric HP-tau species, either as such or as components of larger aggregates, are found in brain extracts from patients affected by AD or frontotemporal dementia and parkinsonism linked to chromosome 17 (FTDP-17) [242]. An ≈180 kDa small oligomer is found in advanced AD patients (Braak stage ≥4), but not in earlier AD stages, or in aged controls [212]. It migrates similarly to synthetic cross-linked tau dimers, and aggregates *in vitro* and *in vivo* into larger, off-pathway oligomers that do not convert into fibrils [212]. Purified tau oligomeric fractions from three AD patients seed *in vitro* tau aggregation and lead to mixed oligomers that impair the synaptic function of hippocampal CA1 slices (synthetic non-seeded tau oligomers are inactive) [251]. Injection of purified AD oligomers in WT mice causes the rapid, transient onset of behavioral deficits with persistent detection of NFTs, suggesting the physiological rescue of deficits by aggregation of neurotoxic oligomers into insoluble aggregates. Accordingly, injection of purified, insoluble tau filaments from AD patients does not cause neurotoxicity [251].

Intracellular tau oligomers, protofibrils, and other on-pathway or off-pathway intermediates to NFTs cause neurotoxicity. Oligomer-specific antibodies determine a constant and significant (up to four times) higher concentration of tau oligomers in AD patients compared with aged controls [252]. The same tau species are found in the extracellular space [252], suggesting a prion-like, infectious mechanism that spreads tauopathies in various CNS compartments [69] through cellular penetration and *cell-to-cell propagation* [253]. The internalization of fibrillized tauRD species in mouse cerebellar neural C17.2 stem cells is followed by seeding (aggregation of endogenous FL tau), and by slow but measurable cell-to-cell transmission [254]. Mouse primary hippocampal and cortical neurons incubated with tau species ranging between monomers and long fibrils show size-selective internalization of oligomers *via* endocytosis [255].

Tau oligomers up to short fibrils co-localize with lysosomes. Monomers and large fibrils appear not to be internalized [255]. Accordingly, internalization-dependent neurotoxicity on cholinergic neurons is observed only with synthetic tau trimers—not with monomers or dimers [241].

In vivo cell-to-cell propagation of tau is proven by treatment of human FL tau-expressing ALZ17 mice (which do not produce NFTs) with brain extracts from TG mice overexpressing human P301S tau [256]. The extracts cause the progressive formation of tau fibrils and NFTs in ALZ17 mice. The fibril spread indicates cell-to-cell propagation among anatomically connected brain regions [256]. Injection of synthetic human P301S tau fibrils in PS19 mice expressing FL human tau causes pathological disturbances at 8 months of age, with a spreading pattern dependent on the injection site [257]. NFTs from injected mice are more AD-like than NFTs from old, non-injected PS19 mice [257]. The characterization of a P301S injection-dependent mouse model proves the transmission of pathological tau species *via* anatomical connectivity, not through physical proximity [258].

Injection-free tau propagation takes place between synaptically connected brain areas of TG mice reversibly overexpressing human P301L tau in the EC (the locus of initial tau pathology appearance in AD) [259,260]. P301L tau aggregates appear in brain regions where the transgene is inactive, and act as seeds for the aggregation of endogenous murine tau. Accordingly, mice show neuronal loss exceeding transgene expression at 24 months of age [260].

Injection of total brain homogenates from patients diagnosed with one of six tauopathies in human tau-expressing ALZ17 mice causes the appearance of aggregates [261]. Their spreading, their morphology and staining, and the presence of human tau isoforms in the aggregates show their human tauopathy-specific structure [261]. A slower but measurable growth and cross-species spreading of tauopathy aggregates after injection of human brain homogenates is observed also in WT mice expressing mouse tau [261].

A recent paper [262], employing extremely sensitive and tau species-selective fluorescent tau probes, supports a molecular mechanism of cell internalization and cell-to-cell propagation in SH-SY5Y neuroblastoma cells (Figure 6.10).

Extracellular tau, either in the monomeric (M) or small oligomeric (O) state, is refractory to aggregation in the inducer-free extracellular space (step 1, Figure 6.10). M [262] and O species are taken up by SH-SY5Y cells (step 2) through macropinocytosis. M crowding in a confined space, and the acidic environment of endocytic vesicles [263] promote nucleation of Ms (step 3), converting them to Os. Exogenous Os are then released from macropinocytic vesicles (step 4), and act as seeds for the formation of intracellular, O-seeded oligomers that capture endogenous tau. Seeding and oligomerization happens either in newly formed lysosomal/endosomal

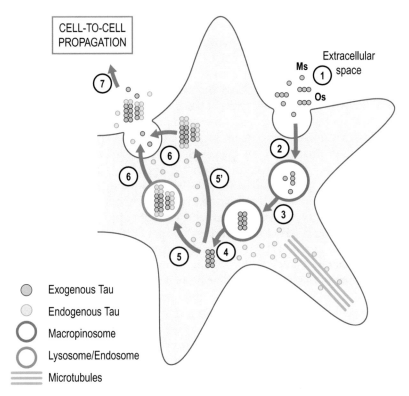

FIGURE 6.10   Cell-to-cell propagation of tau: neuronal spreading of the neurotoxic species.

compartments (step 5) or directly in the cytoplasm (step 5′). Most Os contribute to intracellular aggregation and fibril/NFT formation, but a part of them exits the cell, either as such or after partial fragmentation (step 6). Freshly externalized Os and Ms access the extra cellular space and accelerate cell-to-cell propagation (step 7, Figure 6.10).

The externalization of labeled Os from an SH-SY5Y cell batch and their internalization/seeding into another SH-SY5Y cell culture is experimentally validated (isolation of the Os-containing extracellular medium from the first experiment, incubation of the extracellular medium with a second SH-SY5Y batch, identification of intracellular Os) [262]. It is worth noting that neuronal activity, and in particular presynaptic excitatory activity, triggers the fast release of endogenous tau [264]. The slower kinetics of tau clearance may cause an increase of extracellular tau and, in pathology-prone neuronal environments, may lead to prion-like neuronal propagation of tau [265].

The involvement of heparan sulfate proteoglycans (HSPGs), the cell surface receptors involved in prion infection, is proposed for macropino-cytosis-driven tau propagation *in vitro* and *in vivo* [266]. Macropinosomes are leakier vesicles than endosomes or lysosomes. Tau oligomers could penetrate them by interacting with their lipid bilayer membrane, thus gaining access to the cytosolic space [266].

Interactions between tau species and *lipid membranes* are reported, although less frequently than for Aβ and other amyloid proteins. Cholesterol co-localizes with NFTs in the brains of tauopathy-affected patients [267], and its role as a modulator of tau HP is assessed [268]. Lipid rafts are seldom associated with tau [269], and do not seem to play a relevant role in the aggregation and neurotoxicity of tau species. Similarly, the insertion of amyloidogenic proteins in lipid rafts to form pores as depicted in Figures 6.6 and 6.7 does not apply to tau [270]. NMR shows how positively charged regions in tau3R fragments interact with the polar heads of moderately anionic membranes (composed of a 4:1 neutral/anionic phospholipid mixture) to adopt a β-sheet-driven rigid structure that does not penetrate the lipid membrane body (step 1, Figure 6.11) [270]. A purely anionic phospholipid membrane causes the switch of some tau3R regions to random coil/α-helix, and their internalization in the lipid membrane body (step 2, Figure 6.11) [270].

The membrane–MTBR tau interaction causes membrane leakage, and may have physiological relevance referred to FL human tau and its neurotoxicity. Similar studies with FL tau show the compaction of FL tau incubated with the anionic phospholipid 1,2-dimyristoyl-sn-glycero-3-[phosphorac-(1-glycerol)], and the fast disruption of membrane integrity measured

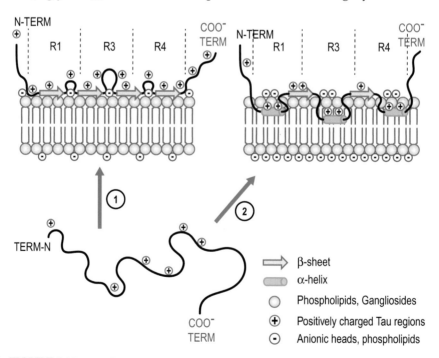

FIGURE 6.11   Membrane–MTBR tau interactions: charge-dependent molecular interactions.

by X-ray and neutron-scattering techniques [189]. The amount of neuronal membrane-bound tau is inversely proportional to its phosphorylation, suggesting a role in membrane trafficking for membrane-bound tau that may be impaired by HP-tau [271].

Rational drug discovery efforts require a validated target, i.e., soluble neurotoxic oligomeric species (and the aggregation/disaggregation steps producing them), or pathology-promoting interactions, to be minimized using immunomodulation or small molecule inhibitors. Classical isolation and purification techniques may threaten the stability of highly reactive and toxic tau oligomers, or of tau-binding partner complexes. Currently available *in vivo* imaging techniques do not yet provide single tau species-directed exquisite specificity. Tau-specific tracers are being characterized, and used in clinical settings [272], to unravel the pathophysiology of tauopathies and to provide further guidance to drug discovery efforts.

6.3 CHAPERONE-DRIVEN DISAGGREGATION OF PROTEIN AGGREGATES

The disassembly of aggregation intermediates and of NFTs happens following two alternative pathways. The first acts on the equilibria ruling the steps leading to NFTs, forcing the processes described in Section 6.2 to proceed backwards towards either tau monomers or non-toxic tau species. Some small molecules listed in Section 6.4 do exactly that, and are described in detail in a chemistry-oriented companion book [95]. The second approach targets an enzyme or an enzyme complex, capable of capturing an advanced aggregation intermediate (in particular tau fibrils or NFTs) and to disassemble it step by step.

Members of the *heat shock chaperone 100 (HSP100) protein family* act in cooperation with co-chaperones to disaggregate protein aggregates with different morphologies [273]. Bacterial ClpB and yeast Hsp104 are stress-activated, essential protein disaggregases for the disassembly/dissolution of large aggregates [274,275]. Prokaryotes and lower eukaryotes rescue misfolded proteins and aggregates to reutilize them, rather than simply disposing of/degrading them [276,277].

Hsp100 family members, including ClpB and Hsp104, are AAA$^+$ (ATPases associated with diverse cellular activities) protein complexes, composed of a hexaprotomeric ring [278]. Hsp104 works at best together with an ATP-dependent chaperone (Hsp70) and two co-chaperones (the J-domain protein Hsp40 and a nucleotide exchange factor (NEF)) [279]. ClpB works with the bacterial counterparts of Hsp70 (DnaK) and Hsp40 (DnaJ) [280]. The Hsp70/DnaK complexes physically bind to Hsp100 disaggregases and regulate–stimulate their activity [281,282].

Each individual unit of ClpB and Hsp104 is identical, but each complex works differently depending on the complexity of the "disaggregating task" (smaller or larger aggregates, tighter interactions between the components of the aggregate, etc.) [283]. ClpB acts in a multiple ATPase, non-cooperative disassemble/unfold/refold mode to dismantle and recycle the molecular components of small disordered aggregates [284] (path A, Figure 6.12). Similar substrates are processed by a single protomer of Hsp104 (path B), although ClpB shows a higher disaggregase activity towards disordered protein aggregates than Hsp104 [285]. Possibly, bacteria need to quickly react to stress-induced protein denaturation and disordered aggregation.

ClpB does not disaggregate amyloidogenic fibrils, due to the high energy needed for their disassembly [285]. Conversely, Hsp104 switches to a cooperative protomer activation to provide the energy needed to unfold the local, stable β-sheet structure domains of amyloids and disaggregate them. Sub-global cooperation is enough for "loose" amyloid proteins (path C: recruitment of a second/step C_1 and third subunit/step C_2, amyloid disaggregation/step C_3). Global/hexameric cooperation is needed for more "organized" amyloid aggregates (path D: sequential recruitment of all subunits/D_1, amyloid disaggregation/step D_2, Figure 6.12) [285].

The amyloid-disaggregating activity of Hsp104 complexes is strongly enhanced by two yeast small chaperones, Hsp26 and Hsp42 [286]. Hsp26 and Hsp42, either alone or in combination, bind to amyloid prion proteins and block their fibrillization. Once incubated with functional Hsp104 complexes in the presence of prion fibrils, they accelerate their disassembly *in vitro* and *ex vivo*. They are equally effective in accelerating the disassembly of NDD-related polyQ and α-synuclein fibers, possibly by binding to the fibers and "loosening" them to facilitate their disassembly by Hsp104 complexes [286].

Metazoan homologs of Hsp104 are unknown, possibly because a strong, amyloid-dismantling disaggregase activity could produce incompletely disassembled, neurotoxic small oligomers—a cure worse than the disease [283]. It is worth mentioning that either yeast Hsp104 as such [287] or rationally designed, more potent Hsp104 mutants [288] are evaluated as putative treatments for proteinopathy-dependent NDDs.

Although autophagy (see Chapters 4 and 5) appears to be the best solution for higher eukaryotes to get rid of insoluble aggregates in general, and amyloid fibrils in particular, the first observations of a mammalian disaggregase activity on several protein aggregates (including the Aβ_{40} peptide [289]) are now attributable to chaperone complexes.

6.3.1 The Target: Hsp110

The Hsp70 chaperone system, and in particular the chaperone–co-chaperone heat shock cognate 70 (Hsc70)/DnaJ protein homolog 1 (Hdj1) complex, shows marginal disaggregating activity on luciferase- and green

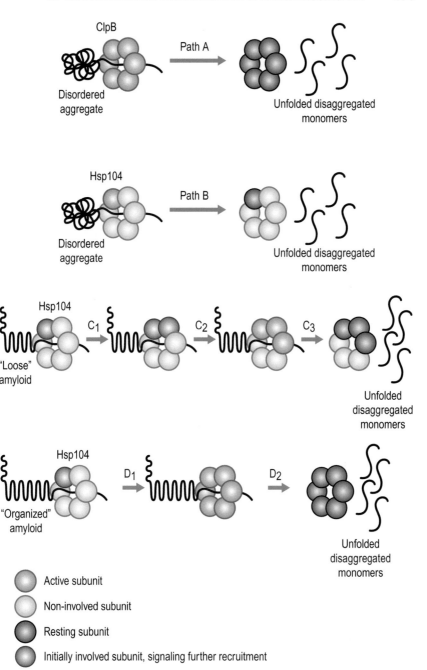

FIGURE 6.12 Disaggregase complexes in bacteria and yeasts: disordered and amyloido-genic substrates.

fluorescent protein (GFP)-based, ≥ 500 kDa disordered aggregates [290]. Rat cytosolic extracts are ≈ 10-fold more active than purified Hsc70/Hdj1 (disaggregation and reactivation of aggregates in ≥ 4 hours), although the yeast Hsp104 complex acts faster (full reactivation in 30–60 minutes). The crude activity is fully recovered when the *constitutive Hsp110 family member* Apg-2 is added to purified Hsc70/Hdj1 [290].

Hsp110 proteins are structurally related to the Hsp70 family. They act either as an ATP-dependent chaperone [291] or as the major NEF co-chaperone for Hsp70 proteins [292]. The binding specificity and kinetics of Hsp70 and Hsp110 proteins differentiate their functions in the PQC machinery [293]. Both chaperones favor binding to hydrophobic, aggregation-prone domains. Hsp70 has a binding preference for aliphatic amino acid-rich sequences, while Hsp110 binds aromatic amino acid-rich sequences. Hsp110 shows even faster binding kinetics than Hsp70, with no significant tighter binding/slower release of the substrate observed for the ADP Hsp70 state [293] (see Chapter 2, Section 2.2.2).

The faster kinetics of Hsp110 favor its binding to substrate proteins, but the higher recurrence of aliphatic/Hsp70-targeted lipophilic stretches in eukaryotic proteins ($\approx 30\%$ *vs.* $\approx 8\%$ for aromatic/Hsp110 targeted) makes the two binding events similarly frequent (top left, Hsp110, and top right, Hsp70, Figure 6.13). Hsp110-aromatic sequence binding is transient, releasing the unmodified misfolded peptide regardless of the ATP/ADP state of Hsp110 (green, middle, Figure 6.13), but is as easily re-established due to its fast kinetics [293]. This gives a prevalent *holdase* function to Hsp110, which prevents the misfolded peptide to aggregate [294].

Conversely, aliphatic sequences remain bound to Hsp70 long enough to hydrolyze ATP, to tighten the ADP Hsp70–substrate interaction, and to use the energy to contribute to the unfolding–refolding of the substrate protein (bottom, Figure 6.13). This mechanism attributes a prevalent *foldase* role to Hsp70, thoroughly described in Chapter 2, Section 2.2.2 [295].

It is conceivable that Hsc70/Apg-2 complexes (either preformed or assembled after the binding of one of the two chaperones to the substrate) bind two neighboring hydrophobic sequences in an aggregation-prone, misfolded protein (Figure 6.13, top middle) [293]. A synergistic effect is observed when equimolar amounts of Hsc70 and Apg-2 cooperate (with the support of Hsp40/Hdj1) to disaggregate and renature luciferase [296]. This indicates an influence of each chaperone on the other chaperone's ATP cycle, assisting the overall unfold/refold/release cycle. A proposed "clamping and walking" mechanism entails one among Hsc70 and Apg-2 tightly bound in its ADP state to the aggregate, while the other is in its ATP-loosely substrate-bound state. Due to the high flexibility of the Hsc70–Apg-2 complex, the latter chaperone "walks" the sequence to find another hydrophobic, misfolded sequence. Then, the new interaction "clamps" the formerly unbound chaperone in a different position on the

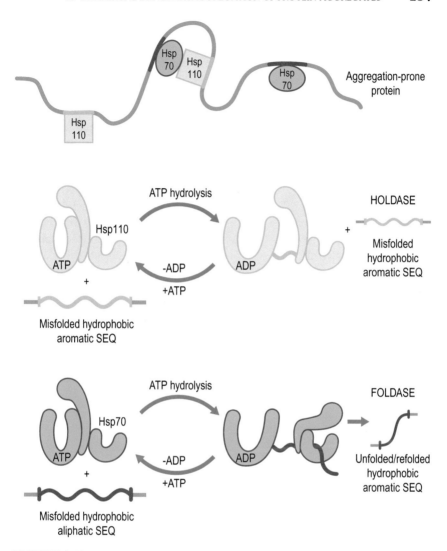

Aggregation-prone protein

Hsp110

ATP hydrolysis

Hsp110

HOLDASE

ATP

-ADP   ADP

+

+ATP

Misfolded hydrophobic aromatic SEQ

Misfolded hydrophobic aromatic SEQ

ATP hydrolysis

Hsp70

FOLDASE

ATP

-ADP   ADP

+

+ATP

Unfolded/refolded hydrophobic aromatic SEQ

Misfolded hydrophobic aliphatic SEQ

FIGURE 6.13   Disaggregases, holdases and foldases: Hsp110 and Hsp70 complexes.

aggregate and starts its ATP cycle to unfold/refold the sequence. Meanwhile the formerly tightly bound chaperone (now back to its ATP state, after releasing the refolded sequence) starts walking around, and continues the cycle [296].

Another report [297] questions the involvement of Hsp110 ATPase activity in the Hsp110–Hsp70–Hsp40 disaggregase cycle, and suggests an NEF-limited role for Hsp110 proteins. It suggests that the Hsp110 concentration should be ≤10% than Hsp70 to achieve the best disaggregase complex efficiency—reportedly superior to previous observations [293]. The disaggregase-specific role of Hsp110/Apg-2 is confirmed, as other NEFs do not assist luciferase disaggregation, and knockdown of Atg-2 causes a significant lifespan reduction in thermally stressed *C. elegans* [297]. ATPase- and efficiency-related discrepancies may derive from different experimental settings, which include more severe aggregation conditions (translating into harder aggregates to disassemble and reactivate), and more "disaggregase power" requested to Hsp110/Apg-2 (translating into activation of its cooperative ATPase activity) [298].

Ternary Hsp110–Hsp70–Hsp40 complexes do not have the amyloid-disaggregating activity of yeast Hsp104 complexes [290]. Addition of the small chaperone HspB5 causes an acceleration of α-synuclein disassembly and renaturation [286], similar to the effect of Hsp26 and Hsp42 on yeast Hsp104 complexes. The disassembly mechanism implies fibers' depolymerization from their ends, and (although accelerated by HspB5) remains extremely slow (≈15 days for 50% disassembly and renaturation) [286]. A slow disaggregase activity may nonetheless be significant to modulate *in vivo* the slow progression of human NDDs.

Interestingly, an Hsp110 KO mouse model shows age-dependent tau HP, NFT accumulation, and behavioral deficits [299]. Molecular abnormalities include sequestration of protein phosphatase 2A (PP2A) in NFTs, and reduced peptidyl prolyl isomerase (PPIase) activity in the brain. Double TG mice (Hsp110 KO and overexpression of mutated amyloid precursor protein (APP)) show significant aggravation of Aβ-related symptoms, and of the neurodegeneration process. Hsp110 is found in close proximity to Aβ senile plaques in AD patients [299].

The physiological and pathological relevance of Hsp110–Hsp70–Hsp40–HspB5-driven disassembly of NDD-relevant amyloid aggregates, and the effect on NDDs of Hsp110 modulators, remains to be determined. Up to now, only the correction of an axonal transport defect in squid axoplasm caused by an ALS-associated mutant SOD1 protein by addition of Apg-1, a human Hsp110 protein family member, is reported in literature [300].

6.4 DISEASE-MODIFYING COMPOUNDS

This chapter deals with neuropathological alterations related to protein misfolding and aggregation in general, and to tau and/or tau-connected events in particular. Sixty-one compounds/scaffolds acting on tau as anti-aggregating and/or as disassembly agents are diffusely covered in the chemistry-oriented companion book [95] devoted to disease-modifying compounds, and are briefly characterized in Table 6.1. Each compound class is numbered as in the companion book, and its chemical core is structurally defined; its mechanism of action and molecular target are

TABLE 6.1   Compounds 6.1a–6.45: Chemical Class, Mechanism, Developing Organization, Development Status

Number	Chemical cpd./class	Target	Organization	Dev. status
6.1a	Methylene blue (MB)	Aβ, tau, huntingtin, TDP-43, prion	TauRX	Ph II
6.1b–d	Azure A–C	Aβ, tau, huntingtin, TDP-43, prion	TauRX	PE
6.3a	LMTX™	Tau, TDP-43, α-synuclein	TauRX	Ph III
6.4	Scyllo-inositol, ELND005, AZD-103	Aβ	Transition Therapeutics— Elan	Ph II
6.5	Alzhemed™, Vivimind™, tramiprosate	Aβ, tau	Neurochem	Ph III; Nu
6.6a,b	Clioquinol (PBT-1), PBT-2	Aβ, tau	Prana	Ph II
6.7	Exebryl-1	Aβ, tau	ProteoTech	Ph I
6.8a,b	EGCG, sunphenon™	Aβ, tau	Taiyo	Ph II, Nu
6.9	GSPE, Meganatural™	Aβ, tau	Polyphenolics	Ph II, Nu
6.10	Curcumin	Aβ, tau	John Douglas French Foundation	Ph II
6.11	Curcumin–Au nanoparticles	Aβ, tau	Natl. Brain Res. Centre, Manesar, India	DD
6.12	Curcumin–phospholipid conjugates	Aβ, tau	Pitiè Salpetriere, Paris	DD
6.13	Amine-containing curcumin analogues	Aβ, tau	Sun-Yat Sen Univ., China	DD
6.14	Sugar–curcumin conjugates	Aβ, tau	University of New York	LO
6.15	Curcumin-derived heterocycles	Aβ, tau	Max Planck, Hamburg	DD
6.16	Oleuropein	Aβ, tau	University of Florence, Italy	LO, Nu
6.17	Myricetin	Aβ, tau, α-synuclein	RIKEN	LO
6.18a,b	Baicalin (6.18a), baicalein (6.18b)	Aβ, α-synuclein	Tongji University, China	LO, Nu
6.19	Tannic acid	Aβ, tau	Tongji University, China	DD
6.20	Exifon	Aβ, tau	Tokyo University	DD

(Continued)

TABLE 6.1   Compounds 6.1a–6.45: Chemical Class, Mechanism, Developing Organization, Development Status *(cont.)*

Number	Chemical cpd./class	Target	Organization	Dev. status
6.21	Emodin	Aβ, tau	Max Planck, Hamburg	DD
6.22a–c	Rheins	Aβ, tau	University of Barcelona	LO
6.23	Tolcapone	Aβ, tau, α-synuclein	Swiss Inst. Technology, Lausanne, CH	DD
6.24	Entacapone	Aβ, tau, α-synuclein	Swiss Inst. Technology, Lausanne, CH	DD
6.25	Memoquin	Aβ	Univ. of Bologna, Italy	PE
6.26	Tetrahydroperforin	Aβ, tau	Univ. Santiago, Chile	LO
6.27a,b	Hemin (6.27a), hematin (6.27b)	Aβ, tau, α-synuclein	Tokyo Institute of Psychiatry	DD
6.28	Phthalocyanine tetrasulfonate	Aβ, tau, α-synuclein, prion	Max Planck Inst., Goettingen and Bonn	DD
6.29a,b	Thioflavin S	Aβ, tau	Cancer Research, UK	LO
6.30a	Thioflavin T	Aβ, tau	University of Pittsburgh	PE
6.30b	Pittsburgh compound B	Aβ, tau	University of Pittsburgh	PE
6.31	Benzothiazolium cyanines	Aβ, tau	Ohio State University	PE
6.32	Multivalent benzothiazole–benzothiazolium cyanines	Aβ, tau	Ohio State University	DD
6.33	BTA-EG$_4$	Aβ, tau	Georgetown University, Washington	LO
6.34	Benzothiazole–metal chelating hybrids	Aβ	Washington Univ., Missouri	DD
6.35	T-284	α-synuclein	Natl. Acad. Sci., Ukraine	DD
6.36	PBB-5	Tau	Mol. Imaging Center, Chiba, Japan	LO
6.37a–c	Rhodanins	Tau	Max Planck, Hamburg	LO
6.38	Thiohydantoins	Tau	Kyoto University	LO
6.39	Benzothiazolyl hydrazides	Tau	Max Planck, Hamburg	LO
6.40a,b	Pyrogalloyl phenylhydrazides	Aβ, tau	Max Planck, Hamburg	DD

TABLE 6.1   Compounds 6.1a–6.45: Chemical Class, Mechanism, Developing Organization, Development Status (*cont.*)

Number	Chemical cpd./class	Target	Organization	Dev. status
6.41	N-Phenylamines	Tau	Max Planck, Hamburg	DD
6.42	Carvedilol	Aβ	Mount Sinai School of Medicine	Ph IV
6.43	2,3-(Difuran-2-yl) quinoxalines	Tau	University of Pennsylvania	DD
6.44a–c	Aminothienopyridazines (ATPZs)	Tau	University of Pennsylvania	PE
6.45	Trehalose	Tau	Max Planck, Hamburg	LO

Not progressed, NP; early discovery, DD; lead optimization, LO; preclinical evaluation, PE; clinical Phase I–II–III, Ph I–Ph III; marketed, MKTD.

mentioned; the public or private laboratory that develops the compound is listed; and the development status—according to publicly available information—is provided.

References

1. Kelly, J. W. The alternative conformations of amyloidogenic proteins and their multi-step assembly pathways. *Curr. Opin. Struct. Biol.* **1998**, *8*, 101–106.
2. Chiti, F.; Dobson, C. M. Protein misfolding, functional amyloid, and human disease. *Annu. Rev. Biochem.* **2006**, *75*, 333–366.
3. Sunde, M.; Blake, C. The structure of amyloid fibrils by electron microscopy and X-ray diffraction. *Adv. Protein Chem.* **1997**, *50*, 123–159.
4. Dobson, C. M. Protein misfolding, evolution and disease. *Trends Biochem. Sci.* **1999**, *24*, 329–332.
5. Sarkar, N.; Dubey, V. K. Exploring critical determinants of protein amyloidogenesis: a review. *J. Pept. Sci.* **2013**, *19*, 529–536.
6. Sipe, J. D.; Benson, M. D.; Buxbaum, J. N.; Ikeda, S.; Merlini, G.; Saraiva, M. J., et al. Amyloid fibril protein nomenclature: 2012 recommendations from the Nomenclature Committee of the International Society of Amyloidosis. *Amyloid* **2012**, *19*, 167–170.
7. Berg, J.M.; Tymoczko, J.L.; Stryer, L. Actin is a polar, self-assembling, dynamic polymer. In *Biochemistry*, 5th edition. Editor: W. H. Freeman, New York, **2002**, 958–960.
8. Morris, A. M.; Watzky, M. A.; Finke, R. G. Protein aggregation kinetics, mechanism, and curve-fitting: a review of the literature. *Biochim. Biophys. Acta* **1794**, *2009*, 375–397.
9. McClellan, A. J.; Tam, S.; Kaganovich, D.; Frydman, J. Protein quality control: chaperones culling corrupt conformations. *Nature Cell Biol.* **2005**, *7*, 736–741.
10. Kaganovich, D.; Kopito, R.; Frydman, J. Misfolded proteins partition between two distinct quality control compartments. *Nature* **2008**, *454*, 1088–1095.
11. Turner, B. J.; Atkin, J. D.; Farg, M. A.; Zang, D. W.; Rembach, A.; Lopes, E. C., et al. Impaired extracellular secretion of mutant superoxide dismutase 1 associates with neurotoxicity in familial amyotrophic lateral sclerosis. *J. Neurosci.* **2005**, *25*, 108–117.

12. Polling, S.; Mok, Y.-F.; Ramdzan, Y. M.; Turner, B. J.; Yerbury, J. J.; Hill, A. F.; Hatters, D. M. Misfolded polyglutamine, polyalanine, and superoxide dismutase 1 aggregate via distinct pathways in the cell. *J. Biol. Chem.* **2014**, *289*, 6669–6680.

13. Hardy, J.; Selkoe, D. J. The amyloid hypothesis of Alzheimer's disease: progress and problems on the road to therapeutics. *Science* **2002**, *297*, 353–356.

14. Prusiner, S. B.; Scott, M. R.; DeArmond, S. J.; Cohen, F. E. Prion protein biology. *Cell* **1998**, *93*, 337–348.

15. Weisberg, S. J.; Lyakhovetsky, R.; Werdiger, A. C.; Gitler, A. D.; Soen, Y.; Kaganovich, D. Compartmentalization of superoxide dismutase 1 (SOD1G93A) aggregates determines their toxicity. *Proc. Natl. Acad. Sci. U.S.A.* **2012**, *109*, 15811–15816.

16. Uversky, V. N. Mysterious oligomerization of the amyloidogenic proteins. *FEBS J.* **2010**, *277*, 2940–2953.

17. Relini, A.; Marano, N.; Gliozzi, A. Misfolding of amyloidogenic proteins and their interactions with membranes. *Biomolecules* **2014**, *4*, 20–55.

18. Rezaei-Ghaleh, N.; Blackledge, M.; Zweckstetter, M. Intrinsically disordered proteins: from sequence and conformational properties toward drug discovery. *ChemBioChem* **2012**, *13*, 230–250.

19. Rosenman, D. J.; Connors, C. R.; Chen, W.; Wang, C.; García, A. E. Aβ monomers transiently sample oligomer and fibril-like configurations: ensemble characterization using a combined MD/NMR approach. *J. Mol. Biol.* **2013**, *425*, 3338–3359.

20. Uversky, V. N. Flexible nets of malleable guardians: Intrinsically disordered chaperones in neurodegenerative diseases. *Chem. Rev.* **2011**, *111*, 1134–1166.

21. Elbaum-Garfinkle, S.; Rhoades, E. Identification of an aggregation-prone structure of tau. *J. Am. Chem. Soc.* **2012**, *134*, 16607–16613.

22. Paz, M. L.; Serrano, L. Sequence determinants of amyloid fibril formation. *Proc. Natl. Acad. Sci. U.S.A.* **2004**, *101*, 87–92.

23. Tenidis, K.; Waldner, M.; Bernhagen, J.; Fischle, W.; Bergmann, M.; Weber, M., et al. Identification of a penta and hexapeptide of islet amyloid polypeptide (IAPP) with amyloidogenic and cytotoxic properties. *J. Mol. Biol.* **2000**, *295*, 1055–1071.

24. Pastor, M. T.; Esteras-Chopo, A.; Serrano, L. Hacking the code of amyloid formation: the amyloid stretch hypothesis. *Prion* **2007**, *1*, 9–14.

25. Gamblin, T. C. Potential structure/function relationships of predicted secondary structural elements of tau. *Biochim. Biophys. Acta* **1739**, *2005*, 140–149.

26. Bolognesi, B.; Cohen, S. I. A.; Aran Terol, P.; Esbjorner, E. K.; Giorgetti, S.; Mossuto, M. F., et al. Single point mutations induce a switch in the molecular mechanism of the aggregation of the Alzheimer's disease associated Aβ$_{42}$ peptide. *ACS Chem. Biol.* **2014**, *9*, 378–382.

27. Li, J.; Uversky, V. N.; Fink, A. L. Effect of familial Parkinson's disease point mutations A30P and A53T on the structural properties, aggregation, and fibrillation of human α-synuclein. *Biochemistry* **2001**, *40*, 11604–11613.

28. Terwel, D.; Lasrado, R.; Snauwaert, J.; Vandeweert, E.; Van Haesendonck, C.; Borghgraef, P.; Van Leuven, F. Changed conformation of mutant tau-P301L underlies the moribund tauopathy, absent in progressive, nonlethal axonopathy of tau-4R/2N transgenic mice. *J. Biol. Chem.* **2005**, *280*, 3963–3973.

29. Almeida, B.; Fernandes, S.; Abreu, I. A.; Macedo-Ribeiro, S. Trinucleotide repeats: a structural perspective. *Front. Neurol.* **2013**, *4*, 76.

30. Scherzinger, E.; Lurz, R.; Turmaine, M.; Mangiarini, L.; Hollenbach, B.; Hasenbank, R., et al. Huntingtin-encoded polyglutamine expansions form amyloid-like protein aggregates in vitro and in vivo. *Cell* **1997**, *90*, 549–558.

31. Albrecht, A.; Mundlos, S. The other trinucleotide repeat: polyalanine expansion disorders. *Curr. Opin. Genet. Devel.* **2005**, *15*, 285–293.

32. Vassar, R.; Bennett, B. D.; Babu-Khan, S.; Kahn, S.; Mendiaz, E. A.; Denis, P., et al. β-Secretase cleavage of Alzheimer's amyloid precursor protein by the transmembrane aspartic protease BACE. *Science* **1999**, *286*, 735–741.

33. Wang, J.-Z.; Xia, Y.-Y.; Grundke-Iqbal, I.; Iqbal, K. Abnormal hyperphosphorylation of tau: sites, regulation, and molecular mechanism of neurofibrillary degeneration. *J. Alzheimer's Dis.* **2013**, *33*, S123–S139.

34. Dear, D. V.; Young, D. S.; Kazlauskaite, J.; Meersman, F.; Oxley, D.; Webster, J., et al. Effects of post-translational modifications on prion protein aggregation and the propagation of scrapie-like characteristics in vitro. *Biochim. Biophys. Acta* **1744**, *2007*, 792–802.

35. Arnaudov, L.; de Vries, R. Thermally induced fibrillar aggregation of hen egg white lysozyme. *Biophys. J.* **2005**, *88*, 515–526.

36. Babenko, V.; Piejko, M.; Wojcik, S.; Mak, P.; Dzwolak, W. Vortex-induced amyloid super-structures of insulin and its component A and B chains. *Langmuir* **2013**, *29*, 5271–5278.

37. Ow, S.; Dunstan, D. E. The effect of concentration, temperature and stirring on hen egg white lysozyme amyloid formation. *Soft Matter* **2013**, *9*, 9692–9701.

38. Nielsen, L.; Khurana, R.; Coats, A.; Frokjaer, S.; Brange, J.; Vyas, S., et al. Effect of environmental factors on the kinetics of insulin fibril formation: elucidation of the molecular mechanism. *Biochemistry* **2001**, *40*, 6036–6046.

39. Leal, S. S.; Cardoso, I.; Valentine, J. S.; Gomes, C. M. Calcium ions promote superoxide dismutase 1 (SOD1) aggregation into non-fibrillar amyloid: a link to toxic effects of calcium overload in amyotrophic lateral sclerosis (ALS)? *J. Biol. Chem.* **2013**, *288*, 25219–25228.

40. Rubin, J.; Khosravi, H.; Bruce, K. L.; Lydon, M. E.; Behrens, S. H.; Chernoff, Y. O.; Bommarius, A. S. Ion-specific effects on prion nucleation and strain formation. *J. Biol. Chem.* **2013**, *288*, 30300–30308.

41. Munishkina, L.; Phelan, C.; Uversky, V.; Fink, A. Conformational behavior and aggregation of alpha-synuclein in organic solvents: modeling the effects of membranes. *Biochemistry* **2003**, *42*, 2720–2730.

42. Kenche, V. B.; Barnham, K. J. Alzheimer's disease & metals: therapeutic opportunities. *Br. J. Pharmacol.* **2011**, *163*, 211–219.

43. Shikama, Y.; Kitazawa, J.; Yagihashi, N.; Uehara, O.; Murata, Y.; Yajima, N., et al. Localized amyloidosis at the site of repeated insulin injection in a diabetic patient. *Intern. Med.* **2010**, *49*, 397–401.

44. Verdone, G.; Corazza, A.; Viglino, P.; Pettirossi, F.; Giorgetti, S.; Mangione, P., et al. The solution structure of human beta2-microglobulin reveals the prodromes of its amyloid transition. *Protein Sci.* **2002**, *11*, 487–499.

45. Uversky, V. N.; Li, J.; Fink, A. L. Pesticides directly accelerate the rate of alpha-synuclein fibril formation: a possible factor in Parkinson's disease. *FEBS Lett.* **2001**, *500*, 105–108.

46. Shacter, E. Quantification and significance of protein oxidation in biological samples. *Drug Metab. Rev.* **2000**, *32*, 307–326.

47. David, D. C. Aging and the aggregating proteome. *Front. Genet.* **2012**, *3*, 347.

48. David, D. C.; Ollikainen, N.; Trinidad, J. C.; Cary, M. P.; Burlingame, A. L.; Kenyon, C. Widespread protein aggregation as an inherent part of aging in C. elegans. *PLoS Biol.* **2010**, e1000450.

49. Keller, J. N.; Hanni, K. B.; Markesbery, W. R. Possible involvement of proteasome inhibition in aging: implications for oxidative stress. *Mech. Ageing Dev.* **2000**, *113*, 61–70.

50. Cuervo, A. M.; Dice, J. F. Age-related decline in chaperone-mediated autophagy. *J. Biol. Chem.* **2000**, *275*, 31505–31513.

51. Brown, M. K.; Naidoo, N. The endoplasmic reticulum stress response in aging and age-related diseases. *Front. Physiol.* **2012**, *3*, 263.

52. Gidalevitz, T.; Kikis, E. A.; Morimoto, R. I. A cellular perspective on conformational disease: the role of genetic background and proteostasis networks. *Curr. Opin. Struct. Biol.* **2010**, *20*, 23–32.

53. Squier, T. C. Oxidative stress and protein aggregation during biological aging. *Exp. Gerontol.* **2001**, *36*, 1539–1550.

54. Ciryam, P.; Tartaglia, G. G.; Morimoto, R. I.; Dobson, C. M.; Vendruscolo, M. Widespread aggregation and neurodegenerative diseases are associated with supersaturated proteins. *Cell Reports* **2013**, *5*, 781–790.
55. Pekar, A. H.; Frank, B. H. Conformation of proinsulin: a comparison of insulin and proinsulin self-association at neutral pH. *Biochemistry* **1972**, *11*, 4013–4016.
56. Alford, J. R.; Kendrick, B. S.; Carpenter, J. F.; Randolph, T. W. High concentration formulations of recombinant human interleukin-1 receptor antagonist: II. Aggregation kinetics. *J. Pharm. Sci.* **2007**, *97*, 3005–3021.
57. Invernizzi, G.; Papaleo, E.; Sabate, R.; Ventura, S. Protein aggregation: mechanisms and functional consequences. *Int. J. Biochem. Cell Biol.* **2012**, *44*, 1541–1554.
58. Roberts, C. J. Non-native protein aggregation kinetics. *Biotech. Bioeng.* **2007**, *5*, 927–938.
59. Gillam, J. E.; MacPhee, C. E. Modelling amyloid fibril formation kinetics: mechanisms of nucleation and growth. *J. Phys.: Condens. Matter* **2013**, *25*, 373101.
60. Paleologou, K. E.; Oueslati, A.; Shakked, G.; Rospigliosi, C. C.; Kim, H.-Y.; Lamberto, G. R., et al. Phosphorylation at S87 is enhanced in synucleinopathies, inhibits a-synuclein oligomerization, and influences synuclein-membrane interactions. *J. Neurosci.* **2010**, *30*, 3184–3198.
61. Guo, W.; Chen, Y.; Zhou, X.; Kar, A.; Ray, P.; Chen, X., et al. An ALS-associated mutation affecting TDP-43 enhances protein aggregation, fibril formation and neurotoxicity. *Nat. Struct. Mol. Biol.* **2011**, *18*, 822–830.
62. Ferrone, F. Analysis of protein aggregation kinetics. *Methods Enzymol.* **1999**, *309*, 256–274.
63. Bhak, G.; Choe, Y. J.; Paik, S. R. Mechanism of amyloidogenesis: nucleation-dependent fibrillation versus double-concerted fibrillation. *BMB Reports* **2009**, *42*, 541–551.
64. Aggeli, A.; Nyrkova, I. A.; Bell, M.; Harding, R.; Carrick, L.; McLeish, T. C., et al. Hierarchical self-assembly of chiral rod-like molecules as a model for peptide beta-sheet tapes, ribbons, fibrils, and fibers. *Proc. Natl. Acad. Sci. U.S.A.* **2001**, *98*, 11857–11862.
65. Kodali, R.; Wetzel, R. Polymorphism in the intermediates and products of amyloid assembly. *Curr. Opin. Struct. Biol.* **2007**, *17*, 48–57.
66. Sabate, R.; Gallardo, M.; Estelrich, J. An autocatalytic reaction as a model for the kinetics of the aggregation of beta-amyloid. *Biopolymers* **2003**, *71*, 190–195.
67. Cohen, S. I. A.; Vendruscolo, M.; Welland, M. E.; Dobson, C. M.; Terentjev, E. M.; Knowles, T. P. J. Nucleated polymerization with secondary pathways. I. Time evolution of the principal moments. *J. Chem. Phys.* **2011**, *135*, 065105.
68. Tanaka, M.; Chien, P.; Yonekura, K.; Weissman, J. S. Mechanism of cross-species prion transmission: an infectious conformation compatible with two highly divergent yeast prion proteins. *Cell* **2005**, *121*, 49–62.
69. Jucker, M.; Walker, L. C. Self-propagation of pathogenic protein aggregates in neurodegenerative diseases. *Nature* **2013**, *501*, 45–51.
70. Raj, A.; Kuceyeski, A.; Weiner, M. A network diffusion model of disease progression in dementia. *Neuron* **2012**, *73*, 1204–1215.
71. Zhou, J.; Gennatas, E. D.; Kramer, J. H.; Miller, B. L.; Seeley, W. W. Predicting regional neurodegeneration from the healthy brain functional connectome. *Neuron* **2012**, *73*, 1216–1227.
72. Andersen, C. B.; Yagi, H.; Manno, M.; Martorana, V.; Ban, T.; Christiansen, G., et al. Branching in amyloid fibril growth. *Biophys. J.* **2009**, *96*, 1529–1536.
73. Gorbenko, G.; Trusova, V. Protein aggregation in a membrane environment. *Adv. Prot. Chem. Struct. Biol.* **2011**, *84*, 113–142.
74. Relini, A.; Canale, C.; de Stefano, S.; Rolandi, R.; Giorgetti, S.; Stoppini, M., et al. Collagen plays an active role in the aggregation of β2-microglobulin under physiopathological conditions of dialysis-related amyloidosis. *J. Biol. Chem.* **2006**, *281*, 16521–16529.
75. Aguzzi, A.; Calella, A. M. Prions: protein aggregation and infectious diseases. *Physiol. Rev.* **2009**, *89*, 1105–1152.

76. Mahal, S. P.; Baker, C. A.; Demczyk, C. A.; Smith, E. W.; Julius, C.; Weissmann, C. Prion strain discrimination in cell culture: the cell panel assay. *Proc. Natl. Acad. Sci. U.S.A.* **2007**, *104*, 20908–20913.

77. Pedersen, J. T.; Heegaard, N. H. H. Analysis of protein aggregation in neurodegenerative disease. *Anal. Chem.* **2013**, *85*, 4215–4227.

78. Miranda, E.; MacLeod, I.; Davies, M. J.; Perez, J.; Romisch, K.; Crowther, D. C.; Lomas, D. A. The intracellular accumulation of polymeric neuroserpin explains the severity of the dementia FENIB. *Hum. Mol. Genet.* **2008**, *17*, 1527–1539.

79. McDonald, J. M.; Savva, G. M.; Brayne, C.; Welzel, A. T.; Forster, G.; Shankar, G. M., et al. The presence of sodium dodecyl sulphate-stable Abeta dimers is strongly associated with Alzheimer-type dementia. *Brain* **2010**, *133*, 1328–1341.

80. Jan, A.; Gokce, O.; Luthi-Carter, R.; Lashuel, H. A. The ratio of monomeric to aggregated forms of Abeta40 and Abeta42 is an important determinant of amyloid-beta aggregation, fibrillogenesis, and toxicity. *J. Biol. Chem.* **2008**, *283*, 28176–28189.

81. Gabrielson, J. P.; Arthur, K. K.; Stoner, M. R.; Winn, B. C.; Kendrick, B. S.; Razinkov, V., et al. Precision of protein aggregation measurements by sedimentation velocity analytical ultracentrifugation in biopharmaceutical applications. *Anal. Biochem.* **2010**, *396*, 231–241.

82. Murphy, R. M. Peptide aggregation in neurodegenerative disease. *Annu. Rev. Biomed. Eng.* **2002**, *4*, 155–174.

83. Heise, H.; Hoyer, W.; Becker, S.; Andronesi, O. C.; Riedel, D.; Baldus, M. Molecular-level secondary structure, polymorphism, and dynamics of full-length α-synuclein fibrils studied by solid-state NMR. *Proc. Natl. Acad. Sci. U.S.A.* **2005**, *102*, 15871–15876.

84. Flyvbjerg, H.; Jobs, E.; Leibler, S. Kinetics of self-assembling microtubules: an "inverse problem" in biochemistry. *Proc. Natl. Acad. Sci. U.S.A.* **1996**, *93*, 5975–5979.

85. Lomakin, A.; Chung, D. S.; Benedek, G. B.; Kirschner, D. A.; Teplow, D. B. On the nucleation and growth of amyloid beta-protein fibrils: detection of nuclei and quantitation of rate constants. *Proc. Natl. Acad. Sci. U.S.A.* **1996**, *93*, 1125–1129.

86. Bernstein, S. L.; Dupuis, N. F.; Lazo, N. D.; Wyttenbach, T.; Condron, M. M.; Bitan, G., et al. Amyloid-β protein oligomerization and the importance of tetramers and dodecamers in the aetiology of Alzheimer's disease. *Nat. Chem.* **2009**, *1*, 326–331.

87. Heck, A. J.; van den Heuvel, R. H. Investigation of intact protein complexes by mass spectrometry. *Mass Spectrom. Rev.* **2004**, *23*, 368–389.

88. Robinson, J. L.; Geser, F.; Stieber, A.; Umoh, M.; Kwong, L. K.; Van Deerlin, V. M., et al. TDP-43 skeins in amyotrophic lateral sclerosis show properties of amyloids. *Acta Neuropathol.* **2013**, *125*, 121–131.

89. Serem, W. K.; Bett, C. K.; Ngunjiri, J. N.; Garno, J. C. Studies of the growth, evolution, and self-aggregation of β-amyloid fibrils using tapping-mode atomic force microscopy. *Microsc. Res. Tech.* **2011**, *74*, 699–708.

90. Frid, P.; Anisimov, S. V.; Popovic, N. Congo red and protein aggregation in neurodegenerative diseases. *Brain Res. Rev.* **2007**, *53*, 135–160.

91. Groenning, M. J. Binding mode of Thioflavin T and other molecular probes in the context of amyloid fibrils—current status. *Chem. Biol.* **2010**, *3*, 1–18.

92. Ohrfelt, A.; Zetterberg, H.; Andersson, K.; Persson, R.; Secic, D.; Brinkmalm, G., et al. Identification of novel α-synuclein isoforms in human brain tissue by using an online nanoLC-ESI-FTICR-MS method. *Neurochem. Res.* **2011**, *36*, 2029–2042.

93. Brody, D. L.; Holtzman, D. M. Active and passive immunotherapy for neurodegenerative disorders. *Annu. Rev. Neurosci.* **2008**, *31*, 175–193.

94. Ross, C. A.; Poirier, M. A. Protein aggregation and neurodegenerative disease. *Nat. Med.* **2004**, *10*, S10–S17.

95. Seneci, P. Chemical modulators of protein misfolding and neurodegenerative disease. Elsevier, accepted for publication, **2015**.

96. Lotz, G. P.; Legleiter, J. The role of amyloidogenic protein oligomerization in neurodegenerative disease. *J. Mol. Med.* **2013**, *91*, 653–664.

97. Lorenzo, A.; Yankner, B. A. Beta-amyloid neurotoxicity requires fibril formation and is inhibited by Congo Red. *Proc. Natl. Acad. Sci. U.S.A.* **1994**, *91*, 12243–12247.

98. Novitskaya, V.; Bocharova, O. V.; Bronstein, I.; Baskakov, I. V. Amyloid fibrils of mammalian prion protein are highly toxic to cultured cells and primary neurons. *J. Biol. Chem.* **2006**, *281*, 13828–13836.

99. Alonso, A. C.; Li, B.; Grundke-Iqbal, I.; Iqbal, K. Polymerization of hyperphosphorylated tau into filaments eliminates its inhibitory activity. *Proc. Natl. Acad. Sci. U.S.A.* **2006**, *103*, 8864–8869.

100. Saudou, F.; Finkbeiner, S.; Devys, D.; Greenberg, M. E. Huntingtin acts in the nucleus to induce apoptosis but death does not correlate with the formation of intranuclear inclusions. *Cell* **1998**, *95*, 55–66.

101. Katzman, R.; Terry, R.; DeTeresa, R.; Brown, T.; Davies, P.; Fuld, P., et al. Clinical, pathological, and neurochemical changes in dementia: a subgroup with preserved mental status and numerous neocortical plaques. *Ann. Neurol.* **1988**, *23*, 138–144.

102. Braak, H.; Braak, E. Morphological changes in the human cerebral cortex in dementia. *J. Hirnforsch.* **1991**, *32*, 277–282.

103. Tompkins, M. M.; Hill, W. D. Contribution of somal Lewy bodies to neuronal death. *Brain Res.* **1997**, *775*, 24–29.

104. Kuemmerle, S.; Gutekunst, C. A.; Klein, A. M.; Li, X. J.; Li, S. H.; Beal, M. F., et al. Huntingtin aggregates may not predict neuronal death in Huntington's disease. *Ann. Neurol.* **1999**, *46*, 842–849.

105. Ferreira, S. T.; Vieira, M. N.; De Felice, F. G. Soluble protein oligomers as emerging toxins in Alzheimer's and other amyloid diseases. *IUBMB Life* **2007**, *59*, 332–345.

106. Chimon, S.; Shaibat, M. A.; Jones, C. R.; Calero, D. C.; Aizezi, B.; Ishii, Y. Evidence of fibril-like beta-sheet structures in a neurotoxic amyloid intermediate of Alzheimer's beta-amyloid. *Nat. Struct. Mol. Biol.* **2007**, *14*, 1157–1164.

107. Bitan, G.; Kirkitadze, M. D.; Lomakin, A.; Vollers, S. S.; Benedek, G. B.; Teplow, D. B. Amyloid beta-protein (Abeta) assembly: Abeta 40 and Abeta 42 oligomerize through distinct pathways. *Proc. Natl. Acad. Sci. U.S.A.* **2003**, *100*, 330–335.

108. Hoshi, M.; Sato, M.; Matsumoto, S.; Noguchi, A.; Yasutake, K.; Yoshida, N.; Sato, K. Spherical aggregates of beta-amyloid (amylospheroid) show high neurotoxicity and activate tau protein kinase I/glycogen synthase kinase-3beta. *Proc. Natl. Acad. Sci. U.S.A.* **2003**, *100*, 6370–6375.

109. Lesne, S.; Koh, M. T.; Kotilinek, L.; Kayed, R.; Glabe, C. G.; Yang, A., et al. A specific amyloid-beta protein assembly in the brain impairs memory. *Nature* **2006**, *440*, 352–357.

110. Ono, K.; Condron, M. M.; Teplow, D. B. Structure neurotoxicity relationships of amyloid beta-protein oligomers. *Proc. Natl. Acad. Sci. U.S.A.* **2009**, *106*, 14745–14750.

111. Tanaka, M.; Collins, S. R.; Toyama, B. H.; Weissman, J. S. The physical basis of how prion conformations determine strain phenotypes. *Nature* **2006**, *442*, 585–589.

112. Ding, T. T.; Lee, S. J.; Rochet, J. C.; Lansbury, P. T., Jr. Annular alpha-synuclein protofibrils are produced when spherical protofibrils are incubated in solution or bound to brain-derived membranes. *Biochemistry* **2002**, *41*, 10209–10217.

113. Kayed, R.; Head, E.; Thompson, J. L.; McIntire, T. M.; Milton, S. C.; Cotman, C. W.; Glabe, C. Common structure of soluble amyloid oligomers implies common mechanism of pathogenesis. *Science* **2003**, *300*, 486–489.

114. Bolognesi, B.; Kumita, J. R.; Barros, T. P.; Esbjorner, E. K.; Luheshi, L. M.; Crowther, D. C., et al. ANS binding reveals common features of cytotoxic amyloid species. *ACS Chem. Biol.* **2010**, *5*, 735–740.

115. Lotz, G. P.; Legleiter, J.; Aron, R.; Mitchell, E. J.; Huang, S. Y.; Ng, C. P., et al. Hsp70 and Hsp40 functionally interact with soluble mutant huntingtin oligomers in a classic ATP-dependent reaction cycle. *J. Biol. Chem.* **2010**, *285*, 38183–38193.

116. Martinez-Vicente, M.; Talloczy, Z.; Wong, E.; Tang, G.; Koga, H.; Kaushik, S., et al. Cargo recognition failure is responsible for inefficient autophagy in Huntington's disease. *Nat. Neurosci.* **2010**, *13*, 567–574.

117. Bennett, E. J.; Shaler, T. A.; Woodman, B.; Ryu, K. Y.; Zaitseva, T. S.; Becker, C. H., et al. Global changes to the ubiquitin system in Huntington's disease. *Nature* **2007**, *448*, 704–711.

118. Hands, S. L.; Wyttenbach, A. Neurotoxic protein oligomerisation associated with polyglutamine diseases. *Acta Neuropathol.* **2010**, *120*, 419–437.

119. Olzscha, H.; Schermann, S. M.; Woerner, A. C.; Pinkert, S.; Hecht, M. H.; Tartaglia, G. G., et al. Amyloid-like aggregates sequester numerous metastable proteins with essential cellular functions. *Cell* **2011**, *144*, 67–78.

120. Castellani, R. J.; Lee, H.-G.; Siedlak, S. L.; Nunomura, A.; Hayashi, T.; Nakamura, M., et al. Reexamining Alzheimer's disease: evidence for a protective role for amyloid-β protein precursor and amyloid-β. *J. Alzheimer's Dis.* **2009**, *18*, 447–452.

121. Bucciantini, M.; Rigacci, S.; Stefani, M. Amyloid aggregation: role of biological membranes and the aggregate–membrane system. *J. Phys. Chem. Lett.* **2014**, *5*, 517–527.

122. Burke, K. A.; Yates, E. A.; Legleiter, J. Biophysical insights into how surfaces, including lipid membranes, modulate protein aggregation related to neurodegeneration. *Front. Neurol.* **2013**, *4*, 17.

123. Butterfield, S. M.; Lashuel, H. A. Amyloidogenic protein–membrane interactions: mechanistic insight from model systems. *Angew. Chem. Int. Ed.* **2010**, *49*, 5628–5654.

124. Knight, J. D.; Hebda, J. A.; Miranker, A. D. Conserved and cooperative assembly of membrane-bound α-helical states of islet amyloid polypeptide. *Biochemistry* **2006**, *45*, 9496–9508.

125. Zhao, H.; Tuominen, E. K. J.; Kinnunen, P. K. J. Formation of amyloid fibers triggered by phosphatidylserine-containing membranes. *Biochemistry* **2004**, *43*, 10302–10307.

126. Jia, Y.; Qian, Z.; Zhang, Y.; Wei, G. Adsorption and orientation of human islet amyloid polypeptide (hIAPP) monomer at anionic lipid bilayers: implications for membrane-mediated aggregation. *Int. J. Mol. Sci.* **2013**, *14*, 6241–6258.

127. Lee, C.; Sun, Y.; Huang, H. W. Membrane-mediated peptide conformation change from alpha-monomers to β-aggregates. *Biophys. J.* **2010**, *98*, 2236–2245.

128. Korade, Z.; Kenworthy, A. K. Lipid rafts, cholesterol, and the brain. *Neuropharmacology* **2008**, *55*, 1265–1273.

129. Pike, L. J. The challenge of lipid rafts. *J. Lipid Res.* **2008**, *50*, S323–S328.

130. Sonnino, S.; Aureli, M.; Grassi, S.; Mauri, L.; Prioni, S.; Prinetti, A. Lipid rafts in neurodegeneration and neuroprotection. *Mol. Neurobiol.* **2013**, *48*, doi: 10.1007/s12035-013-8614-4.

131. Marin, R.; Rojo, J. A.; Fabelo, N.; Fernandez, C. E.; Diaz, M. Lipid raft disarrangement as a result of neuropathological progresses: a novel strategy for early diagnosis? *Neuroscience* **2013**, *245*, 26–39.

132. Prinetti, A.; Chigorno, V.; Tettamanti, G.; Sonnino, S. Sphingolipid-enriched membrane domains from rat cerebellar granule cells differentiated in culture. A compositional study. *J. Biol. Chem.* **2000**, *275*, 11658–11665.

133. Pfrieger, F. W.; Ungerer, N. Cholesterol metabolism in neurons and astrocytes. *Prog. Lipid Res.* **2011**, *50*, 357–371.

134. Wu, G.; Lu, Z. H.; Kulkarni, N.; Amin, R.; Ledeen, R. W. Mice lacking major brain gangliosides develop parkinsonism. *Neurochem. Res.* **2011**, *36*, 1706–1714.

135. Kracun, I.; Rosner, H.; Drnovsek, V.; Heffer-Lauc, M.; Cosovic, C.; Lauc, G. Human brain gangliosides in development, aging and disease. *Int. J. Dev. Biol.* **1991**, *35*, 289–295.

136. Mori, K.; Mahmood, M. I.; Neya, S.; Matsuzaki, K.; Hoshino, T. Formation of GM1 ganglioside clusters on the lipid membrane containing sphingomyeline and cholesterol. *J. Phys. Chem. B* **2012**, *116*, 5111–5121.

137. Sanghera, N.; Correia, B. E.; Correia, J. R.; Ludwig, C.; Agarwal, S.; Nakamura, H. K., et al. Deciphering the molecular details for the binding of the prion protein to main ganglioside GM1 of neuronal membranes. *Chem. Biol.* **2011**, *18*, 1422–1431.

138. Bucciantini, M.; Nosi, D.; Forzan, M.; Russo, E.; Calamai, M.; Pieri, L., et al. Toxic effects of amyloid fibrils on cell membranes: the importance of ganglioside GM1. *FASEB J.* **2012**, *26*, 818–831.

139. Evangelisti, E.; Cecchi, C.; Cascella, R.; Sgromo, C.; Becatti, M.; Dobson, C. M., et al. Membrane lipid composition and its physicochemical properties define cell vulnerability to aberrant protein oligomers. *J. Cell Sci.* **2012**, *125*, 2416–2427.

140. Matsuzaki, K.; Kato, K.; Yanagisawa, K. Aβ polymerization through interaction with membrane gangliosides. *Biochim. Biophys. Acta* **1801**, *2010*, 868–877.

141. Molander-Melin, M.; Blennow, K.; Bogdanovic, N.; Dellheden, B.; Mansson, J. E.; Fredman, P. Structural membrane alterations in Alzheimer brains found to be associated with regional disease development; increased density of gangliosides GM1 and GM2 and loss of cholesterol in detergent-resistant membrane domains. *J. Neurochem.* **2005**, *92*, 171–182.

142. Okada, T.; Ikeda, K.; Wakabayashi, M.; Ogawa, M.; Matsuzaki, K. Formation of toxic Aβ(1-40) fibrils on GM1 ganglioside-containing membranes mimicking lipid rafts: polymorphisms in Aβ(1-40) fibrils. *J. Mol. Biol.* **2008**, *382*, 1066–1074.

143. Di Scala, C.; Yahi, N.; Lelièvre, C.; Garmy, N.; Chahinian, H.; Fantini, J. Biochemical identification of a linear cholesterol-binding domain within Alzheimer's β amyloid peptide. *ACS Chem. Neurosci.* **2013**, *4*, 509–517.

144. Fantini, J.; Yahi, N.; Garmy, N. Cholesterol accelerates the binding of Alzheimer's β-amyloid peptide to ganglioside GM1 through a universal hydrogen-bond-dependent sterol tuning of glycolipid conformation. *Front. Physiol.* **2013**, *4*, 120.

145. Lingwood, D.; Binnington, B.; Róg, T.; Vattulainen, I.; Grzybek, M.; Coskun, U., et al. Cholesterol modulates glycolipid conformation and receptor activity. *Nat. Chem. Biol.* **2011**, *7*, 260–262.

146. Leduc, V.; Jasmin-Belanger, S.; Poirier, J. APOE and cholesterol homeostasis in Alzheimer's disease. *Trends Mol. Med.* **2010**, *16*, 469–477.

147. Hooff, G. P.; Peters, I.; Wood, W. G.; Muller, W. E.; Eckert, G. P. Modulation of cholesterol, farnesylpyrophosphate, and geranylgeranylpyrophosphate in neuroblastoma SH-SY5Y-APP695 cells: impact on amyloid β-protein production. *Mol. Neurobiol.* **2010**, *41*, 341–350.

148. Kolsch, H.; Heun, R.; Jessen, F.; Popp, J.; Hentschel, F.; Maier, W.; Lutjohann, D. Alterations of cholesterol precursor levels in Alzheimer's disease. *Biochim. Biophys. Acta* **1801**, *2010*, 945–950.

149. Arispe, N.; Pollard, H. B.; Rojas, E. Giant multilevel cation channels formed by Alzheimer disease amyloid β-protein Aβ P-(1-40) in bilayer membranes. *Proc. Natl. Acad. Sci. U.S.A.* **1993**, *90*, 10573–10577.

150. Lashuel, H. A.; Lansbury, P. T., Jr. Are amyloid diseases caused by protein aggregates that mimic bacterial pore-forming toxins? *Q. Rev. Biophys.* **2006**, *39*, 167–201.

151. Pellistri, F.; Bucciantini, M.; Invernizzi, G.; Gatta, E.; Penco, A.; Frana, A. M., et al. Different ataxin-3 amyloid aggregates induce intracellular Ca2+ deregulation by different mechanisms in cerebellar granule cells. *BBA-Mol. Cell Res.* **1833**, *2013*, 3155–3165.

152. Malisauskas, M.; Zamotin, V.; Jass, J.; Noppe, W.; Dobson, C.; Morozova-Roche, L. Amyloid protofilaments from the calcium-binding protein equine lysozyme: formation of ring and linear structures depends on pH and metal ion concentration. *J. Mol. Biol.* **2003**, *330*, 879–890.

153. Kayed, R.; Pensalfini, A.; Margol, L.; Sokolov, Y.; Sarsoza, F.; Head, E., et al. Annular protofibrils are a structurally and functionally distinct type of amyloid oligomer. *J. Biol. Chem.* **2009**, *284*, 4230–4237.

154. Quist, A.; Doudevski, L.; Lin, H.; Azimova, R.; Ng, D.; Frangione, B., et al. Amyloid ion channels: a common structural link for protein-misfolding disease. *Proc. Natl. Acad. Sci. U.S.A.* **2005**, *102*, 10427–10432.
155. Lin, H.; Bhatia, R.; Lal, R. Amyloid beta protein forms ion channels: implications for Alzheimer's disease pathophysiology. *FASEB J.* **2001**, *15*, 2433–2444.
156. Bischofberger, M.; Gonzalez, M. R.; van der Goot, F. G. Membrane injury by pore-forming proteins. *Curr. Opin. Cell Biol.* **2009**, *21*, 589–595.
157. Prangkio, P.; Yusko, E. C.; Sept, D.; Yang, J.; Mayer, M. Multivariate analyses of amyloid-β oligomer populations indicate a connection between pore formation and cytotoxicity. *PLoS One* **2012**, *7*, e47261.
158. Jang, H.; Zheng, J.; Lal, R.; Nussinov, R. New structures help the modeling of toxic amyloid beta ion channels. *Trends Biochem. Sci.* **2008**, *33*, 91–100.
159. Volles, M. J.; Lansbury, P. T., Jr. Vesicle permeabilization by protofibrillar alpha-synuclein is sensitive to Parkinson's disease-linked mutations and occurs by a pore-like mechanism. *Biochemistry* **2002**, *41*, 4595–4602.
160. Pollard, H. B.; Rojas, E.; Arispe, N. A new hypothesis for the mechanism of amyloid toxicity, based on the calcium channel activity of amyloid beta protein (A beta P) in phospholipid bilayer membranes. *Ann. N. Y. Acad. Sci.* **1993**, *695*, 165–168.
161. Lashuel, H. A.; Hartley, D.; Petre, B. M.; Walz, T.; Lansbury, P. T., Jr. Amyloid pores from pathogenic mutations. *Nature* **2002**, *418*, 291.
162. Kourie, J. I.; Farrelly, P. V.; Henry, C. L. Channel activity of deamidated isoforms of prion protein fragment 106-126 in planar lipid bilayers. *J. Neurosci. Res.* **2001**, *66*, 214–220.
163. Kim, H.-Y.; Cho, M.-K.; Kumar, A.; Maier, E.; Siebenhaar, C.; Becker, S., et al. Structural properties of pore-forming oligomers of alpha-synuclein. *J. Am. Chem. Soc.* **2009**, *131*, 17482–17489.
164. Zakharov, S. D.; Hulleman, J. D.; Dutseva, E. A.; Antonenko, Y. N.; Rochet, J.-C.; Cramer, W. A. Helical alpha-synuclein forms highly conductive ion channels. *Biochemistry* **2007**, *46*, 14369–14379.
165. Monoi, H.; Futaki, S.; Kugimiya, S.; Minakata, H.; Yoshihara, K. Poly-L-glutamine forms cation channels: relevance to the pathogenesis of the polyglutamine diseases. *Biophys. J.* **2000**, *78*, 2892–2899.
166. Hirakura, Y.; Lin, M. C.; Kagan, B. L. Alzheimer amyloid Abeta1-42 channels: effects of solvent, pH, and Congo Red. *J. Neurosci. Res.* **1999**, *57*, 458–466.
167. Arispe, N.; Rojas, E.; Pollard, H. B. Alzheimer disease amyloid beta protein forms calcium channels in bilayer membranes: blockade by tromethamine and aluminum. *Proc. Natl. Acad. Sci. U.S.A.* **1993**, *90*, 567–571.
168. Arispe, N.; Pollard, H. B.; Rojas, E. Zn^{2+} interaction with Alzheimer amyloid beta protein calcium channels. *Proc. Natl. Acad. Sci. U.S.A.* **1996**, *93*, 1710–1715.
169. Inoue, S. In situ Abeta pores in AD brain are cylindrical assembly of Abeta protofilaments. *Amyloid* **2008**, *15*, 223–233.
170. Pountney, D. L.; Lowe, R.; Quilty, M.; Vickers, J. C.; Voelcker, N. H.; Gai, W. P. Annular α-synuclein species from purified multiple system atrophy inclusions. *J. Neurochem.* **2004**, *90*, 502–512.
171. Mak, D. O.; Webb, W. W. Two classes of alamethicin transmembrane channels: molecular models from single-channel properties. *Biophys. J.* **1995**, *69*, 2323–2336.
172. Yang, L.; Harroun, T. A.; Weiss, T. M.; Ding, L.; Huang, H. W. Barrel-stave model or toroidal model? A case study on melittin pores. *Biophys. J.* **2001**, *81*, 1475–1485.
173. Smith, P. E.; Brender, J. R.; Ramamoorthy, A. Induction of negative curvature as a mechanism of cell toxicity by amyloidogenic peptides: the case of islet amyloid polypeptide. *J. Am. Chem. Soc.* **2009**, *131*, 4470–4478.
174. Rausch, J. M.; Marks, J. R.; Rathinakumar, R.; Wimley, W. C. Beta-sheet pore-forming peptides selected from a rational combinatorial library: mechanism of pore

formation in lipid vesicles and activity in biological membranes. *Biochemistry* **2007**, *46*, 12124–12139.

175. Bechinger, B.; Lohner, K. Detergent-like actions of linear amphipathic cationic antimicrobial peptides. *Biochim. Biophys. Acta Biomembr.* **1758**, *2006*, 1529–1539.

176. Relini, A.; Cavalleri, O.; Rolandi, R.; Gliozzi, A. The two-fold aspect of the interplay of amyloidogenic proteins with lipid membranes. *Chem. Phys. Lipids* **2009**, *158*, 1–9.

177. Canale, C.; Torrassa, S.; Rispoli, P.; Relini, A.; Rolandi, R.; Bucciantini, M., et al. Natively folded HypF-N and its early amyloid aggregates interact with phospholipid monolayers and destabilize supported phospholipid bilayers. *Biophys. J.* **2006**, *91*, 4575–4588.

178. Kayed, R.; Sokolov, Y.; Edmonds, B.; McIntire, T. M.; Milton, S. C.; Hall, J. E.; Glabe, C. G. Permeabilization of lipid bilayers is a common conformation-dependent activity of soluble amyloid oligomers in protein misfolding diseases. *J. Biol. Chem.* **2004**, *279*, 46363–46366.

179. Sokolov, Y.; Kozak, J. A.; Kayed, R.; Chanturiya, A.; Glabe, C.; Hall, J. E. Soluble amyloid oligomers increase bilayer conductance by altering dielectric structure. *J. Gen. Physiol.* **2006**, *128*, 637–647.

180. Brender, J. R.; Heyl, D. L.; Samisetti, S.; Kotler, S. A.; Osborne, J. M.; Pesaru, R. R.; Ramamoorthy, A. Membrane disordering is not sufficient for membrane permeabilization by islet amyloid polypeptide: studies of IAPP(20-29) fragments. *Phys. Chem. Chem. Phys.* **2013**, *15*, 8908–8915.

181. Engel, M.; Yigittop, H.; Elgersma, R.; Rijkers, D.; Liskamp, R.; de Kruijff, B., et al. Islet amyloid polypeptide inserts into phospholipid monolayers as monomer. *J. Mol. Biol.* **2006**, *356*, 783–789.

182. Engel, M. F. M.; Khemtémourian, L.; Kleijer, C. C.; Meeldijk, H. J. D.; Jacobs, J.; Verkleij, A. J., et al. Membrane damage by human islet amyloid polypeptide through fibril growth at the membrane. *Proc. Natl. Acad. Sci. U.S.A.* **2008**, *105*, 6033–6038.

183. Cheng, B.; Gong, H.; Xiao, H.; Petersen, R. B.; Zheng, L.; Huang, K. Inhibiting toxic aggregation of amyloidogenic proteins: a therapeutic strategy for protein misfolding diseases. *Biochim. Biophys. Acta* **1830**, *2013*, 4860–4871.

184. Wang, Q.; Yu, X.; Li, L.; Zheng, J. Inhibition of amyloid-β aggregation in Alzheimer's disease. *Curr. Pharm. Des.* **2014**, *20*, 1223–1243.

185. Bulic, B.; Pickhardt, M.; Mandelkow, E. Progress and developments in tau aggregation inhibitors for Alzheimer disease. *J. Med. Chem.* **2013**, *56*, 4135–4155.

186. Wischik, C. M.; Harrington, C. R.; Storey, J. M. D. Tau-aggregation inhibitor therapy for Alzheimer's disease. *Biochem. Pharmacol.* **2014**, *88*, 529–539.

187. Mandelkow, E. M.; Mandelkow, E. Biochemistry and cell biology of tau protein in neurofibrillary degeneration. *Cold Spring Harbor Perspect. Med.* **2012**, a006247.

188. Mylonas, E.; Hascher, A.; Bernado, P.; Blackledge, M.; Mandelkow, E.; Svergun, D. I. Domain conformation of tau protein studied by solution small-angle X-ray scattering. *Biochemistry* **2008**, *47*, 10345–10353.

189. Jones, E. M.; Dubey, M.; Camp, P. J.; Vernon, B. C.; Biernat, J.; Mandelkow, E., et al. Interaction of tau protein with model lipid membranes induces tau structural compaction and membrane disruption. *Biochemistry* **2012**, *51*, 2539–2550.

190. Martin, L.; Latypova, X.; Terro, F. Post-translational modifications of tau protein: Implications for Alzheimer's disease. *Neurochem. Int.* **2011**, *58*, 458–471.

191. Combs, B.; Gamblin, C. T. FTDP-17 tau mutations induce distinct effects on aggregation and microtubule interactions. *Biochemistry* **2012**, *51*, 8597–8607.

192. Götz, J.; Deters, N.; Doldissen, A.; Bokhari, L.; Ke, Y.; Wiesner, A., et al. A decade of tau transgenic animal models and beyond. *Brain Pathol.* **2007**, *17*, 91–103.

193. Crowther, R. A. Straight and paired helical filaments in Alzheimer disease have a common structural unit. *Proc. Natl. Acad. Sci. U.S.A.* **1991**, *88*, 2288–2292.

194. Moreno-Herrero, F.; Perez, M.; Baro, A. M.; Avila, J. Characterization by atomic force microscopy of Alzheimer paired helical filaments under physiological conditions. *Biophys. J.* **2004**, *86*, 517–525.

195. Wegmann, S.; Jung, Y. J.; Chinnathambi, S.; Mandelkow, E. M.; Mandelkow, E.; Muller, D. J. Human tau isoforms assemble into ribbon-like fibrils that display polymorphic structure and stability. *J. Biol. Chem.* **2010**, *285*, 27302–27313.

196. Moore, C. L.; Huang, M. H.; Robbennolt, S. A.; Voss, K. R.; Combs, B.; Gamblin, T. C.; Goux, W. J. Secondary nucleating sequences affect kinetics and thermodynamics of tau aggregation. *Biochemistry* **2011**, *50*, 10876–10886.

197. Li, W.; Lee, M.-Y. Characterization of two VQIXXK motifs for tau fibrillization in vitro. *Biochemistry* **2006**, *45*, 15692–15701.

198. Mukrasch, M. D.; von Bergen, M.; Biernat, J.; Fischer, D.; Griesinger, C.; Mandelkow, E.; Zweckstetter, M. The "jaws" of the tau–microtubule interaction. *J. Biol. Chem.* **2007**, *282*, 12230–12239.

199. Jeganathan, S.; von Bergen, M.; Srutlach, H.; Steinhoff, H.-J.; Mandelkow, E. Global hairpin folding of tau in solution. *Biochemistry* **2006**, *45*, 2283–2293.

200. Wischik, C. M.; Novak, M.; Edwards, P. C.; Klug, A.; Tichelaar, W.; Crowther, R. A. Structural characterization of the core of the paired helical filament of Alzheimer disease. *Proc. Natl. Acad. Sci. U.S.A.* **1988**, *85*, 4884–4888.

201. Wegmann, S.; Medalsy, I. D.; Mandelkow, E.; Müller, D. J. The fuzzy coat of pathological human Tau fibrils is a two-layered polyelectrolyte brush. *Proc. Natl. Acad. Sci. U.S.A.* **2013**, E313–E321.

202. Barrantes, A.; Sotres, J.; Hernando-Perez, M.; Benitez, M. J.; de Pablo, P. J.; Baro, A. M., et al. Tau aggregation followed by atomic force microscopy and surface plasmon resonance, and single molecule tau-tau interaction probed by atomic force spectroscopy. *J. Alzheimer's Dis.* **2009**, *18*, 141–151.

203. Kuret, J.; Congdon, E. E.; Li, G.; Yin, H.; Yu, X.; Zhong, Q. Evaluating triggers and enhancers of tau fibrillization. *Microsc. Res. Tech.* **2005**, *67*, 141–155.

204. Sibille, N.; Sillen, A.; Leroy, A.; Wieruszeski, J. M.; Mulloy, B.; Landrieu, I.; Lippens, G. Structural impact of heparin binding to full-length tau as studied by NMR spectroscopy. *Biochemistry* **2006**, *45*, 12560–12572.

205. Zhu, H. L.; Fernandez, C.; Fan, J. B.; Shewmaker, F.; Chen, J.; Minton, A. P.; Liang, Y. Quantitative characterization of heparin binding to tau protein: implication for inducer-mediated tau filament formation. *J. Biol. Chem.* **2010**, *285*, 3592–3599.

206. Friedhoff, P.; von Bergen, M.; Mandelkow, E. M.; Davies, P.; Mandelkow, E. A nucleated assembly mechanism of Alzheimer paired helical filaments. *Proc. Natl. Acad. Sci. U.S.A.* **1998**, *95*, 15712–15717.

207. Carlson, S. W.; Branden, M.; Voss, K.; Sun, Q.; Rankin, C. A.; Gamblin, T. C. A complex mechanism for inducer mediated tau polymerization. *Biochemistry* **2007**, *46*, 8838–8849.

208. Ramachandran, G.; Udgaonkar, J. B. Understanding the kinetic roles of the inducer heparin and of rod-like protofibrils during amyloid fibril formation by tau protein. *J. Biol. Chem.* **2011**, *286*, 38948–38959.

209. Ramachandran, G.; Udgaonkar, J. B. Evidence for the existence of a secondary pathway for fibril growth during the aggregation of tau. *J. Mol. Biol.* **2012**, *421*, 296–314.

210. Caughey, B.; Lansbury, P. T. Protofibrils, pores, fibrils, and neurodegeneration: separating the responsible protein aggregates from the innocent bystanders. *Annu. Rev. Neurosci.* **2003**, *26*, 267–298.

211. Maeda, S.; Sahara, N.; Saito, Y.; Murayama, M.; Yoshiike, Y.; Kim, H., et al. Granular tau oligomers as intermediates of tau filaments. *Biochemistry* **2010**, *46*, 3856–3861.

212. Patterson, K. R.; Remmers, C.; Fu, Y.; Brooker, S.; Kanaan, N. M.; Vana, L., et al. Characterization of prefibrillar tau oligomers in vitro and in Alzheimer disease. *J. Biol. Chem.* **2011**, *286*, 23063–23076.

213. Morozova, O. A.; March, Z. M.; Robinson, A. S.; Colby, D. W. Conformational features of tau fibrils from Alzheimer's disease brain are faithfully propagated by unmodified recombinant protein. *Biochemistry* **2013**, *52*, 6960–6967.

214. Bulic, B.; Pickhardt, M.; Schmidt, B.; Mandelkow, E.-M.; Waldmann, H.; Mandelkow, E. Development of tau aggregation inhibitors for Alzheimer's disease. *Angew. Chem. Int. Ed.* **2009**, *48*, 1740–1752.
215. Mena, R.; Edwards, P. C.; Harrington, C. R.; Mukaetova-Ladinska, E. B.; Wischik, C. M. Staging the pathological assembly of truncated tau protein into paired helical filaments in Alzheimer's disease. *Acta Neuropathol.* **1996**, *91*, 633–641.
216. Galvan, M.; David, J. P.; Delacourte, A.; Luna, J.; Mena, R. Sequence of neurofibrillary changes in aging and Alzheimer's disease: a confocal study with phospho-tau antibody, AD2. *J. Alzheimer's Dis.* **2001**, *3*, 417–425.
217. Iqbal, K.; Zaidi, T.; Thompson, C. H.; Merz, P. A.; Wisniewski, H. M. Alzheimer paired helical filaments: bulk isolation, solubility, and protein composition. *Acta Neuropathol.* **1984**, *62*, 167–177.
218. Arriagada, P. V.; Growdon, J. H.; Hedley-Whyte, E. T.; Hyman, B. T. Neurofibrillary tangles but not senile plaques parallel duration and severity of Alzheimer's disease. *Neurology* **1992**, *42*, 631–639.
219. Ramachandran, G.; Udgaonkar, J. B. Mechanistic studies unravel the complexity inherent in tau aggregation leading to Alzheimer's disease and the tauopathies. *Biochemistry* **2013**, *52*, 4107–4126.
220. Messing, L.; Decker, J. M.; Joseph, M.; Mandelkow, E.; Mandelkow, E. M. Cascade of tau toxicity in inducible hippocampal brain slices and prevention by aggregation inhibitors. *Neurobiol. Aging* **2013**, *34*, 1343–1354.
221. Götz, J.; Xia, D.; Leinenga, G.; Chew, Y. L.; Nicholas, H. R. What renders TAU toxic. *Front. Neurol.* **2013**, *4*, 72.
222. Morris, M.; Maeda, S.; Vossel, K.; Mucke, L. The many faces of tau. *Neuron* **2011**, *70*, 410–426.
223. Butner, K. A.; Kirschner, M. W. Tau protein binds to microtubules through a flexible array of distributed weak sites. *J. Cell Biol.* **1991**, *115*, 717–730.
224. Dawson, H. N.; Cantillana, V.; Jansen, M.; Wang, H.; Vitek, M. P.; Wilcock, D. M., et al. Loss of tau elicits axonal degeneration in a mouse model of Alzheimer's disease. *Neuroscience* **2010**, *169*, 516–531.
225. Roberson, E. D.; Scearce-Levie, K.; Palop, J. J.; Yan, F.; Cheng, I. H.; Wu, T., et al. Reducing endogenous tau ameliorates amyloid β-induced deficits in an Alzheimer's disease mouse model. *Science* **2007**, *316*, 750–754.
226. Shipton, O. A.; Leitz, J. R.; Dworzak, J.; Acton, C. E. J.; Tunbridge, E. M.; Denk, F., et al. Tau protein is required for amyloid β-induced impairment of hippocampal long-term potentiation. *J. Neurosci.* **2011**, *31*, 1688–1692.
227. Roberson, E. D.; Halabisky, B.; Yoo, J. W.; Yao, J.; Chin, J.; Yan, F., et al. Amyloid-β/Fyn-induced synaptic, network, and cognitive impairments depend on tau levels in multiple mouse models of Alzheimer's disease. *J. Neurosci.* **2011**, *31*, 700–711.
228. King, M. E.; Kan, H. M.; Baas, P. W.; Erisir, A.; Glabe, C. G.; Bloom, G. S. Tau-dependent microtubule disassembly initiated by prefibrillar β-amyloid. *J. Cell Biol.* **2006**, *175*, 541–546.
229. Stamer, K.; Vogel, R.; Thies, E.; Mandelkow, E.; Mandelkow, E. M. Tau blocks traffic of organelles, neurofilaments, and APP vesicles in neurons and enhances oxidative stress. *J. Cell Biol.* **2002**, *156*, 1051–1063.
230. Kuret, J.; Chirita, C. N.; Congdon, E. E.; Kannanayakal, T.; Li, G.; Necula, M., et al. Pathways of tau fibrillization. *Biochim. Biophys. Acta* 1739, **2005**, 167–178.
231. Giannakopoulos, P.; Herrmann, F. R.; Bussiere, T.; Bouras, C.; Kovari, E.; Perl, D. P., et al. Tangle and neuron numbers, but not amyloid load, predict cognitive status in Alzheimer's disease. *Neurology* **2003**, *60*, 1495–1500.
232. Gomez-Isla, T.; Hollister, R.; West, H.; Mui, S.; Growdon, J. H.; Petersen, R. C., et al. Neuronal loss correlates with but exceeds neurofibrillary tangles in Alzheimer's disease. *Ann. Neurol.* **1997**, *41*, 17–24.

233. van de Nes, J. A.; Nafe, R.; Schlote, W. Non-tau based neuronal degeneration in Alzheimer's disease—an immunocytochemical and quantitative study in the supragranular layers of the middle temporal neocortex. *Brain Res.* **2008**, *1213*, 152–165.
234. Santacruz, K.; Lewis, J.; Spires, T.; Paulson, J.; Kotilinek, L.; Ingelsson, M., et al. Tau suppression in a neurodegenerative mouse model improves memory function. *Science* **2005**, *309*, 476–481.
235. Mocanu, M. M.; Nissen, A.; Eckermann, K.; Khlistunova, I.; Biernat, J.; Drexler, D., et al. The potential for beta-structure in the repeat domain of tau protein determines aggregation, synaptic decay, neuronal loss, and coassembly with endogenous tau in inducible mouse models of tauopathy. *J. Neurosci.* **2008**, *28*, 737–748.
236. Sydow, A.; Van der Jeugd, A.; Zheng, F.; Ahmed, T.; Balschun, D.; Petrova, O., et al. Tau-induced defects in synaptic plasticity, learning, and memory are reversible in transgenic mice after switching off the toxic tau mutant. *J. Neurosci.* **2011**, *31*, 2511–2525.
237. de Calignon, A.; Fox, L. M.; Pitstick, R.; Carlson, G. A.; Bacskai, B. J.; Spires-Jones, T. L.; Hyman, B. T. Caspase activation precedes and leads to tangles. *Nature* **2010**, *464*, 1201–1204.
238. Kuchibhotla, K. V.; Wegmann, S.; Kopeikina, K. J.; Hawkes, J.; Rudinskiy, N.; Andermann, M. L., et al. Neurofibrillary tangle-bearing neurons are functionally integrated in cortical circuits in vivo. *Proc. Natl. Acad. Sci. U.S.A.* **2014**, *111*, 510–514.
239. Spires-Jones, T. L.; Kopeikina, K. J.; Koffie, R. M.; de Calignon, A.; Hyman, B. T. Are tangles as toxic as they look? *J. Mol. Neurosci.* **2011**, *45*, 438–444.
240. Andorfer, C.; Acker, C. M.; Kress, Y.; Hof, P. R.; Duff, K.; Davies, P. Cell-cycle reentry and cell death in transgenic mice expressing nonmutant human tau isoforms. *J. Neurosci.* **2005**, *25*, 5446–5454.
241. Tian, H.; Davidowitz, E.; Lopez, P.; Emadi, S.; Moe, J.; Sierks, M. Trimeric tau is toxic to human neuronal cells at low nanomolar concentrations. *Int. J. Cell Biol.* **2013**, 260787.
242. Berger, Z.; Roder, H.; Hanna, A.; Carlson, A.; Rangachari, V.; Yue, M., et al. Accumulation of pathological tau species and memory loss in a conditional model of tauopathy. *J. Neurosci.* **2007**, *27*, 3650–3662.
243. Sahara, N.; DeTure, M.; Ren, Y.; Ebrahim, A.-S.; Kang, D.; Knight, J., et al. Characteristics of TBS-extractable hyperphosphorylated tau species: aggregation intermediates in rTg4510 mouse brain. *J. Alzheimer's Dis.* **2013**, *33*, 249–263.
244. Polydoro, M.; Dzhala, V. I.; Pooler, A. M.; Nicholls, S. B.; McKinney, A. P.; Sanchez, L., et al. Soluble pathological tau in the entorhinal cortex leads to presynaptic deficits in an early Alzheimer's disease model. *Acta Neuropathol.* **2014**, *127*, 257–270.
245. Mellone, M.; Kestoras, D.; Andrews, M. R.; Dassie, E.; Crowther, R. A.; Stokin, G. B., et al. Tau pathology is present *in vivo* and develops *in vitro* in sensory neurons from human P301S tau transgenic mice: a system for screening drugs against tauopathies. *J. Neurosci.* **2013**, *33*, 18175–18189.
246. Kimura, T.; Fukuda, T.; Sahara, N.; Yamashita, S.; Murayama, M.; Mizoroki, T., et al. Aggregation of detergent-insoluble tau is involved in neuronal loss but not in synaptic loss. *J. Biol. Chem.* **2010**, *285*, 38692–38699.
247. Wittmann, C. W.; Wszolek, M. F.; Shulman, J. M.; Salvaterra, P. M.; Lewis, J.; Hutton, M.; Feany, M. B. Tauopathy in drosophila: neurodegeneration without neurofibrillary tangles. *Science* **2001**, *293*, 711–714.
248. Shulman, J. M.; Feany, M. B. Genetic modifiers of tauopathy in drosophila. *Genetics* **2003**, *165*, 1233–1242.
249. Maeda, S.; Sahara, N.; Saito, Y.; Murayama, S.; Ikai, A.; Takashima, A. Increased levels of granular tau oligomers: an early sign of brain aging and Alzheimer's disease. *Neurosci. Res.* **2006**, *54*, 197–201.
250. Sahara, N.; Maeda, S.; Yoshiike, Y.; Mizoroki, T.; Yamashita, S.; Murayama, M., et al. Molecular chaperone-mediated tau protein metabolism counteracts the formation of granular tau oligomers in human brain. *J. Neurosci. Res.* **2007**, *85*, 3098–3108.

251. Lasagna-Reeves, C. A.; Castillo-Carranza, D. L.; Sengupta, U.; Guerrero-Munoz, M. J.; Kiritoshi, T.; Neugebauer, V., et al. Alzheimer brain-derived tau oligomers propagate pathology from endogenous tau. *Sci. Rep.* **2012**, *2*, 700.
252. Lasagna-Reeves, C. A.; Castillo-Carranza, D. L.; Sengupta, U.; Sarmiento, J.; Troncoso, J.; Jackson, G. R.; Kayed, R. Identification of oligomers at early stages of tau aggregation in Alzheimer's disease. *FASEB J.* **2012**, *26*, 1946–1959.
253. Gerson, J. E.; Kayed, R. Formation and propagation of tau oligomeric seeds. *Front. Neu.* **2013**, *4*, 93.
254. Frost, B.; Jacks, R. L.; Diamond, M. I. Propagation of tau misfolding from the outside to the inside of a cell. *J. Biol. Chem.* **2009**, *284*, 12845–12852.
255. Wu, J. W.; Herman, M.; Liu, L.; Simoes, S.; Acker, C. M.; Figueroa, H., et al. Small misfolded Tau species are internalized via bulk endocytosis and anterogradely and retrogradely transported in neurons. *J. Biol. Chem.* **2013**, *288*, 1856–1870.
256. Clavaguera, F.; Bolmont, T.; Crowther, R. A.; Abramowski, D.; Frank, S.; Probst, A., et al. Transmission and spreading of tauopathy in transgenic mouse brain. *Nat. Cell Biol.* **2009**, *11*, 909–913.
257. Iba, M.; Guo, J. L.; McBride, J. D.; Zhang, B.; Trojanowski, J. Q.; Lee, V. M.-Y. Synthetic tau fibrils mediate transmission of neurofibrillary tangles in a transgenic mouse model of Alzheimer's-like tauopathy. *J. Neurosci.* **2013**, *33*, 1024–1037.
258. Ahmed, Z.; Cooper, J.; Murray, T. K.; Garn, K.; McNaughton, E.; Clarke, H., et al. A novel in vivo model of tau propagation with rapid and progressive neurofibrillary tangle pathology: the pattern of spread is determined by connectivity, not proximity. *Acta Neuropathol.* **2014**, *127*, 667–683.
259. Liu, L.; Drouet, V.; Wu, J. W.; Witter, M. P.; Small, S. A.; Clelland, C.; Duff, K. Trans-synaptic spread of tau pathology in vivo. *PLoS One* **2012**, *7*, e31302.
260. de Calignon, A.; Polydoro, M.; Suárez-Calvet, M.; William, C.; Adamowicz, D. H.; Kopeikina, K. J., et al. Propagation of tau pathology in a model of early Alzheimer's disease. *Neuron* **2012**, *73*, 685–697.
261. Clavaguera, F.; Akatsu, H.; Fraser, G.; Crowther, R. A.; Frank, S.; Hench, J., et al. Brain homogenates from human tauopathies induce tau inclusions in mouse brain. *Proc. Natl. Acad. Sci. U.S.A.* **2013**, *110*, 9535–9540.
262. Michel, C. H.; Kumar, S.; Pinotsi, D.; Tunnacliffe, A.; St. George-Hyslop, P.; Mandelkow, E., et al. Extracellular monomeric tau protein is sufficient to initiate the spread of tau protein pathology. *J. Biol. Chem.* **2014**, *289*, 956–967.
263. Wille, H.; Drewes, G.; Biernat, J.; Mandelkow, E. M.; Mandelkow, E. Alzheimer-like paired helical filaments and antiparallel dimers formed from microtubule-associated protein tau *in vitro*. *J. Cell Biol.* **1992**, *118*, 573–584.
264. Pooler, A. M.; Phillips, E. C.; Lau, D. H. W.; Noble, W.; Hanger, D. P. Physiological release of endogenous tau is stimulated by neuronal activity. *EMBO Rep.* **2013**, *14*, 389–394.
265. Yamada, K.; Holth, J. K.; Liao, F.; Stewart, F. R.; Mahan, T. E.; Jiang, H., et al. Neuronal activity regulates extracellular tau in vivo. *J. Exp. Med.* **2014**, *211*, 387–393.
266. Holmes, B. B.; DeVos, S. L.; Kfoury, N.; Li, M.; Jacks, R.; Yanamandra, K., et al. Heparan sulfate proteoglycans mediate internalization and propagation of specific proteopathic seeds. *Proc. Natl. Acad. Sci. U.S.A.* **2013**, E3138–E3147.
267. Distl, R.; Meske, V.; Ohm, T. G. Tangle-bearing neurons contain more free cholesterol than adjacent tangle-free neurons. *Acta Neuropathol.* **2001**, *101*, 547–554.
268. Fan, Q. W.; Yu, W.; Senda, T.; Yanagisawa, K.; Michikawa, M. Cholesterol-dependent modulation of tau phosphorylation in cultured neurons. *J. Neurochem.* **2001**, *76*, 391–400.
269. Hernandez, P.; Lee, G.; Sjoberg, M.; Maccioni, R. B. Tau phosphorylation by cdk5 and fyn in response to amyloid peptide Aβ25-35: involvement of lipid rafts. *J. Alzheimer's Dis.* **2009**, *16*, 149–156.

270. Künze, G.; Barré, P.; Scheidt, H. A.; Thomas, L.; Eliezer, D.; Huster, D. Binding of the three-repeat domain of tau to phospholipid membranes induces an aggregated-like state of the protein. *Biochim. Biophys. Acta* **1818**, *2012*, 2302–2313.

271. Pooler, A. M.; Usardi, A.; Evans, C. J.; Philpott, K. L.; Noble, W.; Hanger, D. P. Dynamic association of tau with neuronal membranes is regulated by phosphorylation. *Neurobiol. Aging* **2012**, *33*, 431.e27–431.e38.

272. Maruyama, M.; Shimada, H.; Suhara, T.; Shinotoh, H.; Ji, B.; Maeda, J., et al. Imaging of tau pathology in a tauopathy mouse model and in Alzheimer patients compared to normal controls. *Neuron* **2013**, *79*, 1094–1108.

273. Doyle, S. M.; Genest, O.; Wickner, S. Protein rescue from aggregates by powerful molecular chaperone machines. *Nat. Rev. Mol. Cell Biol.* **2013**, *14*, 617–629.

274. Mogk, A.; Tomoyasu, T.; Goloubinoff, P.; Rudiger, S.; Röder, D.; Langen, H.; Bukau, B. Identification of thermolabile *Escherichia coli* proteins: prevention and reversion of aggregation by DnaK and ClpB. *EMBO J.* **1999**, *18*, 6934–6949.

275. Parsell, D. A.; Kowal, A. S.; Singer, M. A.; Lindquist, S. Protein disaggregation mediated by heat-shock protein Hsp104. *Nature* **1994**, *372*, 475–478.

276. Weibezahn, J.; Tessarz, P.; Schlieker, C.; Zahn, R.; Maglica, Z.; Lee, S., et al. Thermotolerance requires refolding of aggregated proteins by substrate translocation through the central pore of ClpB. *Cell* **2004**, *119*, 653–665.

277. Tessarz, P.; Mogk, A.; Bukau, B. Substrate threading through the central pore of the Hsp104 chaperone as a common mechanism for protein disaggregation and prion propagation. *Mol. Microbiol.* **2008**, *68*, 87–97.

278. Lee, S.; Sowa, M. E.; Choi, J. M.; Tsai, F. T. The ClpB/Hsp104 molecular chaperone—a protein disaggregating machine. *J. Struct. Biol.* **2004**, *146*, 99–105.

279. Glover, J. R.; Lindquist, S. Hsp104, Hsp70, and Hsp40: a novel chaperone system that rescues previously aggregated proteins. *Cell* **1998**, *94*, 73–82.

280. Zolkiewski, M. ClpB cooperates with DnaK, DnaJ, and GrpE in suppressing protein aggregation. A novel multi-chaperone system from *Escherichia coli*. *J. Biol. Chem.* **1999**, *274*, 28083–28086.

281. Rosenzweig, R.; Moradi, S.; Zarrine-Afsar, A.; Glover, J. R.; Kay, L. E. Unraveling the mechanism of protein disaggregation through a ClpB–DnaK interaction. *Science* **2013**, *339*, 1080–1083.

282. Miot, M.; Reidy, M.; Doyle, S. M.; Hoskins, J. R.; Johnston, D. M.; Genest, O., et al. Species-specific collaboration of heat shock proteins (Hsp) 70 and 100 in thermotolerance and protein disaggregation. *Proc. Natl Acad. Sci. U.S.A.* **2011**, *108*, 6915–6920.

283. Murray, A. N.; Kelly, J. W. Hsp104 gives clients the individual attention they need. *Cell* **2012**, *151*, 695–697.

284. Martin, A.; Baker, T. A.; Sauer, R. T. Rebuilt AAA+ motors reveal operating principles for ATP-fuelled machines. *Nature* **2005**, *437*, 1115–1120.

285. DeSantis, M. E.; Leung, E. H.; Sweeny, E. A.; Jackrel, M. E.; Cushman-Nick, M.; Neuhaus-Follini, A., et al. Operational plasticity enables Hsp104 to disaggregate diverse amyloid and nonamyloid clients. *Cell* **2012**, *151*, 778–793.

286. Duennwald, M. L.; Echeverria, A.; Shorter, J. Small heat shock proteins potentiate amyloid dissolution by protein disaggregases from yeast and humans. *PLoS Biol.* **2012**, *10*, e1001346.

287. Vashist, S.; Cushman, M.; Shorter, J. Applying Hsp104 to protein-misfolding disorders. *Biochem. Cell Biol.* **2010**, *88*, 1–13.

288. Jackrel, M. E.; DeSantis, M. E.; Martinez, B. A.; Castellano, L. M.; Stewart, R. M.; Caldwell, K. A., et al. Potentiated Hsp104 variants antagonize diverse proteotoxic misfolding events. *Cell* **2014**, *156*, 170–182.

289. Murray, A. N.; Solomon, J. P.; Wang, Y. J.; Balch, W. E.; Kelly, J. W. Discovery and characterization of a mammalian amyloid disaggregation activity. *Protein Sci.* **2010**, *19*, 836–846.

290. Shorter, J. The mammalian disaggregase machinery: Hsp110 synergizes with Hsp70 and Hsp40 to catalyse protein disaggregation and reactivation in a cell-free system. *PLoS One* **2011**, *6*, e26319.
291. Hrizo, S. L.; Gusarova, V.; Habiel, D. M.; Goeckeler, J. L.; Fisher, E. A.; Brodsky, J. L. The Hsp110 molecular chaperone stabilizes apolipoprotein B from endoplasmic reticulum-associated degradation (ERAD). *J. Biol. Chem.* **2007**, *282*, 32665–32675.
292. Dragovic, Z.; Broadley, S. A.; Shomura, Y.; Bracher, A.; Hartl, F. U. Molecular chaperones of the Hsp110 family act as nucleotide exchange factors of Hsp70s. *EMBO J.* **2006**, *25*, 2519–2528.
293. Xu, X.; Boateng Sarbeng, E.; Vorvis, C.; Kumar, D. P.; Zhou, L.; Liu, Q. Unique peptide substrate binding properties of 110-kDa heat-shock protein (Hsp110) determine its distinct chaperone activity. *J. Biol. Chem.* **2012**, *287*, 5661–5672.
294. Oh, H. J.; Easton, D.; Murawski, M.; Kaneko, Y.; Subjeck, J. R. The chaperoning activity of hsp110. Identification of functional domains by use of targeted deletions. *J. Biol. Chem.* **1999**, *274*, 15712–15718.
295. Mayer, M. P.; Schroder, H.; Rudiger, S.; Paal, K.; Laufen, T.; Bukau, B. Multistep mechanism of substrate binding determines chaperone activity of Hsp70. *Nat. Struct. Biol.* **2000**, *7*, 586–593.
296. Mattoo, R. U. H.; Sharma, S. K.; Priya, S.; Finka, A.; Goloubinoff, P. Hsp110 is a *bona fide* chaperone using ATP to unfold stable misfolded polypeptides and reciprocally collaborate with Hsp70 to solubilize protein aggregates. *J. Biol. Chem.* **2013**, *288*, 21399–21411.
297. Rampelt, H.; Kirstein-Miles, J.; Nillegoda, N. B.; Chi, K.; Scholz, S. R.; Morimoto, R. I.; Bukau, B. Metazoan Hsp70 machines use Hsp110 to power protein disaggregation. *EMBO J.* **2012**, *31*, 4221–4235.
298. Torrente, M. P.; Shorter, J. The metazoan protein disaggregase and amyloid depolymerase system. *Prion* **2014**, *7*, 457–463.
299. Eroglu, B.; Moskophidis, D.; Mivechi, N. F. Loss of Hsp110 leads to age-dependent tau hyperphosphorylation and early accumulation of insoluble amyloid β. *Mol. Cell. Biol.* **2010**, *30*, 4626–4643.
300. Songa, Y.; Nagy, M.; Ni, W.; Tyagi, N. K.; Fenton, W. A.; López-Giráldez, F., et al. Molecular chaperone Hsp110 rescues a vesicle transport defect produced by an ALS-associated mutant SOD1 protein in squid axoplasm. *Proc. Natl. Acad. Sci. U.S.A.* **2013**, *110*, 5428–5433.

Conclusions

Is this book comprehensive? I guess not. Although I tried to cover as much as possible, considering the planned book length, there's a lot of relevant matters which either you won't find here, or which have been just briefly mentioned. It had to be so, as protein misfolding and neurodegenerative diseases are extremely wide-ranging subjects. Even focusing only on disease-modifying approaches, hundreds of putative targets could—and maybe should—be mentioned as relevant.

I made two major decisions here that drove my selections and defined at least two major absences in this book. I described five major pathways to rescue impaired cells, tissues, and organs residing in the nervous system through proteinopathy-directed approaches—chaperone-driven refolding, ubiquitination and proteasomal degradation, autophagy, aggrephagy, and disaggregation/anti-aggregation of neurotoxic oligomers and protofibrils. Rather than mentioning all the putative targets in each pathway, I delved deep into one or two of them for each pathway. Hopefully, you have agreed with at least some of my choices among relevant targets that, if properly modulated by external agents, could lead to successful therapeutic interventions.

Hits, leads, and candidates are the other major missing elements here. Biological therapeutics are intentionally absent—not because they're not relevant for neurodegenerative proteinopathies, but because I'm a chemist and I'm "addicted" to small molecule modulators. You found small molecule therapeutics—a total close to 300 of them—only at the end of each chapter, listed in a table including their name and molecular target. That's because a second, chemistry-oriented companion book covers them extensively. Its chapters mirror Chapters 2 to 6 here, that is to say that those molecules act on the same molecular targets and mechanisms described here. You may read about their discovery and structural optimization, their development status and biological–pharmacological characterization, their strengths, and weaknesses. If you are looking for an integrated, multidisciplinary description of prospective avenues towards disease-modifying drugs against protein misfolding-dependent neurodegenerative diseases, you should find something useful here and in the forthcoming, chemistry-oriented companion book. The split in a biology-oriented and a chemistry-oriented set of books leaves you the freedom to access only what's closer to your research and development interests. I hope, though, that at least some of you share with me the attraction for both the targets causing neurodegeneration, and the small molecule leads and candidates that may eventually bring us closer to long-awaited, sorely needed disease-modifying drugs reducing the burden of neurodegenerative diseases.

Index

Printed in the United States
By Bookmasters